Modern Theory of
Critical Phenomena

ADVANCED BOOK CLASSICS

David Pines, Series Editor

Anderson, P.W., *Basic Notions of Condensed Matter Physics*
Bethe H. and Jackiw, R., *Intermediate Quantum Mechanics, Third Edition*
Cowan, G. and Pines, D., *Complexity: Metaphors, Models, and Reality*
de Gennes, P.G., *Superconductivity of Metals and Alloys*
d'Espagnat, B., *Conceptual Foundations of Quantum Mechanics, Second Edition*
Feynman, R., *Photon-Hadron Interactions*
Feynman, R., *Quantum Electrodynamics*
Feynman, R., *Statistical Mechanics*
Feynman, R., *The Theory of Fundamental Processes*
Gell-Mann, M. and Ne'eman, Y., *The Eightfold Way*
Khalatnikov, I. M. *An Introduction to the Theory of Superfluidity*
Ma, S-K., *Modern Theory of Critical Phenomena*
Migdal, A. B., *Qualitative Methods in Quantum Theory*
Negele, J. W. and Orland, H., *Quantum Many-Particle Systems*
Nozières, P., *Theory of Interacting Fermi Systems*
Nozières, P. and Pines, D., *The Theory of Quantum Liquids*
Parisi, G., *Statistical Field Theory*
Pines, D., *Elementary Excitations in Solids*
Pines, D., *The Many-Body Problem*
Quigg, C., *Gauge Theories of the Strong, Weak, and Electromagnetic Interactions*
Schrieffer, J.R., *Theory of Superconductivity, Revised*
Schwinger, J., *Particles, Sources, and Fields, Volume I*
Schwinger, J., *Particles, Sources, and Fields, Volume II*
Schwinger, J., *Particles, Sources, and Fields, Volume III*
Schwinger, J., *Quantum Kinematics and Dynamics*
Wyld, H.W., *Mathematical Methods for Physics*

Modern Theory of
Critical Phenomena

Shang-keng Ma
University of California, San Diego

Routledge
Taylor & Francis Group
NEW YORK AND LONDON

First published 1976 by Westview Press

Published 2018 by Routledge
605 Third Avenue, New York, NY 10017
4 Park Square, Milton Park, Abingdon, Oxon OX14 4RN

Routledge is an imprint of the Taylor & Francis Group, an informa business

A CIP catalog record for this book is available from the Library of Congress.

ISBN 13: 978-0-7382-0301-0 (pbk)

Dedicated to my parents

Hsin-yeh and Tsu-wen

CONTENTS

Editor's Foreword

Perseus Publishing's *Frontiers in Physics* series has, since 1961, made it possible for leading physicists to communicate in coherent fashion their views of recent developments in the most exciting and active fields of physics—without having to devote the time and energy required to prepare a formal review or monograph. Indeed, throughout its nearly forty year existence, the series has emphasized informality in both style and content, as well as pedagogical clarity. Over time, it was expected that these informal accounts would be replaced by more formal counterparts—textbooks or monographs—as the cutting-edge topics they treated gradually became integrated into the body of physics knowledge and reader interest dwindled. However, this has not proven to be the case for a number of the volumes in the series: Many works have remained in print on an on-demand basis, while others have such intrinsic value that the physics community has urged us to extend their life span.

The *Advanced Book Classics* series has been designed to meet this demand. It will keep in print those volumes in *Frontiers in Physics* that continue to provide a unique account of a topic of lasting interest. And through a sizable printing, these classics will be made available at a comparatively modest cost to the reader.

The lectures contained in the late Shang Ma's lecture-note volume, *Modern Theory of Critical Phenomena,* describe the remarkable flowering of this field in the 1960's and early 1970's. Ma's deep understanding of the field, combined with his lucent writing and attention to pedagogical detail, made his book an instant classic, in great demand by graduate students and experienced researchers alike. This has continued to be the case for the last twenty-five years. I am accordingly very pleased that their publication in the *Advanced Book Classics* series will continue to make the lectures readily available for future generations of scientists interested in understanding and extending our knowledge of critical phenomena.

David Pines
Cambridge, England
May, 2000

PREFACE

This book is an introduction to the modern theory of critical phenomena. Research in this field has been extensive over the past few years and the theory has undergone rapid development following the pioneering work of Wilson (1971) on the renormalization group (abbreviated as RG) approach to scaling.

This book is intended for use by the graduate student in physical sciences or engineering who has no previous knowledge of critical phenomena. Nor does the reader need any background in group theory or any advanced mathematics. Elementary statistical mechanics is the only prerequisite.

The first six chapters can be read without going

through any tedious calculation and can be used as an introductory text. They cover the outstanding features of critical phenomena and basic ideas of the RG. The rest of the book treats more advanced topics, and can be used in an advanced course in statistical physics.

I want to stress the distinction between the following two approaches to complex physical problems:

(i) Direct solution approach. This means calculation of physical quantities of interest in terms of parameters given in the particular model — in other words, solving the model. The calculation may be done analytically or numerically, exactly or approximately.

(ii) Exploiting symmetries. This approach does not attempt to solve the model. It considers how parameters in the model change under certain symmetry transformations. From various symmetry properties, one deduces some characteristics of physical quantities. These characteristics are generally independent of the quantitative values of the parameters. By symmetry transformations I mean those which are relatively simple, like reflection, translation, or rotation. I would not call a complete solution of a complicated model a symmetry property of that model.

Approach (ii) is not a substitute for approach (i). Experience tells us that one should try (ii) as far as one can before attempting (i), since (i) is often a very difficult task. Results of (ii) may simplify this task greatly. Outstanding examples of this may be found in the study of rotations in atomic physics, translations in solid state physics, and isotopic spin rotations in nuclear physics. A great deal can be learned from (ii) even without attempting (i).

To a large extent, the traditional effort in the theory of critical phenomena has taken approach (i). The mean field theory is an example of an approximate solution. Onsager's theory of the Ising model is an example of an exact solution. There are many numerical solutions of various models. While the mean field theory often seems too crude, the exact solutions are too complicated. A peculiar feature of critical phenomena is that there is very little one can do to improve the mean field theory substantially without solving the problem exactly. This makes the theory of critical phenomena a very difficult field. Many of its contributors have been mathematical talents.

The new renormalization group theory takes approach (ii). The renormalization group is a set of symmetry transformations. It tells a great deal when applied to critical phenomena although it is not a substitute for a complete solution. I would say that its role in critical phenomena is as important as the role of the rotation group in atomic physics. Although it is not as simply defined as rotations, it is not too complicated either. The fact that it is accessible to mathematically less sophisticated people like myself is an important reason for the recent rapid advances in critical phenomena. The field is now less exclusive, so that many can now understand and contribute to it.

The purpose of this volume is to introduce this new approach, beginning at a very elementary level, and to present a few selected topics in some detail. Some technical points which are often taken for granted in the literature are elaborated. This volume is not intended to be a review of the vast new field, but rather to serve as a text for those who want to learn the basic material and to equip themselves for more advanced readings and contributions.

In spite of its great success, the new renormalization group approach to the theory of critical phenomena still lacks a firm mathematical foundation. Many conclusions still remain tentative and much has not been understood. It is very important to distinguish between plausible hypotheses and established facts. Often the suspicious beginner sees this distinction very clearly. However, after he enters the field, he is overwhelmed by jargon and blinded by the successes reported in the literature. In this volume the reader will encounter frequent emphasis upon ambiguities and uncertainties. These emphases must not be interpreted as discouraging notes, but are there simply to remind the reader of some of the questions which need to be resolved and must not be ignored.

The book is roughly divided into two parts. The first part is devoted to the elaboration of basic ideas following a brief survey of some observed critical phenomena. The second part gives selected applications and discussion of some more technical points.

Very little will be said about the vast literature concerned with approach (i) mentioned above, since there are already many books and reviews available and our main concern is approach (ii) via the RG. However, the mean field theory and the closely related Gaussian approximation will be discussed in detail (Chapter III) because they are very simple and illustrative.

Kadanoff's idea of block construction (1966) will be introduced at an early stage (Chapter II), as it is an essential ingredient of RG theory. The scaling hypothesis will be introduced as a purely phenomenological hypothesis (Chapter IV). The idea of scale transformations is also fundamental to the RG. The definitions of the RG, the idea of fixed points, and connection to critical exponent will be examined in Chapters V and VI.

The basic abstract ideas of the RG are easy to understand, but to carry out these ideas and verify them explicitly turns out to be difficult. Even the simplest examples of the realization of the RG are rather complicated. Several examples, including Wilson's approximate recursion formula, the case of small ϵ, and some two-dimensional numerical calculations, will be presented, and some fundamental difficulties and uncertainties discussed (Chapters VII, VIII).

The very successful technique of the ϵ expansion and the $1/n$ expansion will be developed and illustrated with simple calculations. The basic assumptions behind these expansions are emphasized (Chapter IX). The effect of impurities on critical behaviors will be discussed at length, followed by a study of the self-avoiding random walk problem (Chapter X).

The material in the first ten chapters concerns static (time-averaged) critical phenomena. The remaining four chapters will be devoted to dynamic (time-varying) critical phenomena. Mode-mode coupling, relaxation times, the generalization of the RG ideas to dynamics, etc. will be explained (Chapters XI, XII). A few simple dynamic models are then discussed as illustrations of the application of the RG ideas (Chapter XIII). Finally the perturbation expansion in dynamics is developed and some technical points are elaborated (Chapter XIV).

The material presented in this volume covers only a small fraction of the new developments in critical phenomena over the past four years. Instead of briefly discussing many topics, I have chosen to discuss a few topics in some depth. Since I came to the study of critical phenomena and the RG only quite recently, I remember well the questions a beginner asks, and have tried throughout this volume to bring up such questions and to provide answers to them. Many of the questions brought up, however, still have no answer.

My knowledge in this field owes much to my collaboration and conversations with several colleagues, A. Aharony, M. E. Fisher, B. I. Halperin, P. C. Hohenberg, Y. Imry, T. C. Lubensky, G. F. Mazenko, M. Nauenberg, B. G. Nickel, P. Pfeuty, J. C. Wheeler, K. G. Wilson, and Y. Yang. I am very grateful to K. Friedman, H. Gould, G. F. Mazenko, W. L. McMillan, J. Rehr, A. Aharony, K. Elinger and J. C. Wheeler for their valuable comments on the manuscript.

Special thanks are due to D. Pines, without whose constant encouragement and fruitful suggestions this book would not have been written.

The support of an Alfred P. Sloan Foundation Fellowship and a grant from the National Science

Foundation helped to make this book possible.

Finally, it is my great pleasure to acknowledge the skillful assistance of Ms. Annetta Whiteman in typing this book.

Shang-keng Ma

I. INTRODUCTION

SUMMARY

We review briefly representative empirical data of critical phenomena. Definitions of the critical point, order parameter, critical exponents, etc. are introduced. Qualitative features of critical behavior are summarized. A discussion of mean field theory is included.

1. CRITICAL POINTS AND ORDER PARAMETERS

In describing the macroscopic properties of a piece of material, we are concerned with quantities such as total mass, total energy, total magnetic moment, and other totals of the constituent's particles. For homogeneous materials, it is convenient to divide these quantities by the volume V of the material to obtain the mass density, the energy density, the magnetization, etc., which we shall subsequently refer to as mechanical variables. There are other examples of important mechanical variables which are less familiar and not so easily defined or visualized as those just mentioned. Some important examples are quantum amplitudes of Bose fluids, staggered magnetization of antiferromagnets, Fourier components of atom density in a crystal, etc.

There are also quantities such as the applied pressure p, the temperature T, and the magnetic field h. These are "applied fields." They characterize the environment, or the "reservoir" with which the material is in contact. In most cases, the values of mechanical variables are uniquely determined if the value of external fields are specified.

There are some remarkable cases where a certain mechanical variable is not uniquely determined, but has choices, for special values of applied fields. For example, at $T = 373°K$ and $p = 1$ atm, the density ρ of H_2O is not fixed but has the choice of a high value (water) or low value (steam). This happens whenever (T, p) is on a curve as shown in Figure 1. 1a. This curve terminates at the point

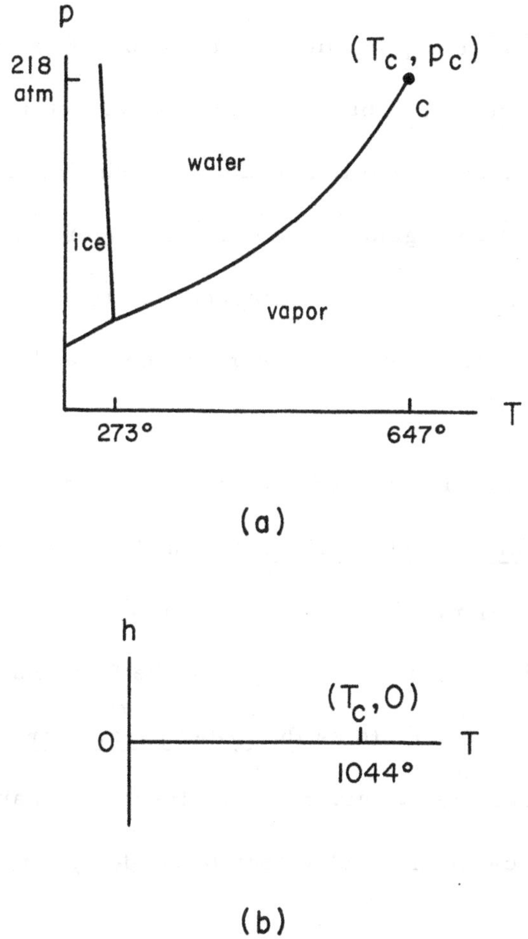

Figure 1. 1. (a) The liquid-gas critical point of H_2O: $T_c = 647\,°K$, $p_c = 218$ atm. (b) The ferro-magnetic point of Fe: $T_c = 1044\,°K$, $h_c = 0$.

(T_c, p_c). This point is a liquid-gas critical point, above which the choice ceases.

Another example is the ferromagnetic state of materials like iron and nickel. The magnetization vector m is not fixed when the applied field h is zero. It can point in different directions. This free choice of directions ceases when $T > T_c$, the Curie temperature. The line oc in Figure 1.1b is analogous to the curve in Figure 1.1a. The point $(T = T_c, h = 0)$ is a ferromagnetic critical point. For $T > T_c$, the material becomes paramagnetic and $m = 0$ when $h = 0$.

Phenomena observed near a critical point are referred to as critical phenomena. A mechanical variable which is undetermined, namely the density ρ in the liquid-gas case and the magnetization m in the ferromagnetic case, will be referred to as the order parameter.*

Besides the two given above, there are many other kinds of critical points and associated order parameters.

---

*This is not the most precise definition, since there may be more than one undetermined mechanical variable. Further criteria may be needed to narrow down the choice of an order parameter.

Interesting examples are the superfluid critical point (the λ point) of liquid helium, and the superconductivity critical points of many metals and alloys. Their associated order parameters are the quantum amplitude of helium atoms and that of electron pairs, respectively. In either case, the order parameter is a complex number. More examples are given in Table 1.1. It turns out that critical phenomena observed in many different materials near various kinds of critical points have quite a few features in common. These common features will be the subject matter of all subsequent discussions. However, to discuss the subject in very general terms would not be the most instructive. We shall instead discuss mainly the ferromagnetic critical point. Generalization to other critical points will be made as we proceed. The ferromagnetic critical phenomena are most easily visualizable and suitable for introductory discussion.

2. QUALITATIVE PICTURE

Ferromagnetism has long been studied extensively. There is a vast literature on this subject. Here we shall be content with the following qualitative information:

Table 1.1

Examples of critical points and their order parameters

Critical Point	Order Parameter	Example	T_c (°K)
Liquid-gas	Density	H_2O	647.05[3]
Ferro-magnetic	Magnetization	Fe	1044.0[1]
Antiferro-magnetic	Sublattice magnetization	FeF_2	78.26[2]
λ-line	He^4-amplitude	He^4	1.8-2.1[4]
Super-conductivity	Electron pair amplitude	Pb	7.19[5]
Binary fluid mixture	Concentration of one fluid	CCl_4-C_7F_{14}	301.78[6]
Binary alloy	Density of one kind on a sub-lattice	Cu-Zn	739[7]
Ferroelectric	Polarization	Triglycine sulfate	322.5[8]

(1) Kadanoff et al. (1967).

(2) Ahlers et al. (1974).

(3) J. M. H. Levelt Sengers (1974).

(4) G. Ahlers (1973).

(5) P. Heller (1967).

(6) P. Heller (1967).

(7) J. Als-Nielsen and O. Dietrich (1966).

(8) J. A. Gonzalo (1966).

(a) The source of magnetization is the spin of electrons in the incomplete atomic shells, typically the d and f shells of transition metal atoms like Fe, Ni, and Co. Each electron spin carries one Bohr-magneton of magnetic moment. Orbital angular momenta do not contribute.

(b) The electron spins will have lower energy when they are parallel because of the "exchange effect." This effect is a combined result of Coulomb repulsion between electrons and the Pauli exclusion principle. The latter keeps electrons with parallel spins apart and thereby reduces the Coulomb energy.

There are many other complicated effects due to atomic and crystal structures of the materials. For example, the spins may prefer to line up along a particular crystalline axis, or in a particular plane. These are, respectively, the "uniaxial" or "planar" ferromagnetics. If there is no preferred direction, then they are "isotropic." In any case, the energy is lower when more spins agree, or line up, in ferromagnetic materials.

At zero temperature, the system is in a lowest energy state and all spins point in the same direction, but this direction is not unique. There are different choices. In the isotropic case, for example, any direction is possible as long as all spins point in the same direction. There is thus a finite magnetization. The material is ferromagnetic. As the temperature T is increased from zero, thermal noise randomizes the spins. If the temperature is not too high, there will still be a net fraction of spins pointing in the same direction at all times. As the temperature is increased, this fraction is reduced. When T reaches T_c, the critical temperature, and beyond, this fraction vanishes. The material becomes paramagnetic. Transitions like this are often called "second order" for reasons which are not relevant to our study here. For T near T_c, the tendency of order (lining up spins by losing energy) and the tendency of disorder (randomizing spins by thermal noise) nearly balance each other. When $T < T_c$, order wins and the ferromagnetic state is often called the "ordered phase," while for $T > T_c$, disorder wins, hence the paramagnetic state is called the "disordered phase."

It is quite plausible that, for T just above T_c, there must be large regions (much larger than crystal unit cells) in which a net fraction of spins are lined up. This is because the ordering tendency can almost, but not quite, succeed. It achieves order in big patches but cannot make a finite fraction of all patches agree. For T just below T_c, the ordering barely makes it. Most patches, except a small, but finite, fraction have spins pointing in random directions.

When the size of the patches becomes large, the time required for ordering or disordering becomes long. The exchange effect, which is the ordering mechanism, operates on neighboring spins. The thermal noise turns spins randomly, without coherence over space or time. For these short range effects to create, destroy, or turn around large patches of lined up spins would take a long time. The "relaxation time," i.e., the time needed to approach thermal equilibrium after a disturbance, is therefore very long when T is near T_c. This in fact makes experimental work difficult. The experimenters must wait longer.

The qualitative picture described above is very sketchy. It may not be plausible at all to a reader who has not been exposed to this subject. The available experimental and theoretical data seem to indicate that many of the peculiar features of critical phenomena are manifestations of the large sizes of spin patches and long relaxation times.

Customarily, critical phenomena are classified into two categories, static and dynamic. Static phenomena concern equilibrium properties such as magnetization, susceptibility, specific heat, the probability distribution of spin configurations, and the average size of the spin patches. Dynamic phenomena concern time dependent phenomena, such as relaxation times, heat diffusion, and spin wave propagation. Up until now static phenomena have been much better understood than dynamic phenomena.

We shall study statics first and devote most of our space to it. Dynamics will be discussed later. We need to know a great deal about static phenomena in order to study the dynamics.

3. THERMODYNAMIC PROPERTIES AND EXPONENTS

Let us review briefly some observations of static phenomena near ferromagnetic critical points.

(a) Order parameter m as a function of T, the exponent β

When the external field h vanishes, the magnetization m below T_c is a decreasing function of T and vanishes at T_c. (See Figure 1.2.) For T very close to T_c, the power law behavior

$$m \propto (T_c - T)^{\beta} \tag{1.1}$$

is a common feature, where β is called a critical exponent. The observed values of β are shown in Table 1.2

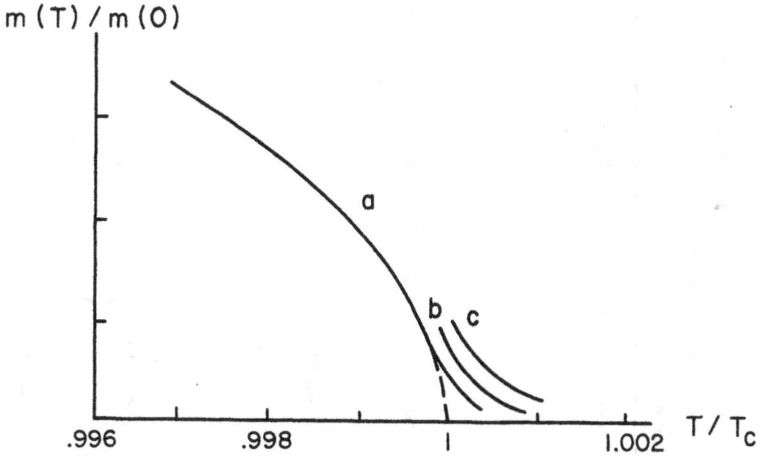

Figure 1.2. Magnetization vs temperature of single crystal $YFeO_3$ for T very close to T_c. A small field h is applied. Curves a, b, c are taken at h = 38 Oe, 210 Oe and 460 Oe, respectively. The broken line is the extrapolation to h = 0. (Data taken from G. Gorodebsky et al. (1966) Fig. 1.)

Table 1.2

Exponents of ferromagnetic critical points

Material	Symmetry	T_c (°K)	α, α'	β	γ, γ'	δ	η
Fe	Isotropic	1044.0[2]	$\alpha = \alpha' = -0.120 \pm 0.01$[1]	0.34 ± 0.02[2]	$\gamma = 1.333 \pm 0.015$[2]		0.07 ± 0.07[2]
Ni	Isotropic	631.58[1]	$\alpha = \alpha' = -0.10 \pm 0.03$[2]	0.33 ± 0.03[2]	$\gamma = 1.32 \pm 0.02$[2]	4.2 ± 0.1[2]	
EuO	Isotropic	69.33[1]	$\alpha = \alpha' = -0.09 \pm 0.01$[1]				
$YFeO_3$	Uniaxial	643[2]		0.354 ± 0.005[2]	$\gamma = 1.33 \pm 0.04$[2] $\gamma' = 0.7 \pm 0.1$		
Gd	Anisotropic	292.5[2]			$\gamma = 1.33$[2]	4.0 ± 0.1[2]	

(1) Lederman et al. (1974).

(2) Kadanoff et al. (1967).

for several materials. The exponent β does not appear to be an integer.

(b) Order parameter as a function of h at T_c the exponent δ

Figure 1.3 shows that m is not a smooth function of the external field h when $T = T_c$. One observes, for

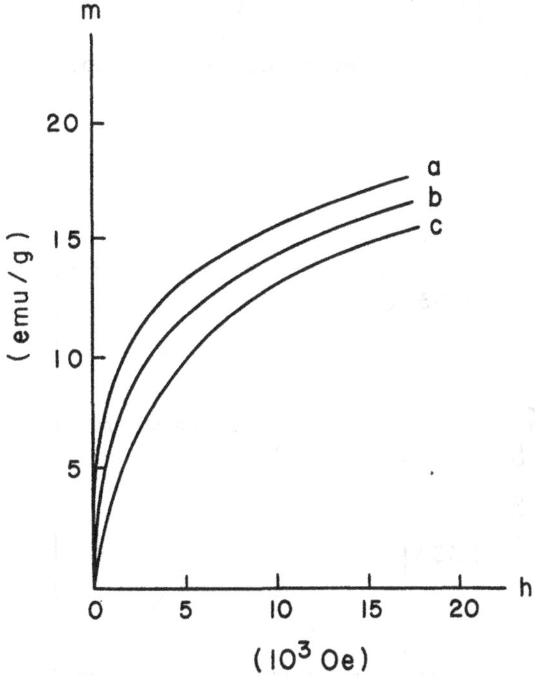

Figure 1.3. Magnetization vs applied field (corrected for demagnetization) at T very near T_c (≈ 630 °K). The curves a, b, c are taken respectively at T = 627.56°, 629.43°. 631.30°. (Taken from J. S. Kouvel and M. E. Fisher (1964), Fig. 1.)

very small h ,

$$m \propto h^{1/\delta} \tag{1.2}$$

where δ is also a critical exponent. Observed values are listed in Table 1.2.

(c) <u>The magnetic susceptibility $\chi = (\partial m/\partial h)_T$, for $h = 0$ as a function of T, the exponent γ</u>

As T approaches T_c, χ is seen to diverge, as shown in Figure 1.4. The divergence is characterized by the exponents γ and γ':

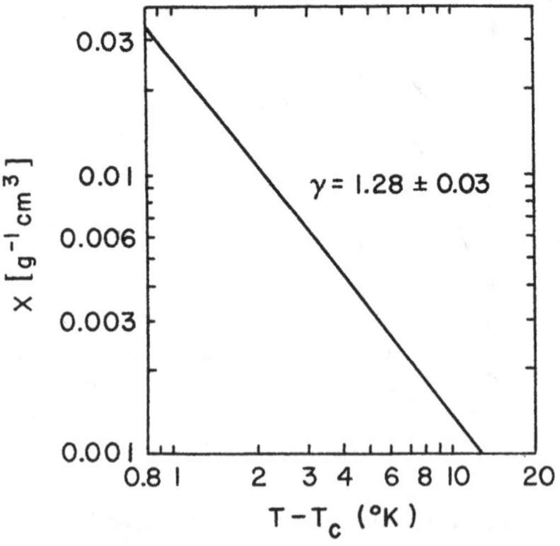

Figure 1.4. Temperature dependence of the magnetic susceptibility of nickel. (Taken from S. Arajs (1965), Fig. 3.)

$$\chi \propto (T - T_c)^{-\gamma}, \quad T > T_c,$$

$$\propto (T_c - T)^{-\gamma'}, \quad T < T_c. \qquad (1.3)$$

Data are consistent with $\gamma = \gamma'$. The proportionality constants in the two cases are not the same. Some measured values are listed in Table 1.2.

(d) Specific heat and the exponent α

The specific heat C at $h = 0$ is observed to have a singularity at T_c as Figure 1.5 shows. This singularity is characterized by the exponents α, α':

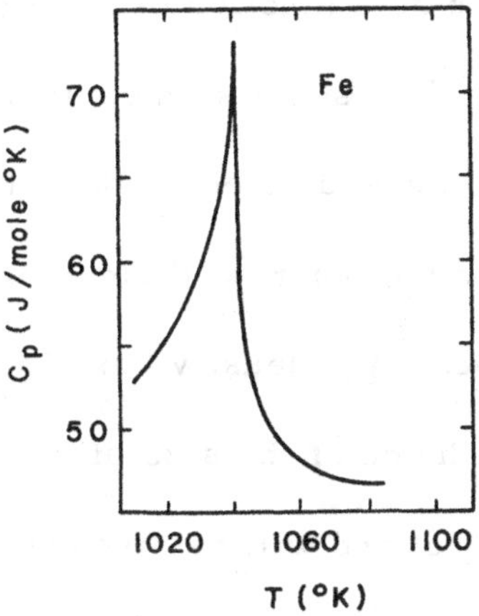

Figure 1.5. Specific heat of iron vs temperature. (Taken from Ledeman et al. (1974), Fig. 2.)

$$C \propto (T - T_c)^{-\alpha}, \quad T > T_c,$$

$$\propto (T_c - T)^{-\alpha'}, \quad T < T_c . \tag{1.4}$$

The observed α and α' are also listed in Table 1.2. The proportionality constants in the two cases are different. One also observes that $\alpha = \alpha'$ within experimental error.

4. FLUCTUATIONS OF THE ORDER PARAMETER, SCATTERING EXPERIMENTS, THE EXPONENT η

If we probe into the details over a scale small compared to the size of the whole material, we see the variation of electron spins as a function of position and time. If we take a "snapshot," we see a spin configuration $\sigma(x)$. The quantity $\sigma(x)$ may be defined as the total spin in a small volume around the point x divided by that small volume, i.e., a "local spin density" or "local order parameter." (The choice of the size of that small volume is an important subject and will be discussed in chapter II.) The spin configuration changes as a result of thermal agitation. At thermal equilibrium the spin configuration follows a probability distribution dictated by the laws of statistical mechanics.

The spin configuration can be probed by scattering experiments, especially neutron scattering. In other words, we can take a snapshot by shining on the material a burst of neutrons. The neutrons are scattered via the interaction between neutron and electron magnetic moments. This interaction is weak and the material is almost transparent to neutrons. The potential felt by a neutron is proportional to the electron spin density $\sigma(x)$. The scattering rate, or cross section, Γ_{fi} for neutrons of initial incoming momentum p_i and final outgoing momentum p_f is proportional to the matrix element of $\sigma(x)$ between these states squared:

$$\Gamma_{fi} \propto \left\langle \left| \int d^3x \; e^{-ip_f \cdot x} \, \sigma(x) \, e^{ip_i \cdot x} \right|^2 \right\rangle \tag{1.5}$$

in the Born approximation. The notation $\langle \ldots \rangle$ indicates statistical averaging over all possible spin configurations.

Let us define the Fourier components σ_k of the spin configuration $\sigma(x)$ as

$$\sigma_k = V^{-1/2} \int d^3x \; e^{-ik \cdot x} \, \sigma(x) \tag{1.6}$$

$$\sigma(x) = V^{-1/2} \sum_k e^{ik \cdot x} \, \sigma_k \; . \tag{1.7}$$

Here V is the volume of the material. We imagine that the material is of cubic shape and plane waves $V^{-1/2} e^{ik \cdot x}$ form a complete set of orthogonal functions. Each plane wave will be called a <u>mode</u>. The spin configuration is then describable as a superposition of modes; σ_k is the amplitude of the mode k.

Thus (1.5) says that

$$\Gamma_{fi} \propto \langle |\sigma_k|^2 \rangle V \tag{1.8}$$

where $k = p_f - p_i$ is the momentum transfer in the neutron scattering process.

The cross section for $k \to 0$ (forward scattering) is observed to diverge as $T \to T_c$. It is found that, for very small k,

$$\Gamma_{fi} \propto k^{-2+\eta} V \text{, at } T_c \tag{1.9}$$

where η is still another critical exponent. (See Figure 1.6 and Table 1.2.)

We pause to introduce more terminology. We define G(k) as

$$G(k) = \int d^3x \, \langle (\sigma(x) - \langle \sigma \rangle)(\sigma(0) - \langle \sigma \rangle) \rangle \, e^{-ik \cdot x} . \tag{1.10}$$

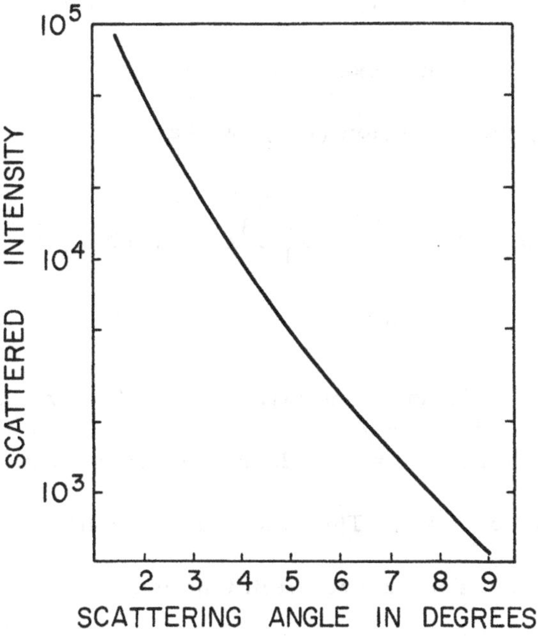

Figure 1.6. Neutron scattering intensity vs scattering angle (which is proportional to the momentum transfer) in Fe at T slightly above T_c. (Taken from Passell et al. (1965), Fig. 9.)

The average $\langle \sigma \rangle$ is the magnetization, which is nonzero if $T < T_c$, or $h \neq 0$. $\sigma(x) - \langle \sigma \rangle$ is thus the deviation from average or the spin fluctuation. The $\langle \ldots \rangle$ in (1.10) is often called the correlation function of spin fluctuations. Thus $G(k)$ is simply the Fourier transform of this correlation function. We shall also call $G(k)$ the correlation function. It will appear very often in our discussions.

Since $\langle \sigma \rangle$ is independent of x, the subtraction of $\langle \sigma \rangle$ makes no difference except at $k = 0$.

By the definition (1.6), we have

$$\langle |\sigma_k|^2 \rangle = V^{-1} \int d^3x_1 \, d^3x_2 \langle \sigma(x_1) \sigma(x_2) \rangle e^{-ik \cdot (x_1 - x_2)}$$

$$= G(k) \tag{1.11}$$

as long as k is not identically zero. We have assumed that the volume V is very large and $\langle \sigma(x_1) \sigma(x_2) \rangle$ depends only on $x = x_1 - x_2$. Thus the neutron scattering experiments measure the correlation function.

The cross section and hence G(k), diverges as $k \to 0$ and $T \to T_c$. What does that imply? In the limit $k \to 0$, (1.10) is

$$G(0) = \int d^3x \, \langle (\sigma(x) - \langle \sigma \rangle)(\sigma(0) - \langle \sigma \rangle) \rangle . \tag{1.12}$$

Since each electron carries one spin, $\sigma(x)$ is bounded by the electron density and cannot diverge. The cause of the divergence of G(0) is that both $\sigma(x) - \langle \sigma \rangle$ and $\sigma(0) - \langle \sigma \rangle$ remain in the same direction over a very large region so that the integral (1.12) is large. Another way of saying this is that spin fluctuations are correlated over a long range. As $T \to T_c$, the range becomes infinite. This observation supports the picture discussed earlier of large patches within each of which a net fraction of spins line up. The average size of the patches becomes very large when T is close to T_c.

An important identity relates the susceptibility χ to the zero k limit of the correlation function (1.12). It is

$$\chi/T = G(0) . \tag{1.13}$$

Thus the susceptibility can also be deduced from neutron scattering data.

5. OBSERVATIONS ON OTHER KINDS OF CRITICAL POINTS

A remarkable fact of nature is that the kind of singular behavior found in ferromagnets also appears at other types of critical points. The definitions of critical exponents given above can be generalized to include many other types of critical points in a natural way as illustrated below.

(a) Liquid-gas critical points

Let ρ_c denote the density at the critical point (see Figure 1.1a), and define $\rho - \rho_c$ as the order parameter.

Along the curve in Figure 1.1a, the value of $\rho - \rho_c$ has the choices $\rho_L - \rho_c$ or $\rho_G - \rho_c$. Observations show that for very small $T_c - T$

$$\begin{aligned} \rho_L - \rho_c &\propto (T_c - T)^\beta \\ \rho_c - \rho_G &\propto (T_c - T)^\beta \; . \end{aligned} \tag{1.14}$$

The two proportionality constants are generally the same. We use the same symbol β since (1.14) is analogous to (1.1). The observed values of β for some liquid-gas critical points are listed in Table 1.3.

If $\rho - \rho_c$ is measured along a line through the critical point at $T = T_c$, one finds

$$\rho - \rho_c \propto (p - p_c)^{1/\delta} \; . \tag{1.15}$$

The symbol δ is used because (1.15) is analogous to (1.2). Some measured values are in Table 1.3.

The compressibility $K = (\partial\rho/\partial p)_T$ plays the role of the susceptibility. It is found to diverge as T approaches T_c. Again the symbol γ is used to characterize the divergence (analogous to (1.3)),

$$K \propto (T - T_c)^{-\gamma}, \quad T > T_c$$
$$\propto (T_c - T)^{-\gamma'}, \quad T < T_c \quad . \tag{1.16}$$

(See Table 1. 3.)

The specific heat at constant density C_V is found to diverge also at liquid-gas critical points with exponents α and α' given in Table 1. 3.*

The fluctuation of the density can be measured by light scattering experiments like the measurements of spin fluctuations in magnetic systems sketched before. If the fluid atoms carry no magnetic moments, then the density fluctuation can be probed by neutron scattering. The potential seen by a neutron is proportional to the density of atomic nuclei. The exponent η can be introduced to characterize the observed divergence of the scattering cross section at T_c as the momentum transfer k approaches zero, in analogy to Eq. (1.9). (See Table 1. 3.)

So far experimental data on liquid-gas critical

---

*The specific heat at constant pressure C_p at a liquid-gas critical point diverges like the compressibility K, i.e., $|T - T_c|^{-\gamma}$.

Table 1.3

Exponents for various critical points

Critical Points	Material		T_c(°K)	α, α'	β	γ, γ'	δ	η
Antiferro-magnetic	$CoCl_2 \cdot 6H_2O$	Uniaxial	2.29[4]	$\alpha \leqslant 0.11$[4] $\alpha' \leqslant 0.19$[4]	0.23 $\pm$ 0.02[4]			
	FeF_2	Uniaxial	78.26[3]	$\alpha = \alpha' = 0.112$ $+ 0.044$[3]				
	$RbMnF_3$	Isotropic	83.05[1]	$\alpha = \alpha' = -0.139$ $+ 0.007$[1]	0.316 $\pm$ 0.008[2]	$\gamma = 1.397$ $\pm$ 0.034[2]		0.067 $\pm$ 0.01[2]
Liquid-gas	CO_2	n = 1	304.16[5]	$\alpha \sim 1/8$[4]	0.3447 $\pm$ 0.0007[5]	$\gamma = \gamma' = 1.20$ $\pm$ 0.02[5]	4.2[4]	
	Xe		289.74[5]	$\alpha = \alpha' = 0.08$ $\pm$ 0.02[6]	0.344 $\pm$ 0.003[5]	$\gamma = \gamma' = 1.203$ $\pm$ 0.002	4.4 $\pm$ 0.4[4]	
	He^3		3.3105[8]	$\alpha \leqslant 0.3$[4] $\alpha' \leqslant 0.2$[4]	0.361 $\pm$ 0.001[5]	$\gamma = \gamma' = 1.15$ $\pm$ 0.03[5]		
	He^4		5.1885[5]	$\alpha = 0.127$[7] $\alpha' = 0.159$	0.3554 $\pm$ 0.0028[5]	$\gamma = \gamma' = 1.17$ $\pm$ 0.0005[5]		

γ	He^4		1.8-2.1[9]	$-0.04 \leqslant \alpha = \alpha' < 0$				
Binary mixture	CCl_4-C_7F_{14}	n = 1	301.78[10]		0.335 ± 0.02[10]	$\gamma = 1.2$[10]	~ 4[10]	
Binary alloy	Co-Zn	n = 1	739[11]		0.305 ± 0.005[11]	$\gamma = 1.25 \pm 0.02$[11]		
Ferro-electric	Triglycine sulfate	n = 1	322.6[12]			$\gamma = \gamma' = 1.00 \pm 0.05$[12]		

(1) Lederman et al. (1974).

(2) Corliss et al. (1969).

(3) Ahlers et al. (1974).

(4) Kadanoff et al. (1967).

(5) J. M. H. Levelt Sengers (1974).

(6) C. Edwards et al. (1968).

(7) M. R. Moldorev (1969).

(8) B. Wallace and H. Meyer (1970).

(9) G. Ahlers (1973).

(10) P. Heller (1973).

(11) J. Als-Nielsen and O. Dietrich (1966).

(12) J. A. Gonzalo (1966).

points have been more abundant than those on other kinds of critical points.

(b) Antiferromagnetic critical points

In an antiferromagnetic crystal (imagine a cubic crystal for simplicity), the interaction between electron spins is such that spins in nearest neighboring cells tend to point in opposite directions. As a result spin direction becomes ordered and alternates from one cell to the next at low temperatures. We can imagine two interpenetrating sublattices. There is an average magnetization m on one sublattice and -m on the other. There is a critical temperature T_c (called the Neél temperature) above which m vanishes. m is usually referred to as the staggered magnetization and plays the role of the order parameter. There are uniaxial antiferromagnets with m along or opposite to one special axis, planar antiferromagnets with m in any direction in a plane, and isotropic antiferromagnets with m in any direction. m can be measured by neutron scattering. The alternating spin directions give rise to a large scattering cross section proportional to $(mV)^2$ for a momentum transfer $k_o + K$, with

$$k_o = (\pi/a, \pi/a, \pi/a) \tag{1.17}$$

and K being any vector in the reciprocal lattice. Here a is the lattice constant, and V is the volume of the crystal probed by the neutron beam.* This $(mV)^2$ dependence of the cross section is evident in view of (1.5) and the fact that the spin density is periodic with period 2a. Note that if the momentum transfer vector k is not equal to $k_o + K$ given above, then the cross section would be proportional to V, not V^2.

The measured m for small $T_c - T$ follows the power law

$$m \propto (T_c - T)^\beta \tag{1.18}$$

where the symbol β is used again as in (1.14) and (1.1). Some observed values are in Table 1.3.

The quantity that is analogous to h would be an external "staggered" magnetic field whose direction alternates from one cell to the next. Such a field is not experimentally

*We assume that the whole volume of the crystal is traversed by the beam.

accessible. Consequently, the analog of δ in (1.2) and that of γ in (1.3) are not measurable directly.

We can define the local staggered magnetization $m(x)$ through

$$m(x) = V^{-1/2} \sum_k m_k e^{ik \cdot x} = \sigma(x) e^{-ik_o \cdot x}$$
$$m_k = \sigma_{k+k_o} \qquad (1.19)$$

where the σ_k are the Fourier components of the electron spin configuration. The neutron scattering cross section of momentum transfer $k + k_o$ (with $k \neq 0$) thus measures the correlation function

$$G(k) = \langle |m_k|^2 \rangle \qquad (1.20)$$

apart from a factor proportional to V (not V^2). In the limit $V \to \infty$ one finds the behavior

$$G(k) \propto k^{-2+\eta} \qquad (1.21)$$

at T_c like the ferromagnetic and liquid-gas cases. In the limit $k \to 0$ for small but nonzero $T - T_c$ we can use the exponent γ to specify the behavior of G:

$$\lim_{k \to 0} \langle |m_k|^2 \rangle \equiv G(0) \propto (T - T_c)^{-\gamma}, \quad T > T_c$$
$$\propto (T_c - T)^{-\gamma'}, \quad T < T_c \qquad (1.22)$$

The use of γ follows from an identity similar to (1.13) which allows us to define the susceptibility as $G(0)T$ for the staggered magnetization even though the staggered field is not realizable by experiments. Some observed values of exponents are given in Table 1.3. It should be clear by now that exponents can be defined in a natural way as soon as the order parameter is defined.

(c) Binary alloy critical points

Ordered binary alloys are very closely analogous to antiferromagnets. A classic example is the β-brass, a cubic crystal made of 50% Zn and 50% Cu. At low temperatures, the nearest neighbors of Zn atoms are predominantly Cu atoms, making a configuration of alternating Zn and Cu atoms. Again we can imagine two interpenetrating sublattices. Let $\Delta\rho$ = (density of Cu) - (density of Zn) on one sublattice. Then $\Delta\rho$ has the opposite value on the other sublattice. The value of $\Delta\rho$ has two choices of sign. Its magnitude decreases as T increases and vanishes if T

exceeds the critical temperature. Thus $\Delta\rho$ is naturally identified as the order parameter and the definitions of exponents follow. (See Table 1.3.)

(d) The λ point of liquid He^4 and critical points of superconductivity

Liquid He^4 is a system of Bosons (He^4 atoms) which is found to remain a liquid down to nearly 0°K at normal pressure. In quantum mechanics it can be described by a complex Bose field. The field amplitude $\psi(x)$ is analogous to the spin density $\sigma(x)$ in ferromagnets, the local density $\rho(x)$ in a liquid or gas, and local staggered magnetization $m(x)$ in an antiferromagnet. Although $\psi(x)$ is not directly measurable, many theoretical arguments and indirect observations indicate that the λ point of liquid He^4 is a critical point with

$$\bar{\psi} \equiv V^{-1} \int d^3x \; \psi(x) \qquad (1.23)$$

as the order parameter. Below the λ point, i.e., for $T < T_c$ ($T_c \approx 2$°K and varies slightly with pressure), $\bar{\psi}$ assumes a nonzero magnitude but its direction in the complex plane is not fixed. This behavior is analogous to planar ferromagnets and planar antiferromagnets. Having identified the order parameter, we can proceed to define the exponents as we did before. However, since ψ is not directly measurable, since the external field analogous to h does not exist in the laboratory, and since there is no scattering experiment measuring the correlation of ψ, the exponents β, γ, δ, η cannot be observed. Of course, the specific heat is measurable. A logarithmic divergence $-\ln|T - T_c|$ has been found. Such a divergence can be regarded as a very small α but a large coefficient proportional to $1/\alpha$:

$$-\ln|T - T_c| = \lim_{\alpha \to 0} \frac{1}{\alpha}\left(|T - T_c|^{-\alpha} - 1\right) . \qquad (1.24)$$

Various properties of superconductors are attributed to the nonzero value of an order parameter $\bar{\Delta}$, which plays the same role as the $\bar{\psi}$ in liquid He^4:

$$\bar{\Delta} = V^{-1} \int d^3x \; \Delta(x) \qquad (1.25)$$

Here $\Delta(x)$ is the complex field amplitude of "Cooper pair Bosons." Each Boson is now a pair of electrons with nearly opposite momenta and opposite spins. The order parameter $\bar{\Delta}$ vanishes when $T \geq T_c$ and the superconductivity ceases. Again β, γ, δ, η are not measurable directly. The specific heat of various superconductors shows a discontinuity at T_c (see Figure 1.7). This is not a power law behavior as observed at other kinds of critical points. We shall argue later that the experimentally attained values of $T - T_c$ have not been small enough for the power law divergence to show up (see Sec. III.6).

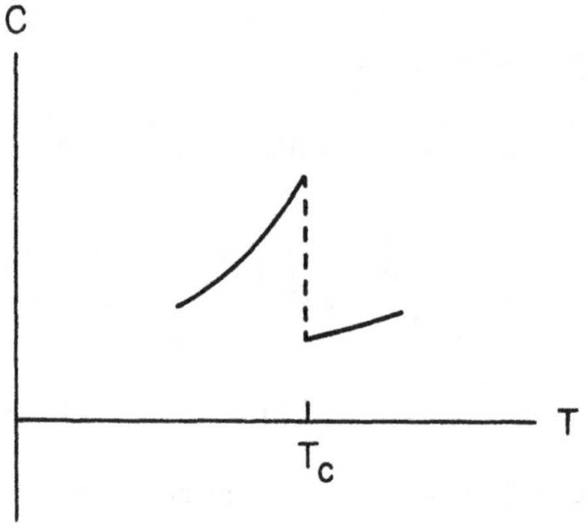

Figure 1.7. Specific heat of a superconductor near T_c.

6. SUMMARY OF QUALITATIVE FEATURES OF STATIC PHENOMENA

The most outstanding features revealed by the data reviewed above are (in the language of ferromagnetism):

(a) Non-uniqueness of the order parameter below T_c

There are different directions which the magnetization m can assume when $h = 0$, $T < T_c$. For $T > T_c$, m vanishes.

(b) Singular behavior

Many thermodynamic quantities are singular functions of $T - T_c$ and h. The correlation function at the critical point is a singular function of k.

(c) Universality of critical exponents

The singularities are characterized by nonintegral powers of $|T - T_c|$, k, or h. These powers, the critical exponents, are universal in the sense that they are the same for many different materials. The symmetry properties of materials do seem to make some difference. The exponents for uniaxial ferromagnets differ from those for isotropic ferromagnets, for example. There are also other mechanisms such as long range forces which affect the exponents such as the dipolar interaction in some ferromagnets.

In the above brief review of data, we have not touched upon any details of the various techniques of measurement. These techniques are by themselves interesting and important, but an adequate review would be beyond the scope of this book.

To the terms specified by the exponents introduced above, there are correction terms (higher powers of $|T - T_c|$, h, and k) which are negligible only if $|T - T_c|$, h, and k are sufficiently small. How small is "sufficiently small?" This is a very difficult question, whose answer depends on the details of each material under observation. It must be answered in order to interpret experimental data properly. We shall examine this question later on. (See Sec. VI. 4.)

7. MEAN FIELD THEORY

The outstanding features of the empirical data have been reviewed. Now we need a theory for a qualitative understanding of the major mechanisms behind these features, and for a basis of quantitative calculations. A completely satisfactory theory has not been established, but there has been considerable progress toward its establishment. Starting in the next chapter, we shall carefully explore theoretical advances in detail. We devote the remainder of this chapter to a brief review of the mean field theory, which is the simplest and oldest theoretical attempt. Unsatisfactory in many respects, it still captures a few important features of critical phenomena. Its simplicity makes it a very valuable tool for a rough analysis.

Consider a ferromagnetic model. Each electron spin is in a local magnetic field h', which is the external field h (which we assume to be very small), plus the field provided by the neighboring spins. The average value m of a spin in the field h' should follow a Curie law, i.e., should be proportional to h' and inversely proportional to T:

$$m = ch'/T , \tag{1.26}$$

where c is a constant. The mean field theory assumes that the field due to the neighboring spins is a function of the average of all spins, namely, m. If m is very small, this field is linear in m. Thus we have

$$h' = h + am \tag{1.27}$$

where a is a constant. We can combine (1.27) and (1.26) to obtain

$$m = hc/(T - T_c) \,, \tag{1.28}$$

$$T_c = ca \,,$$

$$\chi = \partial m/\partial h \propto 1/(T - T_c) \,, \tag{1.29}$$

for $T > T_c$. Equation (1.29) is the Curie-Weiss law. As T approaches T_c, m diverges. For $T < T_c$, Eqs. (1.27) and (1.26) do not have a meaningful solution [m would point opposite to h according to (1.28)]. When m is not very small, we need to keep higher order terms in m in (1.27) for the total field. Including the next power, we have

$$h' = h + (a - bm^2)\,m \,, \tag{1.30}$$

where b is another constant. This gives, instead of (1.28),

$$m(1 - T_c/T + cbm^2/T) = ch/T \,. \tag{1.31}$$

For $T > T_c$ we get the same answer for χ. For $T = T_c$ (1.31) gives

$$m \propto h^{1/3} \,. \tag{1.32}$$

Note that b must be positive for (1.32) to make sense.

For $T < T_c$, the solution for m is nonzero and not unique for $h = 0$:

$$m^2 = (T_c - T)/cb, \quad T < T_c \ . \tag{1.33}$$

The magnitude of m is fixed, but its direction is not. The susceptibility χ can also be worked out easily. One finds

$$\chi \propto 1/(T_c - T), \quad T < T_c \ . \tag{1.34}$$

The magnetic energy can be estimated by $-h' \cdot m$. For $T > T_c$ there is no contribution when $h = 0$. For $T < T_c$ we have $h' \neq 0$ even when $h = 0$; hence

$$E \propto -m^2 \propto T - T_c \quad \text{for} \quad T < T_c , \tag{1.35}$$

and

$$E = 0 \quad \text{for} \quad T > T_c \ . \tag{1.36}$$

It follows that the specific heat $C = \partial E/\partial T$ at $h = 0$ is discontinuous at T_c.

Comparing (1.29), (1.32), (1.33) and (1.34) to (1.1), (1.2), and (1.3), we obtain the exponents predicted by the mean field theory:

$$\beta = \frac{1}{2}, \quad \delta = 3, \qquad\qquad (1.37)$$
$$\gamma = \gamma' = 1 \quad .$$

These values are often referred to in literature as "classical exponents" or "mean field exponents." They do not agree very well with those listed in Tables 1.2 and 1.3. However, in view of how little we put in, the theory is remarkably successful. It shows that the field provided by neighboring spins is responsible for generating a nonzero magnetization below T_c. The theory also exhibits a divergent susceptibility and exponents which are independent of details, i.e., independent of the constants, a, b and c. Furthermore, it is easy to generalize the mean field theory to describe other kinds of critical points, owing to its simplicity and the transparent role played by the order parameter. It is not difficult to convince oneself that when the above steps are repeated for antiferromagnetic, liquid-gas, binary alloy, and other critical points, the same exponents (1.37) and a discontinuity in specific heat will be found. In other words, complete universality of exponents is implied by the mean field theory.

The major unrealistic approximation in the mean field theory is that nonuniform spin configurations have been excluded. The effect of fluctuations has thus been ignored. One trivial consequence is that no statement concerning the correlation function can be made. The role of the spin patches mentioned earlier cannot be accounted for in this simple theory. It turns out to be extremely difficult to understand and to analyze mathematically the effect of fluctuations. All subsequent chapters are directly or indirectly devoted to this problem.

II. MODELS AND BASIC CONCEPTS

SUMMARY

We introduce in this chapter several important concepts. Among them is the block Hamiltonian, which will play an important part in later discussions of the renormalization group. The Ginzburg-Landau Hamiltonian is introduced as a crude form of a block Hamiltonian. To make the explanation of basic ideas concrete and simple, it is convenient to introduce a few well known models, namely, the Ising, XY, Heisenberg, and general n-vector models.

1. SEQUENCE OF MODELS

If we can show quantitatively as well as qualitatively how critical phenomena can be derived from microscopic models via first principles, then we have a complete theory. However, before attempting such a task, we must first examine the merits of various models. In the course of our examination, we shall bring out some fundamental ideas.

The criteria for a microscopic model are not rigid and depend on the phenomena of interest. This is illustrated by the following sequence of models from which critical phenomena can be derived. To be specific, we shall always restrict the discussion to the ferromagnetic critical phenomena in a given crystal.

Model (1): Electrons and atomic nuclei interacting via Coulomb force.

This is certainly a microscopic model from which almost all phenomena, including critical phenomena, can be derived. But clearly this model is not practical as a starting point in studying critical phenomena.

Model (2): Electrons in a prescribed crystal lattice with an effective interaction.

The crystal lattice is now assumed to be known. We take for granted the parameters specifying the electron-electron interaction, the band structure, crystal fields, etc.

furnished by experts in atomic and solid state physics who started from Model (1).

This model is more suitable for analyzing critical phenomena than Model (1), since the formation of the crystal lattice and inner atomic shells are very remote from critical phenomena. We are willing to take them as given.

Since critical phenomena are expected to result from large scale collective behavior of electron spins, we probably do not need to know the band structure and many other details except for their combined effect on the interaction among electron spins. This model may then be further simplified.

Model (3): Classical spins, one in each unit cell of the given crystal lattice, with spin-spin interaction specified.

Here the quantum nature, the electron motion, and many details of Model (2) are ignored. The spin-spin interactions are given by parameters which are so adjusted as to simulate, as nearly as possible, what Model (2) would imply. The art of such simulation is not trivial [see Mattis (1969) for example], and the most commonly studied versions of this model, such as the Ising and Heisenberg models (which we shall define later), are very crude. However, we expect that the important physics lies in how a large number of electron spins behave together. Being a bit crude on the unit cell scale does not matter much. This model is just as microscopic as (1) and (2) as far as critical phenomena are concerned.

Model (4): Classical spins, one in each block of $2 \times 2 \times 2$ unit cells with the spin-spin interaction specified.

This is one step further than Model (3) in eliminating details. Each "spin" here is the net of 8 spins in Model (3). Again as far as large scale behavior is concerned, we really don't care about the details in each block, apart from the combined effect of these details on the interaction of the net spins on blocks. This model is clearly no less microscopic than Model (3).

Model (5): Spins on larger blocks.

Instead of $2 \times 2 \times 2$ cells per block we can take $3 \times 3 \times 3$ or even $10 \times 10 \times 10$ per block. How far can we go in making the blocks bigger and still claim a microscopic model? There is no clear-cut answer. But qualitatively, the answer is that the block size must be much less than the characteristic length of critical phenomena, i.e., the average size of spin patches mentioned in Chapter I. If we take for granted that the patches become larger as T gets closer to T_c, then the block size can be taken to be very large when T is sufficiently close to T_c. The idea and application of block construction was presented by Kadanoff (1966, 1967).

Note that in the sequence of the above models, the details which we expect to be irrelevant to critical phenomena are successively eliminated. When such an elimination process is being carried out, simplification is a practical necessity to avoid excessive mathematical complication.

Most often, experimental data are used to determine parameters in models, as calculation is impractical. Needless to say, the electronic charge, for example, in Model (1) was experimentally determined. The knowledge of crystal structure in Model (2) can be obtained through X-ray scattering. For Models (3), (4) and (5), we can fix the parameters by fitting experimental data obtained at temperatures not close to T_c.

Traditionally most theoretical studies in critical phenomena started with various versions of Model (3). We shall mainly examine Model (5). The Ginzburg-Landau model can be regarded as the crudest version of Model (5). However, it is instructive to introduce Model (3) first and show how Models (4) and (5) can be derived. This is the program for this chapter.

2. CLASSICAL MODELS OF THE CELL HAMILTONIAN

A. The Ising Model

The Ising model is a simulation of a uniaxial ferromagnet. Imagine a cubic crystal. Let each cubic unit cell be labeled by the position vector c of the center of the cell. Let the lattice constant (the length of a side of a unit cell) be 1A for convenience and A will be used as the unit of length. Let the whole crystal be a cube of volume L^3.

In each cell, there is one spin variable σ_c which measures the total spin in the cell c. There are L^3 cells, and thus L^3 spin variables. These variables will be called <u>cell spins</u>. The energy of these spins is a function $H[\sigma]$ of these L^3 cell spins. It is the Hamiltonian for the cell spins. Let us call it the cell Hamiltonian.

The model designed by Ising has a cell Hamiltonian of the form

$$\hat{H}[\sigma] = \frac{1}{2} J \sum_c \sum_r{}' (\sigma_c - \sigma_{c+r})^2 , \tag{2.1}$$

where the primed sum r is taken over only the nearest neighbor cells of c. The spin variables are restricted to two values $\sigma_c = \pm 1$. Equation (2.1) is the simplest way of saying that the energy is smaller if the spins agree with their neighbors than if they are opposite. The constant J can be estimated by the "exchange energy" between a pair of neighboring spins in the uniaxial ferromagnet which (2.1) is supposed to simulate.

To make our later discussion easier, we generalize (2.1) slightly to

$$\hat{H}[\sigma] = \frac{1}{2} J \sum_c \sum_r{}' (\sigma_c - \sigma_{c+r})^2 + \sum_c U(\sigma_c^2) , \tag{2.2}$$

and regard each σ_c as a continuous variable. The additional energy $U(\sigma_c^2)$ is large except near $\sigma_c = \pm 1$, as shown in Figure 2.1. By adjusting the shape of $U(\sigma_c^2)$, we can account for the energy increase when the electron spreads over into neighboring cells. Further generalizations to include next nearest neighbor interaction and other effects can be made, but we shall not discuss them till later.

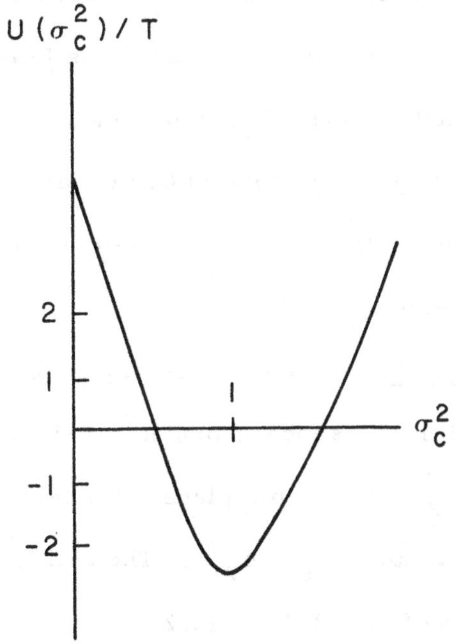

Figure 2.1. A sharp minimum of $U(\sigma_c^2)/T$ at 1 implies that the magnitude of σ_c is effectively restricted to nearly 1.

B. The XY Model and the Heisenberg Model

If the spins are not restricted to pointing along one axis, but allowed to point in any direction, we need to describe each spin by a vector

$$\sigma_c = (\sigma_{1c}, \sigma_{2c}, \sigma_{3c}) \ . \tag{2.3}$$

Introducing the usual dot product of vectors and the notation $\sigma_c^2 = \sigma_c \cdot \sigma_c$, (2.1) or (2.2) (with σ_c interpreted as a vector) becomes the cell Hamiltonian of the <u>Heisenberg model</u>. It simulates isotropic ferromagnets. It is unchanged when all spins are rotated by the same amount. Note that we are treating σ_c as a classical vector and not as Pauli spin matrices.

The <u>XY model</u> is the case intermediate between the Ising model and the Heisenberg model. It simulates magnets with spins mostly pointing in a plane. We can use σ_c as a two-component vector $(\sigma_{1c}, \sigma_{2c})$. The cell Hamiltonian will still have the form (2.1) or (2.2).

C. The Space Dimension d and Spin Dimension n

We introduce the symbol n for the number of components of the spin vector σ_c. Thus for the Ising, XY, and Heisenberg models we have $n = 1$, 2, and 3, respectively. The symbol d will be used for the dimension of the space which the crystal is occupying. Usually, $d = 3$. For a monolayer of spins, $d = 2$; for a chain of spins, $d = 1$.

In later discussions we shall have the opportunity to consider models with unphysical values of d and n.

Clearly, the total number of spin variables is nL^d since there are n components for each spin vector. In the subsequent discussion we shall use n and d as parameters so that statements concerning Ising, XY, or Heisenberg models can be obtained by assigning the appropriate value for n and the appropriate value for d.

The above models are of course only the simplest of many classical models which have been investigated. Some other models will be mentioned as we proceed.

3. STATISTICAL MECHANICS

In the language of probability theory, each of the nL^d spin variables is a random variable. The joint probability distribution P for these random variables is furnished by the law of statistical mechanics:

$$P = Z^{-1} e^{-\hat{H}[\sigma]/T} . \qquad (2.4)$$

Here Z is a normalization constant which is called the partition function:

$$Z = \int e^{-\hat{H}[\sigma]/T} \prod_{i,c} d\sigma_{ic} , \qquad (2.5)$$

where the product is taken over all components and cells. The integral is thus a multiple integral of nL^d variables.

The average value of any physical quantity which is a function of the spin variables can be calculated by integrating over P. For example,

$$\langle \sigma_{ia} \sigma_{ja'} \rangle = \int \sigma_{ia} \sigma_{ja'} P \prod_{i,c} d\sigma_{ic} . \qquad (2.6)$$

The Fourier components σ_k of the spin vector configuration will play an important role in our subsequent discussions:

$$\sigma_c = L^{-d/2} \sum_k e^{ik \cdot c} \sigma_k$$

$$\sigma_k = L^{-d/2} \sum_c e^{ik \cdot c} \sigma_c \tag{2.7}$$

The sum over wave vectors k is taken over the L^d discrete points in the first Brillouin zone. The density of points in the k space, $L^d(2\pi)^{-d}$, is very large since L is a very large number. We can regard the cell Hamiltonian $\hat{H}[\sigma]$ either as a function of the σ_c's or as a function of σ_k's. Equation (2.7) simply defines a change of variables.

As we noted in Chapter 1 (see (1.8)), the scattering experiments measure the correlation function

$$\langle |\sigma_{ik}|^2 \rangle \equiv G(k) \tag{2.8}$$

which is independent of i if the cell Hamiltonian is isotropic in the spin variables.

It must be noted that the cell Hamiltonian describes the spin configuration down to a minimum distance, i.e., the unit cell size. The variation of spins within a unit cell is beyond the description of the cell Hamiltonian. The σ_k

defined by (2.7), which are the same as those in (1.6), are sufficient to specify the spin variations over a scale larger than a unit cell and are the only random variables in the cell Hamiltonian.

Let us add a term

$$-h \sum_c \sigma_{1c} \tag{2.9}$$

to the Hamiltonian, i.e., turn on a magnetic field h in the 1 direction. Let the free energy per unit volume F be defined via (2.5)

$$Z = e^{-FL^d/T} \quad . \tag{2.10}$$

F is then a function of h, T and the parameters specifying the cell Hamiltonian. Thermodynamic properties follow once F is known. We list a few formulas:

Entropy	$S = -\partial F/\partial T$
Specific heat	$C = T\partial S/\partial T = -T\partial^2 F/\partial T^2$
Magnetization	$m = -(\partial F/\partial h)_T$
Susceptibility	$\chi = (\partial m/\partial h)_T = -(\partial^2 F/\partial h^2)_T$ (2.11)

Note that (1.13) follows from (1.12) and the last equation of

(2.11). These formulas, together with integrals like (2.5) and (2.6), form a well defined mathematical framework for calculations. The calculated physical quantities are functions of T and h. If this mathematical framework does describe nature, the calculated physical quantities must agree with the observed ones. In particular, the observed singularities discussed in Chapter I must appear in the calculated results.

The cell Hamiltonian $\hat{H}$ is a smooth function of T, h, and other parameters; and integrations over $\exp(-\hat{H}/T)$ are well defined. How could singularities ever come out of these formulas? Indeed they never would if there were a finite number of spins. They come out only in the limit when the number of spins approaches infinity. Real systems are macroscopic and the number of spins is practically infinite. It is not easy to see how this limit produces the singularities. In fact this is the basic task of a theory of critical phenomena, the task of explaining how singularities come out of a smooth Hamiltonian in the limit of $L^d \to \infty$.

One approach is to calculate exactly and then take the limit $L^d \to \infty$. This turns out to be very difficult. A large proportion of the theoretical work on critical phenomena has been applied to this task. The earliest and most important achievement in this direction was Onsager's solution of the Ising model of $d = 2$. It is important because it shows explicitly and quantitatively, even though in a very complicated manner, that singular behavior observed near T_c does indeed follow from a non-singular Hamiltonian via the laws of statistical mechanics. Onsager's art of exact evaluation of integrals has been pursued and further developed. Numerical evaluation has also been pursued extensively for models which have defied exact solution. We list some of the results in Table 2.1. Although such mathematical work provides us with exact numerical and analytic information, it does not furnish much physical understanding, and thus offers few hints as to how approximations can be made. The situation is roughly that we get the exact answers by doing a tremendous amount of mathematical work, or else we get very few or bad answers. There is no scheme for getting approximate answers by doing less work. This situation is not unexpected. The quantities which we want to calculate, such as the specific heat and susceptibility, are

Table 2.1

Critical exponents for model systems calculated from series expansions

	Spatial Dimension d	Dimension of Order Parameter n	α	β	γ	δ	η	ν
Ising	2	1	0^{*} (log)	0.125^{*}	1.75^{*}	15.04 ± 0.07	0.25^{*}	1^{*}
Heisenberg	2	3			2.5 ± 0.1			
Ising	3	1	0.013 ± 0.01	$0.312^{+0.002}_{-0.005}$	1.250 ± 0.002	5.0 ± 0.05		$0.638^{+0.002}_{-0.001}$
XY, spin 1/2	3	2			1.35 ± 0.03			
XY	3	2	-0.02 ± 0.03		1.318 ± 0.010			0.670 ± 0.006
Heisenberg, spin 1/2	3	3	$-0.20 \pm 0.08^{**}$		1.43 ± 0.01			0.70 ± 0.03
Heisenberg	3	3	-0.14 ± 0.06	0.38 ± 0.03	$1.375^{+0.020}_{-0.010}$			0.703 ± 0.010
Spherical	3	∞	-1^{*}	0.5^{*}	2^{*}	5^{*}	0^{*}	1^{*}

*Exact

Table adopted from Wortis (1973) [also **Baker et al. (1967)].

singular functions of temperature. Experience tells us that approximation methods usually do not work well for singular functions. Furthermore, the singular functions have certain universal behavior, as observations suggested (see Sec. I. 5). Because the answers to not depend on the details, making approximations for the Hamiltonian would not make the problem easier. We shall not attempt to explore the method of exact evaluation of free energy or other quantities, which has been treated extensively in books and papers.

In subsequent discussions, classical models will serve only as a convenient basis for studying qualitative behavior and for the renormalization group approach in Chapter V.

4. BLOCK HAMILTONIANS AND KADANOFF TRANSFORMATIONS

The cell Hamiltonians discussed above describe interactions between cell spins. The parameters in a cell Hamiltonian sum up the relevant effects of the details within a scale smaller than a unit cell. Evidently, we should be able to construct from a cell Hamiltonian a block

Hamiltonian describing interactions between block spins. Here we divide the crystal into cubic blocks, each of which consists of b^d unit cells; $b = 2$, or 5, for example. Let us define a block spin for each block as the sum of the b^d cell spins in the block divided by b^d, namely, the mean of the cell spins in the block. The parameters in the block Hamiltonian sum up the relevant details within a scale of b lattice constants.

How do we construct a block Hamiltonian given a cell Hamiltonian? In principle it is straightforward. Before writing down the answer, let us remember a simple rule in probability theory illustrated by the following example. If $P(q_1, q_2)$ is the probability distribution for the two random variables q_1, q_2, let us define

$$q = \frac{1}{2}(q_1 + q_2) \quad . \tag{2.12}$$

How do we construct the probability distribution for q? The rule is

$$\begin{aligned} P'(q) &= \int dq_1 dq_2 \; \delta\left(q - \frac{1}{2}(q_1 + q_2)\right) P(q_1, q_2) \\ &= \left\langle \delta\left(q - \frac{1}{2}(q_1 + q_2)\right) \right\rangle_P \quad . \end{aligned} \tag{2.13}$$

$P'(q)$ does exactly the same job as $P(q_1, q_2)$, as far as the average values involving q are concerned. For example

$$\langle q^2 \rangle = \int dq\, q^2\, P'(q)$$
$$= \langle (q_1 + q_2)^2/4 \rangle_P \; . \qquad (2.14)$$

The q integral removes the δ function and sets q equal to $(q_1 + q_2)/2$.

Let us label the blocks by the position vectors x of the centers of the blocks. As mentioned above, a block spin is defined as the net spin in a block divided by the number of cells per block:

$$\sigma_x = b^{-d} \sum_c^x \sigma_c \; , \qquad (2.15)$$

where the superscript x on the summation sign denotes the sum over b^d cell spins in the block. The components of σ_x are σ_{ix}, i runs from 1 to n as before. There are L^d/b^d block spins. Thus σ_x is simply the "mean" of the cell spins within the block labeled by x. Clearly, if $b = 1$, the block spins are the same as cell spins. If $b = L$, then the block spin would be the mean of all the cell spins in the system. Note that we use the word "mean" instead of using the word "average" here for the sum of spins divided by the number of spins. The word "average" will be exclusively used for statistical average over the canonical ensemble. In the limit of infinite block size, and only in this limit, we expect the probability distribution for the block spin to be infinitely sharply peaked at the average value.

Following the rule illustrated by (2.13), we write down the probability distribution $P'[\sigma]$ for the block spins:

$$P'[\sigma] = \left\langle \prod_{i,x} \delta\left(\sigma_{ix} - b^{-d} \sum_c^x \sigma_{ic}\right)\right\rangle_P$$

$$\propto \int e^{-\hat{H}[\sigma]/T} \prod_{i,x} \delta\left(\sigma_{ix} - b^{-d} \sum_c^x \sigma_{ic}\right) \prod_{j,c} d\sigma_{jc}$$

$$\equiv e^{-H[\sigma]/T} , \qquad (2.16)$$

where the index c runs over all cells and i, j from 1 to n. The function $H[\sigma]$ of the block spins σ_x is the desired block Hamiltonian. It would be the total free energy were it not for the constraints imposed by the δ functions. In view of the definitions (2.5) and (2.16), we have

$$Z = \int e^{-H[\sigma]/T} \prod_{i,x} d\sigma_{ix} , \tag{2.17}$$

where Z is defined by (2.5).

The block spins σ_x can account for variations of spins over a minimum length of b, i.e., the spatial resolution of the block Hamiltonian H is b, whereas the spatial resolution of the cell Hamiltonian $\hat{H}$ is 1. The block Hamiltonian has a poorer spatial resolution. But if we are not interested in variations of spin over a distance smaller than b, then the block Hamiltonian is equivalent to the cell Hamiltonian in the sense that they produce the same average of interest. In our study of critical phenomena we are only interested in the variations of spins of long wavelengths, much longer than 1 lattice constant.

We can define another block Hamiltonian which is very similar but not identical to the one defined above. We take the cell Hamiltonian $\hat{H}[\sigma]$ and write it as a function of the Fourier components σ_k via (2.7). This is simply a change of variables. Then $\exp(-\hat{H}/T)$ is the probability distribution for the nL^d new random variables σ_{ik}. Again let us remember a simple rule of probability theory which is essentially the same one as illustrated by (2.13). Consider a probability distribution of two random variables $P(q_1, q_2)$. If we are only interested in q_1, then we obtain the probability distribution $P'(q_1)$ for q_1 alone by integrating out q_2:

$$P'(q_1) = \int dq_2 \, P(q_1, q_2) \quad . \tag{2.18}$$

Thus if we only interested in σ_k with k smaller than a specific magnitude Λ, we can integrate out all those σ_{ik} with $k > \Lambda$ from $\exp(-\hat{H}/T)$:

$$\begin{aligned} P' &\propto \int e^{-\hat{H}[\sigma]/T} \prod_{i,k>\Lambda} d\sigma_{ik} \\ &\equiv e^{-H[\sigma]/T} \quad . \end{aligned} \tag{2.19}$$

Now P' is the probability distribution for σ_{ik} with those k vectors within the sphere of radius Λ. The function $H[\sigma]$ of these variables is a block Hamiltonian in the following sense. Without the Fourier components of $k > \Lambda$, variations of spins over a scale shorter than $\sim 2\pi\Lambda^{-1}$ cannot be specified. That is, the spatial resolution of $H[\sigma]$ is $2\pi\Lambda^{-1}$. If we identify $2\pi\Lambda^{-1}$ as a block size b, then $H[\sigma]$

in (2.19) has an interpretation similar to that of $H[\sigma]$ defined by (2.16). In (2.16) we integrate out finer details within each block to downgrade the spatial resolution to a size b. In (2.19) we achieve the same effect by integrating out higher Fourier components. Quantitatively the two block Hamiltonians are not identical. In our qualitative discussions we shall ignore the difference.

Similar to the block spin (2.15), we define

$$\sigma(x) = L^{-d/2} \sum_{k<\Lambda} \sigma_k e^{ik \cdot x} , \tag{2.20}$$

which specifies the spin configuration down to a distance $b \sim \Lambda^{-1}$. Qualitatively, $\sigma(x)$ is the same as σ_x, which is the mean spin over a block.

The procedure (2.16) or (2.19) of obtaining the block Hamiltonian $H[\sigma]$ from the cell Hamiltonian $\hat{H}[\sigma]$ will be referred to as a Kadanoff transformation. We shall denote this procedure symbolically as

$$H[\sigma]/T = K_b \hat{H}[\sigma]/T \tag{2.21}$$

where the subscript b indicates that the block size is b times the cell size. We set $K_1 = 1$.

Evidently, given a block Hamiltonian $H[\sigma]$, we can construct another block Hamiltonian $H''[\sigma]$ for block spins defined on still bigger blocks. For example we can combine 2^d blocks to make one new block. The construction of the new block Hamiltonian is the same as that of the old block Hamiltonian from the cell Hamiltonian. This means lowering the spatial resolution by still another factor of 2, i.e., an application of K_2. We shall write, for combining s^d blocks,

$$H''[\sigma]/T = K_s H[\sigma]/T \quad , \tag{2.22}$$

$$e^{-H''[\sigma]/T} = \int e^{-H[\sigma']/T} \prod_{i,x} \delta\left(\sigma_{ix} - s^{-d} \sum_y^x \sigma'_{iy}\right) \prod_{j,y} d\sigma'_{jy} \tag{2.23}$$

where $s^{-d} \sum_y \sigma'_{iy}$ is the mean spin over the s^d old blocks in the new block centered at x. Alternatively K_s can be defined using Fourier components:

$$e^{-H''[\sigma]/T} = \int e^{-H[\sigma]/T} \prod_{i,\ \Lambda>k>\Lambda/s} d\sigma_{ik} \, , \tag{2.24}$$

which follows (2.19) and brings the cutoff wave number down from Λ to Λ/s.

Clearly,

$$K_s H[\sigma]/T = K_s K_b \hat{H}[\sigma]/T$$

$$= K_{sb} \hat{H}[\sigma]/T \qquad (2.25)$$

The last equality is obvious since combining b^d cells for each block and then combining s^d blocks for each big block is the same as combining $(sb)^d$ cells for each big block. In general

$$K_s K_{s'} = K_{ss'} . \qquad (2.26)$$

Kadanoff transformations will play a major role in the construction of the renormalization group in Chapter V.

Let us comment on a couple of technical points.

The Fourier components σ_{ik} are complex. We need to be more careful in defining the integration over $d\sigma_{ik}$ in (2.19) and in (2.24). Note that $\sigma_{ik}^* = \sigma_{i-k}$ since $\sigma_i(x)$ is real. Thus, there are just two real variables involved in σ_{ik} and σ_{i-k}, namely, $\mathrm{Re}\,\sigma_{ik}$ and $\mathrm{Im}\,\sigma_{ik}$, not four. The integrations such as (2.19) and (2.24) must be defined by integrations over $\mathrm{Re}\,\sigma_{ik}$ and $\mathrm{Im}\,\sigma_{ik}$:

$$d\sigma_{ik}\, d\sigma_{i-k} = d(\mathrm{Re}\,\sigma_{ik})\, d(\mathrm{Im}\,\sigma_{ik})$$

for each pair σ_{ik} and σ_{i-k}.

We have been using continuous spin variables instead of discrete ones. The main reason is that discrete spins will become essentially continuous after a Kadanoff transformation. For example, suppose that we have block spins each of which can assume two values, +1 or -1. We combine 8 blocks to form a new block. The new block spin, i.e., the mean of the 8 old spins, can assume one of 9 values, i.e., ±1, ±6/8, ±4/8, ±2/8, 0. Nine values make the new block spin essentially continuous. Coarse graining destroys the discreteness as we expect intuitively. Furthermore, it is easier to discuss Fourier transforms if we use continuous variables.

However, it is possible to modify the definition of the new block spins to preserve the 2-valueness at the expense of the precise physical meaning of the new spin variables. We shall discuss this point in Chapter VIII.

In summary, the block Hamiltonian is obtained from the cell Hamiltonian by a smearing or coarse graining process. It specifies the interaction over a scale $b \sim \Lambda^{-1}$ and sums up gross features of b^d cell spins. In view of (2.16) or (2.19), it depends on T, as well as other parameters specifying the cell Hamiltonian, like J and h in (2.1) and (2.9). Such dependence is expected to be non-singular for the following reason. The interaction between block spins is short ranged since it originates from the interaction between cell spins which is assumed to be of short range. Thus the interaction between block spins depends on the behavior of a finite number, i.e., $\sim b^d$, of cell spins in several blocks. As we remarked before there can be no singular behavior from a finite number of cell spins. Thus the block Hamiltonian is a smooth function of T and other parameters of the cell Hamiltonian.

We have illustrated the relationship between Model (3) (the cell Hamiltonian), and Models (4) and (5) (the block Hamiltonian), in the beginning of this chapter. Carrying out the calculation explicitly to obtain either (5) from (4), (4) from (3), (3) from (2), or (2) from (1) is very involved and one has to make proper approximations. Some explicit calculations will be presented later when we discuss the renormalization group.

5. GINZBURG-LANDAU FORM

The Ginzburg-Landau approach has played an important role in the theory of superconductivity and other critical phenomena. It starts by assuming a simple form for the block Hamiltonian. Customarily this simple form is written as

$$H[\sigma]/T = \int d^d x \, [a_0 + a_2 \sigma^2 + a_4 \sigma^4 + c(\nabla\sigma)^2 - h \cdot \sigma] \tag{2.27}$$

where

$$\sigma^2 \equiv \sigma(x) \cdot \sigma(x) \equiv \sum_{i=1}^{n} (\sigma_i(x))^2 ,$$

$$\sigma^4 \equiv (\sigma^2)^2 ,$$

$$(\nabla\sigma)^2 \equiv \sum_{\alpha=1}^{d} \sum_{i=1}^{n} \left(\frac{\partial \sigma_i}{\partial x_\alpha} \right)^2 . \tag{2.28}$$

The coefficients a_0, a_2, a_4, c are functions of T, and h is the applied magnetic field divided by T. We assume the system is isotropic in spin space apart from the effect of the applied magnetic field and is also effectively isotropic in coordinate space. The block spin $\sigma(x)$ is defined by (2.20). It specifies the spin configuration down to a distance $b \sim \Lambda^{-1}$. This block Hamiltonian is really a function of the discrete set of Fourier components σ_k:

$$H[\sigma]/T = a_0 L^d + \sum_{k<\Lambda} \sigma_k \cdot \sigma_{-k} (a_2 + ck^2)$$

$$+ L^{-d} \sum_{kk'k''<\Lambda} a_4 (\sigma_k \cdot \sigma_{k'})(\sigma_{k''} \cdot \sigma_{-k-k'-k''})$$

$$- L^{d/2} \sigma_0 \cdot h \ , \qquad (2.29)$$

and $P \propto e^{-H[\sigma]/T}$ is the probability distribution for σ_k, $k < \Lambda$. We can write a similar block Hamiltonian in terms of σ_x:

$$H[\sigma]/T = b^d \sum_x \Big[a_0 + a_2 \sigma_x^2 + a_4 \sigma_x^4$$

$$+ c \frac{b^{-2}}{2} \sum_y{}' (\sigma_x - \sigma_{x+y})^2 - h \cdot \sigma_x \Big] \ . \qquad (2.30)$$

Here σ_x are block spins defined by (2.15). The sum over

y is taken over the 2d nearest neighbors of the block x. The quantities $(\sigma_{x+y} - \sigma_x)/b$ can be identified to a first approximation as the gradient $\nabla\sigma$ in (2.27).

The meaning of the various terms in the Ginzburg-Landau form of the block Hamiltonian can be understood easily through (2.30). The meaning of the $-h\cdot\sigma_x$ term is transparent, and let us forget about it for simplicity. If we drop the gradient term, namely, the $(\sigma_x - \sigma_{x+r})^2$ term, then (2.30) becomes a sum

$$b^d \sum_x U(\sigma_x) , \tag{2.31}$$

with

$$U(\sigma_x) = a_0 + a_2\sigma_x^2 + a_4\sigma_x^4 . \tag{2.32}$$

Each of the terms in the sum (2.31) depends only on σ_x of one block. That is, each block spin is statistically independent of other block spins, since the probability distribution is a product $\prod_x \exp(-U(\sigma_x))$. We then have a system of L^d/b^d non-interacting blocks. Each block has an effective Hamiltonian $b^d U(\sigma_x)$, which is the free energy of the block with mean spin constrained at σ_x. This free energy must be a smooth function of σ_x, T, and other parameters since there is only a finite number of cell spins in a block, as we remarked before. Thus (2.32) is just the first few terms of the power series expansion of this free energy. The coefficients a_0, a_2, a_4, and higher coefficients not included must all be smooth functions of T and other parameters. Since only terms up to the fourth power of σ_x are kept, we expect that the Ginzburg-Landau form might not be a good approximation for large σ_x. It will certainly make no sense if a_4 is negative because $U(\sigma_x)$ would then approach $-\infty$ as $\sigma_x \to \infty$ and the probability distribution $P \propto \exp(-H[\sigma]/T)$ would blow up.

When the gradient term $(\sigma_x - \sigma_{x+y})^2$ is included in (2.30), the block spins are no longer independent. Thus the Ginsburg-Landau form describes the interaction between neighboring blocks by this simple gradient term. For a ferromagnet we expect that the interaction will tend to make a block spin parallel to its neighboring block spins. The gradient term describes this tendency with a strength specified by the coefficient c. This term vanishes only if all block spins have the same value. The greater the difference among the block spins, the larger H/T becomes and hence the smaller the probability. We expect that the interaction between block spins is not completely accounted for by the gradient term alone. There should be higher power products of neighboring block spins.

To sum up, the Ginzburg-Landau form of block Hamiltonian is the simplest but not the most general form. The most general form would involve all powers of products of block spins. The Ginzburg-Landau form keeps only the lowest powers. Of course, it can be simplified further by dropping the σ_x^4 term. However, that would be too serious a simplification, as will be clear in Chapter III.

Note that we have excluded fractional powers or any other kind of singular function of σ_x from the block Hamiltonian. This exclusion is based on the assumption that the interaction between spins is short ranged so that the terms in the block Hamiltonian describe "local properties" pertaining to $\sim b^d$ spins, and therefore must be smooth functions of σ_x, h, and T.

III. THE GAUSSIAN APPROXIMATION

SUMMARY

We solve the Ginzburg-Landau model in the Gaussian approximation. In this approximation the most probable spin configuration is obtained by minimizing the block Hamiltonian. Fluctuations around the most probable configuration are approximately treated as independent modes with Gaussian distributions. The exponents thus obtained are $\alpha = 2 - d/2$, $\beta = 1/2$, $\gamma = 1$, $\delta = 3$, $\eta = 0$. The Ginzburg criterion is discussed. Although the Gaussian approximation is an oversimplification, it illustrates many important features of critical phenomena. We also discuss briefly strong effects of fluctuations for $d \le 2$, including the Hohenberg-Mermin-Wagner theorem.

1. MOST PROBABLE VALUE AND GAUSSIAN APPROXIMATION

Consider a probability distribution $P(q)$. If it is largest at $q = \tilde{q}$, then $\tilde{q}$ is the "most probable value" of q. The average value $\langle q \rangle$ is not necessarily the same as $\tilde{q}$, but is close to it if the maximum is sharply peaked. In this case we can fit the peak of $P(q)$ with a Gaussian

$$e^{-(q-\tilde{q})/2\lambda^2}$$

where λ measures the half width of the peak. Such a fit for $P(q)$ is our Gaussian approximation. The advantage of a Gaussian is that it has many convenient mathematical properties.

We now illustrate the application of this approximation. Consider a system with one degree of freedom. There is one coordinate q and a Hamiltonian $H(q)$, which is just the potential energy (we neglect the kinetic degree of freedom). Let the position of mechanical equilibrium be $\tilde{q}$, where $H(q)$ is minimum. This is the most probable value since $P \propto \exp(-H(q)/T)$. Near $\tilde{q}$ the potential is approximately $H(\tilde{q})$ plus a harmonic oscillator potential

$$H(q)/T = H(\tilde{q})/T + \frac{1}{2}(q - \tilde{q})^2/\lambda^2 + \cdots \tag{3.1}$$

$$\lambda^{-2} \equiv T^{-1}(\partial^2 H/\partial q^2)_{q=\tilde{q}} \ . \tag{3.2}$$

The probability distribution is then approximately a Gaussian if higher powers of $(q - \tilde{q})$ can be neglected:

$$\begin{aligned} P &\propto e^{-H(q)/T} \\ &\propto e^{-(q-\tilde{q})^2/2\lambda^2} \end{aligned} \tag{3.3}$$

Clearly the average of q is then the same as the most probable value $\tilde{q}$:

$$\langle q \rangle = \tilde{q} \tag{3.4}$$

and the fluctuation of q from the average is measured by λ ·

$$\langle (q - \tilde{q})^2 \rangle = \lambda^2 \ . \tag{3.5}$$

In order to justify dropping higher orders of $(q - \tilde{q})$, λ^2 must not be too large. In other words, the fluctuation of q must be sufficiently small, since (3.1) is a Taylor series expansion valid only for small $(q - \tilde{q})$.

Under the Gaussian approximation, the free energy

f is just $H(\tilde{q})$ plus the free energy for a harmonic oscillator

$$e^{-f/T} = \int dq \; e^{-H(\tilde{q})/T - (q-\tilde{q})^2/2\lambda^2}$$

$$f = H(\tilde{q}) - \frac{1}{2} T \ln (2\pi\lambda^2) \; . \qquad (3.6)$$

The generalization to more than one degree of freedom is straightforward. Let there be N degrees of freedom, q_α, $\alpha = 1, 2, \ldots, N$, and write $q = (q_1, \ldots, q_N)$. Let the minimum of $H[q]$ be at $\tilde{q}$. As we learned from mechanics, the potential near the minimum can be approximated by that of N independent harmonic oscillators, the normal modes:

$$H[q]/T \approx H[\tilde{q}]/T + \frac{1}{2} \sum_{\ell=1}^{N} q_\ell'^2/\lambda_\ell^2 \; , \qquad (3.7)$$

where q_ℓ' are linear combinations of $q_\alpha - \tilde{q}_\alpha$ and are called "normal coordinates" giving the amplitudes of normal modes. The quantities λ_ℓ^{-2} are the eigenvalues of the matrix $T^{-1}(\partial^2 H/\partial q_\alpha \partial q_\beta)_{q=\tilde{q}}$. Thus the generalization to more than one degree of freedom is accomplished simply by adding more independent normal modes, and Eqs. (3.4),

(3.5), and (3.6) are generalized, respectively, to

$$\langle q_\alpha \rangle = \tilde{q}_\alpha , \ \langle q'_\ell \rangle = 0 \ , \tag{3.8}$$

$$\langle q'^2_\ell \rangle = \lambda_\ell^2 \ , \tag{3.9}$$

$$f = H[\tilde{q}] - \frac{1}{2} T \sum_\ell \ln (2\pi \lambda_\ell^2) \ . \tag{3.10}$$

In summary, within the Gaussian approximation the most probable configuration $\tilde{q}$ gives the average of q; and q'_ℓ, the fluctuations of q, are independent modes each following a Gaussian distribution.

2. MINIMUM OF THE GINZBURG-LANDAU HAMILTONIAN, LANDAU THEORY

We proceed to apply the Gaussian approximation to a block Hamiltonian $H[\sigma]$. We shall first determine the minimum of $H[\sigma]$ and then approximate $H[\sigma]$ as that minimum plus a sum of harmonic oscillator terms. To be specific we use the Ginzburg-Landau form (see II.4):

$$H[\sigma]/T = \int d^d x [a_0 + a_2 \sigma^2(x) + a_4 \sigma^4(x) + c(\nabla \sigma(x))^2 - h \cdot \sigma(x)] \ . \tag{3.11}$$

Implicity (3.11) is a function of the Fourier components σ_k with $k < \Lambda$ through the relation

$$\sigma(x) = L^{-d/2} \sum_{k<\Lambda} \sigma_k e^{ik\cdot x} . \qquad (3.12)$$

The explicit expression in terms of σ_k is given by (2.29). The role of the q_α in the discussion below (3.6) is played by σ_{ik} .

The most probable spin configuration $\tilde{\sigma}$, which minimizes $H[\sigma]$, must be uniform, i.e.,

$$\tilde{\sigma}(x) = \bar{\sigma} , \qquad (3.13)$$

so that the gradient term vanishes. This means that all Fourier components for this configuration vanish except for $k = 0$:

$$\tilde{\sigma}_k = 0 , \quad k \neq 0$$

$$\tilde{\sigma}_0 = L^{d/2} \bar{\sigma} . \qquad (3.13')$$

Substituting $\tilde{\sigma}(x) = \bar{\sigma}$ in (3.11), or (3.13′) in (2.29), we get

$$H[\tilde{\sigma}]/T = L^d (a_0 + a_2 \bar{\sigma}^2 + a_4 \bar{\sigma}^4 - h \cdot \bar{\sigma}) . \qquad (3.14)$$

The value of $\bar{\sigma}$ can be found by setting the derivative of (3.14) to zero and solving for $\bar{\sigma}$:

$$2\bar{\sigma}(a_2 + 2a_4\bar{\sigma}^2) - h = 0 \ . \tag{3.15}$$

Clearly $\bar{\sigma}$ must be along the direction of h. In the limit of very small h, the desired $\bar{\sigma}$, which is the magnetization m in this approximation, is

$$\bar{\sigma} = h/2a_2 \, , \qquad a_2 > 0 \, , \tag{3.16}$$

$$\bar{\sigma} = m_o\hat{h} + h/(8m_o^2 a_4) \, , \quad a_2 < 0 \, , \tag{3.17}$$

where $\hat{h}$ is the unit vector along h and

$$m_o = (-a_2/2a_4)^{1/2} \ . \tag{3.18}$$

We plot (3.14) in Figure 3.1a for $a_2 > 0$ and in Figure 3.1b for $a_2 < 0$. Recall that a_4 must be positive. The abscissa in these plots is the component of $\bar{\sigma}$ along the direction of h.

We have no information about a_2, a_4 except that they must be smooth functions of the temperature. It was Landau's observation that interesting consequences appear when a_2 vanishes, and that we should look at the small

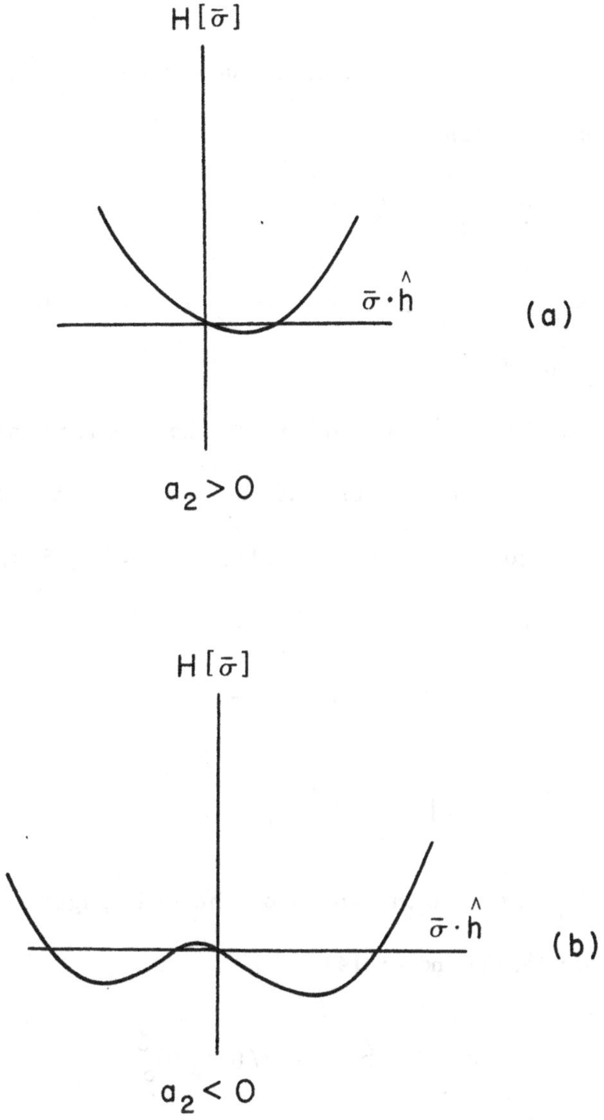

Figure 3.1. Values of the block Hamiltonian evaluated at uniform configurations. A small magnetic field h is assumed to be present.

range of temperatures near T_0, the zero of a_2 (i.e., $a_2 = 0$ if $T = T_0$). Within this small range a_4 may be taken as a constant and

$$a_2 \approx a_2' (T - T_0) \ , \tag{3.19}$$

where a_2' is a constant, since a_2, a_4 are both smooth functions of T.

Assume $a_2' > 0$. When T changes from above T_0 to below, a_2 changes sign. If $T > T_0$, as in the situation shown in Figure 3.1a, Eqs. (3.16) and (3.14) give, to first order in h,

$$m = \bar{\sigma} = h/[2 a_2' (T - T_0)] \tag{3.20}$$

$$H[\tilde{\sigma}]/T = L^d a_0 \ . \tag{3.21}$$

If $T < T_0$, we have the situation shown in Figure 3.1b, thus from (3.17) and (3.14)

$$m = \bar{\sigma} = \hat{h} m_o + h/(8 a_4 m_o^2) \tag{3.22}$$

$$m_o = (a_2'/2 a_4)^{1/2} (T_0 - T)^{1/2} \ , \tag{3.23}$$

$$H[\tilde{\sigma}]/T = L^d [a_0 - (T - T_0)^2 a_2'^2 / 4 a_4] \ . \tag{3.24}$$

It is now very plausible that T_0 can be identified as the critical temperature T_c because, as $h \to 0$, $\bar{\sigma}$ vanishes for $T > T_0$ but not for $T < T_0$. If h is identically zero then $H[\tilde{\sigma}]/T$ is minimum whenever the length of the vector $\bar{\sigma}$ is m_o, regardless of the direction of $\bar{\sigma}$.

So far we have been looking at the most probable spin configuration but have not included any fluctuation of spins. Suppose that we neglect fluctuations altogether. Then we get the "Landau theory of second order phase transition." The free energy FL^d would be simply the minimum of $H[\sigma]$. The susceptibility $\chi = (\partial\bar{\sigma}/\partial h)_T$ is, by (3.20) and (3.22),

$$\chi = [2a_2'(T - T_c)]^{-1}, \quad T > T_c ,$$

$$= [4a_2'(T_c - T)]^{-1}, \quad T < T_c , \qquad (3.25)$$

which diverges at T_c. Compared with (3.21), there is an extra term in (3.24), which gives rise to a discontinuity ΔC of specific heat $C = -T\partial^2 F/\partial T^2$ at T_c:

$$\Delta C = T_c^2 \, a_2'^2 / 2a_4 . \qquad (3.26)$$

For $T = T_c$, $a_2 = 0$, and from (3.15) we have

$$\bar{\sigma}^3 = h/4a_4 \quad . \tag{3.27}$$

These results (3.25) - (3.27) agree with the mean field theory [see (1.26) - (1.37)].

So far we have only looked at the most probable spin configuration. Now we turn to the fluctuations around the most probable configuration in the Gaussian approximation. We shall discuss the cases $T > T_c$ and $T < T_c$ separately.

3. GAUSSIAN APPROXIMATION FOR $T > T_c$

In this case $a_2 > 0$ and we set $h = 0$ for simplicity. The most probable $\sigma(x)$ is zero. The Gaussian approximation is easily read off from (2.29) by keeping terms up to σ^2:

$$H[\sigma]/T \approx a_0 L^d + \sum_{k<\Lambda} \sum_{i=1}^{n} (a_2 + ck^2) \, |\sigma_{ik}|^2, \tag{3.28}$$

which is analogous to (3.7), and we identify

$$\frac{1}{2} \lambda_\ell^{-2} \quad \text{as} \quad (a_2 + ck^2)$$

and σ_{ik} as the normal coordinates q'_ℓ. The results (3.8), (3.9), and (3.10) now take the form

$$\langle \sigma_{ik} \rangle = 0 \,, \tag{3.29}$$

$$G(k) = \langle |\sigma_{ik}|^2 \rangle = \frac{1}{2}(a_2 + ck^2)^{-1} \,, \tag{3.30}$$

$$FL^d = a_0 L^d - \frac{1}{2} T \sum_{k<\Lambda} n \ln[\pi/(a_2 + ck^2)] \,, \tag{3.31}$$

respectively. Here F is the free energy per unit volume.

The quantity $\langle |\sigma_{ik}|^2 \rangle$ is what we called the correlation function $G(k)$. It is proportional to the differential scattering cross section as discussed in Chapter I. (See (1.11).) Since $a_2 = a'_2 (T - T_c)$, Eq. (3.30) gives

$$\lim_{T \to T_c} G(k) \propto k^{-2} \,, \tag{3.32}$$

$$\lim_{k \to 0} G(k) \propto (T - T_c)^{-1} \,, \tag{3.33}$$

where the second quantity is proportional to the magnetic susceptibility χ (see (1.13)).

By the definitions of the critical exponents η and γ (see (1.9) and (1.3)), we have

$$\eta = 0 \ ,$$
$$\gamma = 1 \ , \tag{3.34}$$

in view of (3. 32) and (3. 33).

The specific heat is obtained by differentiating the free energy twice with respect to T. Since a_0 is a smooth function of T, the only possible source of a singular dependence on T is through the vanishing of $a_2 + ck^2$ in the logarithm of (3. 31) when $T - T_c \to 0$ and $k \to 0$. Differentiating (3. 31) twice, we obtain

$$C = -T\partial^2 F/\partial T^2$$
$$= \frac{a_2'^2}{2} T^2 n(2\pi)^{-d} \int d^d k \, (a_2 + ck^2)^{-2} + \ell.s. \ . \tag{3.35}$$

We have replaced the sum over k by an integral over a sphere of radius Λ:

$$\sum_{k<\Lambda} \to L^d (2\pi)^{-d} \int d^d k \quad , \tag{3.36}$$

where the upper limit Λ in the k integral is understood. The symbol "ℓ.s." in (3. 35) means less singular terms, including integrals of $(a_2 + ck)^{-1}$, ln $(a_2 + ck)$, and non-singular terms. Since $a_2 = a_2'(T - T_c)$, a term is more singular if more powers of a_2 appear in the denominator. Note that the $(a_2 + ck^2)^{-1} = G(k)$ measures the fluctuation of the spins. For small $T - T_c$ long wavelength (small k) fluctuations are very large. These large fluctuations are what cause the integral of (3.35) to diverge at $T \to T_c$.

To extract this divergence, let us make the following change of integration variable, from k to k':

$$k = k'/\xi \tag{3.37}$$

$$\xi^{-1} \equiv (a_2/c)^{1/2}$$

$$= (a_2'/c)^{1/2} (T - T_c)^{1/2} \ . \tag{3.38}$$

Then we have

$$C = n \left[\frac{1}{2} (T a_2'/c)^2 (2\pi)^{-d} \int d^d k' (1 + k'^2)^{-2} \right] \xi^{4-d} + \ell.s.$$

$$\equiv C_0 \, \xi^{4-d} + \ell.s. \ . \tag{3.39}$$

The upper limit of the k' integral is $\Lambda\xi$, which goes to infinity as $T - T_c \to 0$ (i.e., $\xi \to \infty$). For $d < 4$, this integral converges and gives a numerical constant.

Restricting ourselves to cases with $d < 4$, we conclude from (3.29) that, for very small $T - T_c$,

$$C \propto (T - T_0)^{-\alpha} + \ell.s. \quad ,$$

$$\alpha = 2 - d/2 \ . \tag{3.40}$$

To sum up, the Gaussian approximation tells us that above the critical temperature T_c the susceptibility (3.23) and the specific heat diverge as T approaches T_c. The critical exponents $\gamma = 1$, $\eta = 0$, $\alpha = 2 - d/2$, are independent of the details of the Hamiltonian, since the parameters a_0, a_2', a_4, and c do not appear in them. The singular behavior does come out of a smooth block Hamiltonian. We shall examine the accuracy of the Gaussian approximation after we discuss the case $T < T_c$.

4. GAUSSIAN APPROXIMATION FOR $T < T_c$

For $T < T_c$ additional features appear as a result of a nonzero $\bar{\sigma}$ even when $h \to 0$. We shall keep a finite but very small h pointing in the 1 direction. Thus $(\tilde{\sigma}_1)_{k=0} = L^{d/2}\bar{\sigma}$, with $\bar{\sigma}$ given by (3.22) and (3.23), and $\tilde{\sigma}_{ik} = 0$ for $k \neq 0$ or $i \neq 1$. Expanding $H[\sigma]/T$ in powers of $\sigma - \tilde{\sigma}$ and keeping all terms up to the second power, we obtain using (2.29) and (3.15),

$$H[\sigma]/T \approx H[\tilde{\sigma}]/T + \sum_{\substack{k<\Lambda \\ k\neq 0}} \Big[(4a_4 m^2 + h/2m + ck^2)|\sigma_{1k}|^2$$

$$+ (h/2m + ck^2) \sum_{i=2}^{n} |\sigma_{ik}|^2 \Big] , \qquad (3.41)$$

where $H[\tilde{\sigma}]/T$ is given by (3.24), and m given by (3.22). Again this expression has the form of (3.7) and we identify $\frac{1}{2}\lambda_\ell^{-1}$ as $h/2m + 4a_4 m^2 + ck^2$ and $h/2m + ck^2$, and q'_ℓ as $\sigma_{ik} - \tilde{\sigma}_{ik}$. The results corresponding to (3.8) - (3.10) are

$$\langle \sigma_{ik} \rangle = \tilde{\sigma}_{ik} , \qquad (3.42)$$

$$G_I(k) \equiv \langle |\sigma_{1k}|^2 \rangle = \frac{1}{2} (2a'_2(T_c - T) + ck^2)^{-1} , \qquad (3.43)$$

$$G_\perp(k) \equiv \langle |\sigma_{ik}|^2 \rangle = \frac{1}{2} (h/2m + ck^2)^{-1} , \quad i = 2, \ldots, n , \qquad (3.44)$$

$$FL^d = H[\sigma]/T - \frac{1}{2} T \sum_{k<\Lambda} \{ \ln[\pi/(2a'_2(T_c - T) + ck^2)]$$

$$+ (n-1) \ln[\pi/ck^2] \} . \qquad (3.45)$$

In (3.43) and (3.45) we have set $h = 0$ and written $2a'_2(T_c - T)$ for $4a_4 m^2$.

We see that σ_{1k}, which will be referred to as the amplitude of the longitudinal mode (parallel to $\bar{\sigma}$), behaves very differently from σ_{ik}, $i = 2, \ldots, n$, which will be referred to as the amplitudes of transverse modes (perpendicular to $\bar{\sigma}$). In the limit $k \to 0$, $G_1(k)$ equals χ (given by (3.25)), but $G_\perp(k)$ becomes m/h, which blows up in the limit $h \to 0$. The result $G_\perp(0) = m/h$ is a consequence of isotropy in spin space, as will be discussed in Sec. IX.7.

The specific heat is again obtained via $C = -T\partial^2 F/\partial T^2$ from (3.45). The first term in (3.45) gives a finite but discontinuous contribution to C as noted before [see (3.26)]. The last term of (3.45), which comes from the transverse modes, does not contribute. The second term, which comes from the longitudinal modes, has the same form as the second term in (3.31) but with $2a_2'(T_c - T)$ replacing $a_2 = a_2'(T - T_c)$. For this term, we can go through the same calculation as (3.35) - (3.39) to obtain

$$C = 2^{\frac{d}{2}-2}\left[\frac{1}{2}(Ta_2'/c)^2 (2\pi)^{-d}\int d^dk'(1+k'^2)^{-2}\, c^{-2}\right]\xi^{4-d} + \ell.s.$$

$$\equiv C_0'\,\xi^{4-d} + \ell.s. \qquad (3.46)$$

Here we have extended (3.38), the definition of ξ, to

$$\xi^{-1} = (a_2'/c)^{1/2} \, |T - T_c|^{1/2} \; . \tag{3.47}$$

Note that what is in the brackets [...] in (3.46) is the same as what is in them in (3.39), but, instead of n in front of the brackets, we have $2^{\frac{d}{2}-2}$ in (3.46).

5. THE CORRELATION LENGTH AND TEMPERATURE DEPENDENCE

The singular temperature dependence of the quantities examined above can be neatly summarized in terms of ξ defined by (3.47). For $T > T_c$, Eqs. (3.30) and (3.39) can be written as

$$G(k) = A\xi^2/(1 + k^2\xi^2) \; ,$$

$$\chi = G(0) = A\xi^2 \; , \tag{3.48}$$

$$A = c/2 \; ,$$

$$C = C_0\,\xi^{4-d} + \ell.s. \tag{3.49}$$

For $T < T_c$ Eqs. (3.43) and (3.46) assume the form

$$G_1(k) = A'\xi^2/(1 + k^2\xi^2/2) \quad ,$$

$$\chi = G_1(0) = A'\xi^2 \quad , \tag{3.50}$$

$$A' = c/4 \quad ,$$

$$C = C_0'\,\xi^{4-d} + \ell.s. \tag{3.51}$$

The constants C_0, C_0' are given by (3.39) and (3.46), respectively.

We shall call ξ the <u>correlation length.</u> It measures the distance over which spin fluctuations are correlated. This is more easily seen in coordinate space. Fourier transforming $G(k)$, we get

$$\langle \sigma(x+r)\,\sigma(x)\rangle = (2\pi)^{-d}\int d^dk\; G(k)\, e^{ik\cdot r}$$

$$= A\xi^{2-d}(2\pi)^{-d}\int d^dk'(1+k'^2)^{-1}\, e^{ik'\cdot r/\xi} \; . \tag{3.52}$$

The upper limit of the k' integral is $\Lambda\xi$ as in (3.39). For $d = 3$, the integral (3.52) is elementary, and we find

$$\langle \sigma(x+r)\,\sigma(x)\rangle \propto \frac{e^{-r/\xi}}{r} \; . \tag{3.53}$$

Thus the correlation of spins persists over a distance ξ. For other values of d, the integral will give other functions, but the same physical picture remains. We can identify ξ as a measure of the average size of the spin patches mentioned in Chapter I. Anticipating subsequent developments, we introduce the exponent ν to characterize the divergence of this size:

$$\xi \propto |T - T_c|^{-\nu} . \tag{3.54}$$

Here in the Gaussian approximation, we have

$$\nu = \frac{1}{2} . \tag{3.55}$$

In (3.48) - (3.51) it is clear that the singular behavior of these quantities for vanishing $|T - T_c|$ and k can be viewed as a result of $\xi \to \infty$. That is, as far as the singular temperature dependence is concerned, ξ is the only relevant length. The role of ξ is evident in the calculation of the specific heat. The change of variable $k = k'/\xi$ in (3.37) enabled us to obtain the singular part of C. The factor ξ^{4-d} in (3.39) follows directly. The $d^d k$ in the free energy per unit volume (i.e., per d powers of length) gives ξ^{-d}.

Differentiating with respect to T twice gives $(\xi^{1/\nu})^2$ since each $T - T_c$ accounts for $\xi^{-1/\nu}$. Thus

$$C \propto \xi^{-d+2/\nu} \propto |T - T_c|^{-(2-d\nu)} \tag{3.56}$$

and

$$\alpha = 2 - d\nu \ . \tag{3.57}$$

Here $\nu = 1/2$. The counting powers of ξ is the basis of the scaling hypothesis which we shall explore in Ch. IV.

6. SUMMARY OF RESULTS AND THE GINZBURG CRITERION

The exponents in the Gaussian approximation can be read off from (3.25) - (3.27), (3.32), (3.39), and (3.46). We have

$$\gamma = \gamma' = 1 \ ,$$

$$\eta = 0 \ ,$$

$$\delta = 3 \ ,$$

$$\beta = 1/2 \ ,$$

$$\alpha = \alpha' = 2 - d/2 \ . \tag{3.58}$$

The ratios of the coefficients

$$A/A' = 2 \,,$$

$$C_0/C_0' = n\,2^{2-d/2} \tag{3.59}$$

are obtained from (3.39), (3.46), (3.48), and (3.50). These results are independent of the specific values of a_2', a_4, and c. To this extent, (3.58) and (3.59) exhibit universality.

The Gaussian approximation gets the same exponents as the mean field theory except for the specific heat. In addition to a discontinuity at T_c there is a divergence for $d < 4$. This divergence is a result of fluctuations of modes of small k. In the mean field theory all modes of $k \neq 0$ are neglected. We can get a rough idea of how important the fluctuations are by comparing the size of the discontinuity ΔC [see (3.26)] to the size of the divergent term given by (3.39) and (3.46). The ratio is

$$C_0 \xi^{4-d} / \Delta C \sim [\zeta_T / |1 - T/T_c|]^{2-d/2} \tag{3.60}$$

where

$$\zeta_T = [(2\pi \xi_o)^{-d}/\Delta C]^{2/(4-d)} \tag{3.61a}$$

$$\sim (a_4/c^{d/2})^{2/(4-d)} \Big/ a_2' T_c \ , \tag{3.61b}$$

$$\xi_o \equiv (c/a_2' T_c)^{1/2} \ . \tag{3.62}$$

Thus when the temperature is close to T_c within a range $\zeta_T T_c$, the fluctuations are expected to be important. The smaller ζ_T is, the smaller this range will be. This qualitative criterion was pointed out by Ginzburg (1960).

The motivation for Ginzburg's study of the fluctuations was to understand why a finite discontinuity in specific heat was observed for some critical points (in superconductors, and in some ferromagnetic and ferroelectric materials) but a singularity was observed in addition to the discontinuity for other critical points. Ginzburg argued that such an observed difference was actually quantitative, not qualitative. When ζ_T happened to be so small that the range $\zeta_T T_c$ could not be resolved experimentally, then the divergence produced by the fluctuations would not be seen.

The length ξ_o defined by (3.62) estimates the correlation length for temperatures far away from T_c [see Eq. (3.47)]. It can be deduced from neutron scattering data, for example. The discontinuity ΔC can be obtained directly from specific heat data. Estimates of ζ_T for various materials can be found in the paper of Ginzburg (1960).

When we apply the above results to a superconductive critical point or to the λ point of He^4, the length ξ_o is not directly measurable but can be estimated. In the former case, ξ_o measures the size of a "Cooper pair" (a few thousand Å), and ΔC is a few joules/ccK°. This makes $\zeta_T \sim 10^{-15}$, which makes the range $\zeta_T T_c \sim 10^{-15}$°K too small to observe at present. In the latter case, ξ_o ($\sim$ a few Å) is the deBroglie wavelength of a helium atom at $T_c \approx 2.2$°K, and ΔC is about a joule/cc°K. Thus $\zeta_T \sim 0.3$ for the λ point.

For antiferromagnetic critical points, the divergence of the specific heat is very pronounced. The length ξ_o is of the order of a lattice spacing and ΔC is not very well defined. The smallness of ξ_o and the pronounced divergence illustrate the importance of fluctuations.

7. FLUCTUATION AND DIMENSION

From the above results for the specific heat $C \propto |T - T_c|^{(d-4)/2}$, it is evident that for $d > 4$, C does not diverge at T_c. This is an indication that 4 is a special dimension. This point will come up again later.

It is also evident that the smaller d, the more serious the divergence. We shall show that for $d \le 2$ the fluctuations become so large that for $n > 1$ a nonzero order parameter $\langle \sigma(x) \rangle$ can no longer exist in the limit of infinite volume and zero external field h. For $d = 1$, $\langle \sigma(x) \rangle$ must vanish in the same limit even for $n = 1$.

Let h point in the 1 direction as before. Then the correlation function of the transverse components σ_i, $i = 2, \ldots, n$ in coordinate space is

$$\langle \sigma_i(x+r)\, \sigma_i(x) \rangle = (2\pi)^{-d} \int d^d k\; G_\perp(k)\, e^{ik\cdot r}$$

$$= (2\pi)^{-d} \int_0^\Lambda dk\; k^{d-1} (h/m + 2ck^2)^{-1}\, e^{ik\cdot r} . \tag{3.63}$$

We have used (3.44) for $G_\perp$. Since the magnitude of $\sigma_i(x)$ is bounded, the integral in (3.63) must be finite for any value of h. In the limit $h \to 0$ and with m assumed finite, (3.63) diverges as $(h/m)^{(d-2)/2}$ for $d < 2$ and diverges like $\ln(h/m)$ for $d = 2$. Thus m cannot be non-zero, i.e., $\langle \sigma(x) \rangle$ must vanish for $n \geq 2$, $d \leq 2$, as $h \to 0$.

Qualitatively the situation is as follows. When $h = 0$, $H[\sigma]/T$ is isotropic in the spin space. The most probable configuration $\tilde{\sigma}$ is uniform with magnitude m, but all directions are equally probable. Now imagine a configuration $\sigma(x)$ with $|\sigma(x)| = m$ but with the direction of $\sigma(x)$ turning gradually as x changes. This configuration would be just as probable as $\tilde{\sigma}$ were it not for the spatial variation of direction which contributes to the $(\nabla\sigma)^2$ term of $H[\sigma]/T$. If the direction of $\sigma(x)$ varies periodically with a very long wavelength, $(\nabla\sigma)^2$ will be very small. For such a configuration the mean spin is zero because of the change of direction. There are many such configurations with very long wavelengths. In the limit of a large volume, the number of these configurations is very large. It is a matter of competition between the number of such long wavelength configurations and the suppression of probability due to $(\nabla\sigma)^2$. For $d > 2$, the latter wins. For $d \le 2$ the former wins, and $\langle \sigma(x) \rangle$ thus vanishes.

Rigorous proofs of the above conclusion have been worked out by Hohenberg (1967), and Mermin and Wagner (1966).

Clearly, the above conclusion applies only for $n \ge 2$ under the condition of continuous rotational invariance in spin space. For $n = 1$, the Ising model of $d = 2$, for example, does have a nonzero $\langle \sigma(x) \rangle$ for $h \to 0$ below T_c.

For $d = 1$ even in the case $n = 1$, $\langle \sigma(x) \rangle$ has to vanish in the limit $h \to 0$ and infinite volume [see Landau and Lifshitz (1953)]. The arguments goes as follows. Figure 3.2a shows the most probable configuration $\tilde{\sigma}(x) = m_o$. For $h \to 0$, $\sigma(x) = -m_o$ becomes equally probable since $H[\sigma]/T$ is unchanged under a change of sign of σ. Figure 3.2b shows a configuration with a "kink." Apart from the region near the kink, $\sigma(x)$ is either m_o or $-m_o$. Thus $H[\sigma]/T$ for such a configuration differs from $H[\tilde{\sigma}]/T$ by an amount

$$c \int (\nabla\sigma)^2 \, dx \equiv K/T \tag{3.64}$$

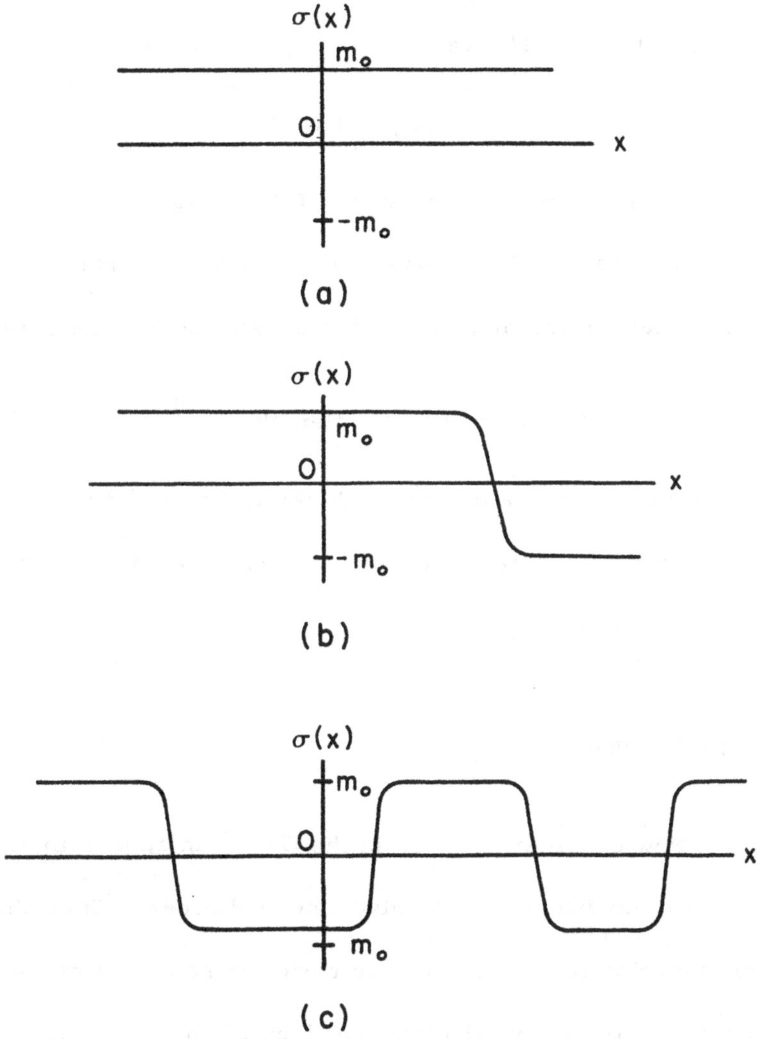

Figure 3.2. (a) The uniform and most probable spin configuration. (b) A configuration with one kink. (c) A configuration with many kinks.

integrated over the region of the kink where $\nabla\sigma \neq 0$. K is thus the "kink energy." Figure 3.2c shows a configuration with many kinks. The value of $H[\sigma]/T$ is then

$$H[\sigma]/T \sim H[\tilde{\sigma}]/T + NK/T \ ,$$

where N is the number of kinks, if the kinks are sufficiently far apart. The locations of the kinks are random. Thus we have a gas of kinks with a density proportional to

$$\text{(number of kinks)/length} \propto e^{-K/T} \ . \tag{3.65}$$

As long as $T \neq 0$ the density of kinks is finite and the sign of $\sigma(x)$ will alternate at random. Thus the average $\sigma(x)$ must be zero.

8. DISCUSSION

Now consider $d > 2$. In the Gaussian approximation, the spin fluctuation is kept only to second order. Thus the approximation is good if the fluctuation is small. Unfortunately, near a critical point the fluctuation is very large and the Gaussian approximation is therefore expected to be poor. In fact the exponents (3.58) do not agree well with the observed ones. One needs to include higher order terms, i.e., to include the "interaction" between the plane wave modes, which are independent modes in the Gaussian approximation. The term responsible for the interaction is the $a_4 \sigma^4$ term in the Ginzburg-Landau Hamiltonian.

Evidently the closer T is to T_c, the larger the fluctuation is and the more important it is to include the effect of interaction among modes. The criterion for how close T must be to T_c for the fluctuation to be large is just the Ginzburg criterion discussed above. That is, the criterion for the breakdown of mean field theory (fluctuations ignored) is the same as for the breakdown of the Gaussian approximation, where fluctuations are included but in a truncated manner. The value of ζ_T given by (3.61) estimates the effect of interaction. The larger a_4 is, the larger ζ_T is.

In Chapter IX we shall examine the details of a perturbation expansion in powers of a_4 for calculating various quantities. It will be shown that the expansion is actually an expansion in powers of

$$a_4 |T - T_c|^{d/2 - 2} \propto [\zeta_T / |1 - T/T_c|]^{2 - d/2}. \qquad (3.66)$$

Therefore, for $d < 4$, such an expansion will not converge for $|T - T_c| \to 0$. It is very difficult to improve the solution beyond the Gaussian approximation without summing the whole perturbation series, i.e., solving the model exactly. This difficulty does not appear for $d > 4$ as (3.66) indicates.

IV. THE SCALING HYPOTHESIS

SUMMARY

We introduce the scaling hypothesis and summarize its consequences with regard to critical exponents. The ideas of scale transformation and scale dimension are discussed.

1. THE CORRELATION LENGTH AND THE SCALING HYPOTHESIS

The scaling hypothesis is a plausible conjecture. It makes no reference to any model and has been very successful in correlating observed data. The basic idea of this hypothesis is that the long range correlation of spin fluctuations near T_c is responsible for all singular behavior.

There are many different ways of formulating this hypothesis mathematically. The following is just one of them.

Recall that a characteristic length ξ, the correlation length, emerged naturally in the Gaussian approximation of Chapter III. It is a measure of the range over which spin fluctuations are correlated, or the average size of a spin patch in which a sizeable fraction of spins point in the same direction. We showed that the singularities in various physical quantities at T_c could be understood as a result of the divergence of ξ at T_c.

A correlation length can be defined without reference to the Gaussian approximation or any specific model. Since $G(k) = \langle |\sigma_k|^2 \rangle$, as a function of k, peaks sharply around $k = 0$, we can identify ξ^{-1} as the width of the peak. $G(k)$ is directly measurable by scattering experiments. More precisely we can define in the absence of an external field,

$$\xi^{-2} = -\frac{1}{2} G^{-1}(0) \left(d^2 G(k)/dk^2\right)_{k=0}, \quad \text{at } h = 0. \tag{4.1}$$

Empirical data shows that ξ diverges at T_c, and we use an exponent ν to characterize this divergence for $|T - T_c| \to 0$:

$$\xi \propto |T - T_c|^{-\nu}, \quad T > T_c$$

$$\propto |T - T_c|^{-\nu'}, \quad T < T_c \tag{4.2}$$

For $T < T_c$ the correlation length must be defined with $G_1(k) = \langle |\sigma_{1k}|^2 \rangle$ if $n > 1$. Here we shall assume that $\nu = \nu'$. This is an additional assumption which will be shown to be plausible later. The proportionality constants in the two cases in (4.2) are, in general, different.

The scaling hypothesis states that the divergence of ξ is responsible for the singular dependence on $T - T_c$ of physical quantities, and, as far as the singular dependence is concerned, ξ is the only relevant length.

This is a very strong hypothesis. It asserts that the large spin patches, but not the details over smaller scales, account for the physics of critical phenomena.

As an illustration of the use of the scaling hypothesis, consider the correlation function $G(k)$. Let $b_1, b_2, \ldots$ be various microscopic lengths, which are much smaller than ξ when $|T - T_c|$ is sufficiently small. Write $G(k)$ as a function of ξk, b_1/ξ, $b_2/\xi, \ldots$:

$$G(k) = f(k\xi\,,\ b_1/\xi\,,\ b_2/\xi\,,\dots)$$

$$= \xi^y\,(g(k\xi) + \text{higher powers of }\ \xi^{-1})$$

$$\approx\ \xi^y\ g(k\xi)\quad , \tag{4.3}$$

where we have expanded f in powers of the second, third, ... arguments and kept only the leading power. Let us clarify (4.3) further. What we did is to keep the first argument of the function $f(\zeta\,,\ \lambda_1\,,\ \lambda_2\,,\dots)$ fixed and consider the limits as $\lambda_1, \lambda_2,\dots$ become very small. This limit of f will be proportional to some powers of $\lambda_1, \lambda_2,\dots$, i.e.,

$$\lim_{\lambda_1,\lambda_2,\dots\to 0} f(\zeta,\ \lambda_1\,,\ \lambda_2\,,\dots) = \tilde{g}(\zeta)\,\lambda_1^{x_1}\,\lambda_2^{x_2}\,\dots$$

where the proportionality constant is called $\tilde{g}(\zeta)$ and the powers $x_1, x_2,\dots$ of course are properties of f. Now write

$$-y = x_1 + x_2 + \dots$$

$$g(\zeta) = \tilde{g}(\zeta)\ b_1^{x_1}\,b_2^{x_2}\,\dots\quad .$$

We then obtain (4.3). The g must contain the right powers of microscopic lengths $b_1, b_2, \ldots$ and other parameters so that the unit of $\xi^y g$ comes out correctly as the unit of the correlation function $G(k)$. Note that k^{-1} is entirely arbitrary and $k\xi$ can assume any value. The scaling hypothesis says that the singular temperature dependence enters through ξ and any other dependence on T implicit in g is smooth and can be regarded as constant over a small temperature range around T_c.

The hypothesis does not give the value of y, nor does it tell what the function g is. But it does relate y to previously defined exponents. Since

$$G(0) = \xi^y g(0) \propto |T - T_c|^{-\nu y} , \tag{4.4}$$

we have

$$\nu y = \gamma , \tag{4.5}$$

by the definition of γ [see Eq. (1.3)] and the fact that $\chi/T = G(0)$. For $|T - T_c| \to 0$, $k \neq 0$, we have, by the definition of η [see Eqs. (1.8), (1.9), and (1.11)], $G(k) \propto k^{-2+\eta}$. This implies that

$$\lim_{k\xi \to \infty} g(k\xi) \propto (k\xi)^{-2+\eta} , \quad (4.6)$$

$$y = 2 - \eta = \gamma/\nu . \quad (4.7)$$

We have thus derived a relationship among the exponents η, γ, and ν. Such a relationship is often referred to as a "scaling law."

2. SCALE TRANSFORMATION AND DIMENSIONAL ANALYSIS

The application of the scaling hypothesis is facilitated by using the ideas of scale transformation and dimensional analysis, because there is only one length ξ to keep track of besides k^{-1} and L, the size of the system.

A scale transformation by a factor $s = 2$ doubles the unit of length by a factor of 2. If $s = 3$ the transformation triples the unit or reference of length, and similarly for other values of s. Under a scale transformation an interval in space Δx is changed to $\Delta x'$:

$$\Delta x \to \Delta x' = s^{-1} \Delta x \quad (4.8)$$

and a wave vector k changes according to

$$k \rightarrow k' = sk \quad . \tag{4.9}$$

In other words, in the new units, a space would appear shrunk to half and a wave vector enlarged to twice as large if $s = 2$. This is similar to a rotation or a translation of reference frame. In the new frame, objects appear rotated or shifted in position.

We define the scale dimension of Δx as -1 and that of k as $+1$, according to the power of s in (4.8) and (4.9). In general if, under a scale transformation

$$A \rightarrow A' = A s^{\lambda} \quad , \tag{4.10}$$

then the scale dimension of A is λ. The usual rules of dimensional analysis apply. For example, the scale dimension of $(\Delta x)^2$ is -2, that of volume is $-d$, and that of $k^{1.05} A$ is $1.05 + \lambda$. Subsequently, we shall use the word "dimension" for "scale dimension," for simplicity.

Clearly the dimension of the correlation length ξ is -1. What is the dimension of $G(k)$? According to (1.10), $G(k)$ is measured in units of (spin density)^2 $\times$ volume. But we have not defined the dimension of spin. On the other hand, (4.3) says that $G(k)$ is proportional to ξ^y times a function of $k\xi$, which has the dimension zero. Thus the dimension of $G(k)$ is $-y$. It follows that the dimension of spin density d_σ satisfies

$$2d_\sigma - d = -y$$

$$d_\sigma = \frac{1}{2}(d - y) = \frac{1}{2}(d - 2 + \eta) . \tag{4.11}$$

We are using the dependence on ξ of G to fix the dimension, since, under the scaling hypothesis, ξ is the only relevant length. Since $G(k) \propto k^{-2+\eta}$ for $\xi \to \infty$, we must have $-2 + \eta$ as the dimension of $G(k)$. Thus $-2 + \eta$ must equal $-y$. Thus the scaling law (4.7) is simply a condition for the consistency in the dimension of $G(k)$.

If the dimension of a quantity is known, the dependence on ξ, and hence on $|T - T_c|$, follows. For example, the total free energy is not changed under a scale transformation, so it has dimension 0. The free energy per unit volume F thus has a dimension d. It follows that as $T \to T_c$,

$$F \propto \xi^{-d} \propto |T - T_c|^{\nu d} \tag{4.12}$$

$$C = -T\partial^2 F/\partial T^2 \propto |T - T_c|^{\nu d - 2} \tag{4.13}$$

$$\alpha = 2 - \nu d \ . \tag{4.14}$$

The last line follows from the definition (1.4) of the specific heat exponent α.

Another example is the magnetization m below T_c. Since m is the average spin density, it has dimension $\frac{1}{2}(d - 2 + \eta)$ according to (4.11). Therefore, for $T \to T_c$,

$$\begin{aligned} m &\propto \xi^{-\frac{1}{2}(d - 2 + \eta)} \\ &\propto |T - T_c|^{\frac{1}{2}\nu(d - 2 + \eta)} , \end{aligned} \tag{4.15}$$

$$\beta = \frac{1}{2}\nu(d - 2 + \eta) , \tag{4.16}$$

by the definition (1.1) of β.

Since, in the presence of an external field h, $m = -\partial F/\partial h$, the dimension of h must be

$$d_h = d - d_\sigma = \frac{1}{2}(d + 2 - \eta) \ . \tag{4.17}$$

For $T = T_c$, $h \neq 0$, we know that m is still finite. It can depend only on some power of h since ξ is infinite. To make the dimension right, we must have

$$m \propto \left(h^{1/d_h}\right)^{d_\sigma} \quad . \tag{4.18}$$

According to the definition of δ (1.2), $m \propto h^{1/\delta}$, we have

$$\delta = d_h/d_\sigma = (d + 2 - \eta)/(d - 2 + \eta) \quad . \tag{4.19}$$

The dimension of $h\xi^{d_h}$ is zero; therefore we have for $T \neq T_c$ and $h \neq 0$,

$$m = \xi^{-d_\sigma} w_\pm\left(h\xi^{d_h}\right) \tag{4.20}$$

where $\pm$ refers to $T > T_c$ or $T < T_c$ and $w_\pm$ are functions which have dimension zero. Similarly, the free energy per volume F should take the form

$$F = \xi^{-d} f_\pm\left(h\xi^{d_h}\right) . \tag{4.21}$$

Now we examine the assumption $\nu = \nu'$ which led to $\alpha = \alpha'$ and $\gamma = \gamma'$. At a fixed nonzero h, we expect that F is a smooth function of $T - T_c$ because the critical point is at $T = T_c$ <u>and</u> $h = 0$. Therefore, we write

$$F = F_o(h) + F_1(h)(T - T_c)$$

$$+ \cdots \quad . \tag{4.22}$$

In order that (4.21) is consistent with (4.22), we must have

$$F = \xi^{-d} f_+\left(h\xi^{d_h}\right) = \xi^{-d}\left(h\xi^{d_h}\right)^{d/d_h}\left[A_+ + B_+\left(h\xi^{d_h}\right)^{-1/d_h \nu} + \cdots\right]$$

$$= h^{d/d_h}\left[A_+ + B_+' \, h^{-1/d_h \nu} (T - T_c) + \cdots\right] \tag{4.23}$$

for $T > T_c$, where $B_+' \propto B_+$, and

$$F = h^{d/d_h}\left[A_- + B_-' \, h^{-1/d_h \nu'} (T - T_c) + \cdots\right] \tag{4.24}$$

for $T < T_c$. Note that

$$T - T_c \propto \xi^{-1/\nu}, \quad \text{for} \quad T > T_c$$

$$\propto -\xi^{-1/\nu'}, \quad \text{for} \quad T < T_c \; . \tag{4.25}$$

If $\nu \neq \nu'$, it would not be possible for the $T - T_c$ terms in (4.23) and (4.24) to agree. Thus the assumption $\nu = \nu'$ is quite reasonable. We summarize the above results in Table 4.1.

Table 4.1

Some consequence of the scaling hypothesis

(a) Scaling Dimensions

ξ	k	σ_k	$\sigma(x)$	h	F
-1	1	$-1+\eta/2$	$\frac{1}{2}(d-2+\eta)$	$\frac{1}{2}(d+2-\eta)$	d

(b) Scaling Laws

$$\nu = \nu' = \gamma/(2-\eta)$$

$$\alpha = \alpha' = 2 - \nu d$$

$$\beta = \frac{1}{2}\nu(d-2+\eta)$$

$$\delta = (d+2-\eta)/(d-2+\eta)$$

3. DISCUSSION

The dimensional analysis given above is very ad hoc. We know that when the unit of length is changed, all lengths, large or small, are affected. Saying that only large lengths are affected is clearly wrong; there must be something else which we have missed if the above arguments are to have meaning. This "something else" will be taken up when we discuss the renormalization group in the subsequent chapters. For the moment however let us just check how the consequences of the scaling hypothesis fit empirical data.

Using the available data listed in Tables 1.2 and 1.3, some nontrivial checks can be made. For example, the first three ferromagnetic materials in Table 1.2 share the same exponents within experimental error. Taking $\gamma = 1.33$, $\eta = 0.07$, we obtain $\nu = 0.69$ using the first relation in Table 4.1b. The rest of Table 4.1b gives $\alpha = -0.07$, $\beta = 0.37$, $\delta = 4.6$, which is in agreement with the data in Table 1.2 to within 10%. From the data for the liquid-gas critical points in Table 1.3 we obtain $\nu = 0.64$ using $\gamma = 1.20$, $\eta = 0.11$. Then Table 4.1b gives $\alpha = 0.08$, $\beta = 0.355$, $\delta = 4.40$. Apart from α the agreement with experiment is excellent.

Such agreement with experimental data must be considered as a great success of the scaling hypothesis, especially in view of the simplicity and the crudeness of the hypothesis. A theory of critical phenomena would not be satisfactory unless it provided a basis for understanding the scaling hypothesis.

V. THE RENORMALIZATION GROUP

SUMMARY

The renormalization group (RG) is defined as a set of symmetry transformations. The motivation for the construction and use of the RG is the same as that for other symmetry transformations like translations and rotations. We define the RG as essentially a combination of the Kadanoff transformation and the scale transformation. The physical basis for the essential ingredients of the RG and alternative definitions are discussed.

1. MOTIVATION

The study of symmetry transformations such as rotations, translations, and isotopic spin rotations, has proven extremely useful in various branches of physics. Many fundamental concepts are associated with the symmetry properties of physical systems — e.g., quantum numbers, selection rules, and multiplicity of degeneracies. Experience shows that a set of symmetry transformations is useful for those particular systems which are at least approximately invariant under symmetry transformations of that set. For example, for an atom, one encounters a Hamiltonian

$$H = H_0 + H_1 \tag{5.1}$$

where H_0 is invariant under rotations around the nucleus and H_1 is a small perturbation. H_1 might be due to an external magnetic field, or a quadruple field. The perturbation is generally not invariant under rotations but often transforms simply and can be classified according to the representations of the rotation group. The machinery of perturbation theory together with systematic symmetry arguments (group theory) determines the general patterns of atomic phenomena almost completely, without the need to dig into the details of H_0 and H_1.

When the detailed dynamics of a physical system is too complicated or unknown, and there is no obvious symmetry, one then tries to construct or guess at some kind of symmetry transformations. Hopefully certain patterns of experimental data can be at least approximately interpreted as consequences of the properties of the physical systems under such symmetry transformations. Examples of the success of such an approach are the isotopic spin rotations and SU(3) in nuclear and particle physics.

Now we face a complicated macroscopic system near its critical point. Naturally we ask: Under what symmetry transformations is such a system invariant, at least approximately? It would be very nice to have a simple set of transformations under which the system at $T = T_c$ is invariant [describable by some Hamiltonian like H_0 of Eq. (5.1)] and if $T \neq T_c$, only a small perturbation proportional to $T - T_c$, although not invariant needs to be considered [like the H_1 term in Eq. (5.1)]. One would then hope that somehow the universal critical exponents would appear as symmetry properties, like the angular momentum quantum numbers that come out of the rotation symmetry.

The fact that microscopic details seem to make very little difference in critical phenomena strongly suggests that the observed general pattern should be interpetable as symmetry properties, even though it is not clear a priori what the desired symmetry transformations are.

There has been some success in constructing the desired transformations, although nothing as neat and simple as the rotation group has been obtained. We shall call the desired transformations the renormalization group, abbreviated as the RG. There is no obvious motivation for this name, but we shall see some justification for it at a technical level later. There are many features of the RG not shared by the familiar symmetry transformations. Its application is not as simple as we suggested above. For example, a system at a critical point will not be simply invariant under the RG but will have certain special properties.

2. DEFINITION OF THE RENORMALIZATION GROUP (RG)

There is room for a great deal of flexibility in defining the RG. There have been no clearcut criteria as to what the optimum definition should be, and there are mathematical ambiguities which have not been completely resolved. So far various applications of the RG have indicated that as long as certain basic ingredients are included in the definition, the physical results derived will not depend on the fine details of the definition. We shall first illustrate these basic ingredients by defining the RG in the simplest manner. Variations and technical points will be discussed later.

Consider a set of probability distributions for the L^d/b^d random variables σ_x, the block spins. Note that, in general, to discuss transformations, we need to consider a sufficiently large set of probability distributions and examine how each of them transforms to another. We need labels, or parameters, to keep track of these probability distributions and to formulate transformations. The choice of parameters is to a great extent arbitrary. For simplicity consider the set of probability distributions of the Ginzburg-Landau form (2.30) with $h = 0$,

$$P \propto e^{-\mathcal{H}} \tag{5.2a}$$

$$\mathcal{H} = b^d \sum_x \left[u_2 \sigma_x^2 + u_4 \sigma_x^4 + \frac{1}{2} b^{-2} \sum_y{}' c(\sigma_x - \sigma_{x+y})^2 \right] . \tag{5.2b}$$

We assume $n = 1$ for simplicity. We can use the triplet of parameters

$$\mu = (u_2, u_4, c) \tag{5.3}$$

to label the probability distributions. Different values of parameters form a three dimensional space which we shall call the <u>parameter space</u>. Every probability distribution of the form (5.2) is represented by a point in this parameter space. Any transformation taking a probability distribution P to another P' is represented by a transformation of a point μ to another point μ'.

Now we define a transformation R_s for any integer $s \geq 1$:

$$\mu' = R_s \mu , \tag{5.4}$$

in the parameter space to represent the transformation of probability distribution $P \propto e^{-\mathcal{H}}$ to $P' \propto e^{-\mathcal{H}'}$ according to the following steps.

Step (i): Apply the Kadanoff transformation K_s to $\mathcal{H}$ to obtain

$$\mathcal{H}''[\sigma] = K_s \mathcal{H}[\sigma] . \tag{5.5}$$

This is a coarse graining procedure as discussed in Chapter II [see Eqs. (2.22)-(2.24)]. Now $P'' \propto e^{-\mathcal{H}''}$ is a probability distribution for $L^d/(sb)^d$ block spins defined on blocks of size (sb). Each new block spin σ_x is the mean of the s^d old block spins and x is the mean of the s^d position vectors of the old blocks within a new block. The Kadanoff transformation (5.5) downgrades the spatial resolution of spin variations to sb. Note that as long as we are not interested in variations within a distance sb, $P'' \propto e^{-\mathcal{H}''}$ is sufficient and is equivalent to P as noted in Chapter II.

Step (ii): Relabel the block spins σ_x in $\mathcal{H}''[\sigma]$ and multiply each of them by a constant λ_s to obtain

$$\mathcal{H}'[\sigma] = (\mathcal{H}''[\sigma])_{\sigma_x \to \lambda_s \sigma_{x'}} , \tag{5.6a}$$

with

$$x' = x/s . \tag{5.6b}$$

Clearly this step shrinks the size of the system by s, and the block size sb is shrunk to b, back to the original size.

The Steps (i) and (ii) can be explicitly written as

$$e^{-\mathcal{H}'[\sigma]} = \int e^{-\mathcal{H}[\sigma'']} \prod_{x'} \delta\left(\lambda_s \sigma_{x'} - s^{-d} \sum_y^x \sigma''_y\right) \prod_y d\sigma''_y , \tag{5.6c}$$

i.e., the same as (2.23) defining the Kadanoff transformation, except that σ_x is renamed $\lambda_s \sigma_{x'}$. Recall that $s^{-d} \sum_y^x \sigma''_y$ is the mean of s^d block spins in a new block centered at x.

Now we write $\mathcal{H}'$ in the Ginzburg-Landau form:

$$\mathcal{H}'[\sigma] = b^d \sum_{x'} \left(\frac{1}{2} c' b^{-2} \sum_{y'} (\sigma_{x'} - \sigma_{x'+y'})^2 + u_2' \sigma_{x'}^2 + u_4' \sigma_{x'}^4 \right). \tag{5.7}$$

The sum over x' is taken over

$$(L'/b)^d = (L/sb)^d \tag{5.8}$$

blocks. The new parameters are identified as

$$\mu' = (u_2', u_4', c') \quad , \tag{5.9}$$

and thus define R_s of (5.4). The set of transformations $\{R_s, s \geq 1\}$ is called RG. It is a semigroup, not a group, since inverse transformations are not defined. It has the property that

$$R_s R_{s'} = R_{ss'} \tag{5.10}$$

only if the factor λ_s in (5.6a) has the s dependence

$$\lambda_s = s^a \quad , \tag{5.11}$$

where a is independent of s. This is because $K_s K_{s'} = K_{ss'}$ [see Eqs. (2.23) - (2.26)], but $\lambda_s \lambda_{s'} = \lambda_{ss'}$ only if (5.11) is satisfied.

In short R_s is essentially a coarse graining followed by a change of scale. There is no net change of the block size b. The two steps are the basic ingredients of the RG. They are suggested by the success of the scaling hypothesis discussed in the previous chapter. We want to construct transformations which hopefully can explain the results of the scaling hypothesis. Therefore we want the RG to resemble a scale transformation. As we mentioned in Chapter IV all lengths, large or small, are affected by a change of scale and ignoring this effect on lengths shorter than ξ was an ad hoc approximation. If we take into account the effect on all lengths, we shall never find invariance under scale transformations even if $\xi \to \infty$. The block size will shrink, in particular. The coarse graining, Step (i), serves the purpose of keeping the block size unchanged. It has the effect of enlarging the block size and balances the effect of shrinkage due to the scale change of Step (ii). The size of the system L changes to $L' = L/s$ in Step (ii). However, since quantities of interest are almost always independent of L for $L \to \infty$, we exclude L from μ. It thus appears possible, but not guaranteed, that there are probability distributions represented by points invariant under R_s in the parameter space. Such points will be called fixed points of R_s and will play a major role in all subsequent discussions. In general one needs to adjust the factor λ_s in such a way that R_s will have a fixed point.

Note that additive constants in $\mathcal{K}$ and $\mathcal{K}'$ have not been kept track of and are not counted as parameters in μ and μ'. As far as the probability distributions are concerned, such constants are of no consequence since the probability distributions have to be normalized anyway. However, if one wants to calculate the free energy, then additive constants do make a big difference and we must keep track of these constants. We shall return to this point later in this chapter.

Before proceeding further we need to answer an important question. That is, whether $\mathcal{H}''$ of (5.5) and hence $\mathcal{H}'$ of (5.6) actually have the Ginzburg-Landau form taken for granted in (5.7). In general the answer is no, as we shall see explicitly in examples in Chapter VII. The Kadanoff transformation generates additional terms proportional to σ_x^6, σ_x^8, $(\sigma_x - \sigma_{x+y})^2 \sigma_x^2$, etc. In other words, the Ginzburg-Landau form is not general enough and the three dimensional parameter space is not adequate to represent probabilities involved in these transformations. We need more parameters, i.e., more entries in μ to specify other kinds of terms in $\mathcal{H}$. How to define the parameters is to a great extent arbitrary and one is guided by considerations of convenience and clarity. As an example we can write

$$\mathcal{H} = b^d \sum_x \Big[u_1\sigma_x + u_2\sigma_x^2 + u_3\sigma_x^3 + \cdots$$

$$+ \frac{1}{2} b^{-2} \sum_y{}' (\sigma_x - \sigma_{x+y})^2 (c + v_1\sigma_x + v_2\sigma_x^2 + \cdots)$$

$$+ \frac{1}{2} b^{-4} \sum_y{}' (2\sigma_x - \sigma_{x+y} - \sigma_{x-y})^2 (w + w_1\sigma_x + w_2\sigma_x^2 + \cdots) \Big]$$

$$+ \cdots \tag{5.12}$$

and define

$$\mu = (u_1, u_2, \ldots ; c, v_1, v_2, \ldots ; w, w_1, w_2, \ldots) . \tag{5.13}$$

Note that odd powers of σ_x will be present when there is no symmetry under the change of sign of σ_x (when there is an external field, for example). Generalization to $n > 1$ is straightforward.

One hopes that in practical calculations with the RG, only several entries in μ will be needed to get a good approximation. For example, in Chapter VII, we shall see that when d is very close to 4, the Ginzburg-Landau form itself is quite adequate.

One must not forget that R_s is simply a symmetry transformation, like a rotation or translation. It represents a change of labeling or references, not any change of physical content. The probability P' represented by $\mu' = R_s\mu$ is equivalent to P represented by μ as far as those spin variations which survived the Kadanoff transformations are concerned. Average values calculated with P' are simply related to those calculated with P. For example

$$\langle \sigma_x \rangle_P = \lambda_s \langle \sigma_{x'} \rangle_{P'}$$

$$\langle \sigma_x \sigma_{x+r} \rangle_P = \lambda_s^2 \langle \sigma_{x/s} \sigma_{(x+r)/s} \rangle_{P'} \, , \quad r > sb \, , \tag{5.14}$$

since the σ_x in P is renamed $\lambda_s \sigma_{x/s}$ in P′. Let us write the correlation function in wave vector space circulated with P as

$$G(k, \mu) = \langle |\sigma_k|^2 \rangle_P \quad . \tag{5.15}$$

It is the Fourier component of $\langle \sigma_x \sigma_{x+r} \rangle_P$ [see Eq. (1.11)]. From (5.14) we obtain

$$G(k, \mu) = \lambda_s^2 s^d \langle |\sigma_{sk}|^2 \rangle_{P'}$$

$$= \lambda_s^2 s^d G(sk, \mu') \quad . \tag{5.16}$$

Note that under the replacement $\sigma_x \rightarrow \lambda_s \sigma_{x'}$, the quantity

$$\sigma_k = L^{-d/2} b^d \sum_x e^{-ik \cdot x} \sigma_x \tag{5.17}$$

is replaced by

$$(sL')^{-d/2} s^d b^d \sum_{x'} e^{-ik \cdot sx'} \lambda_s \sigma_{x'} = s^{d/2} \lambda_s \sigma_{sk} \, . \tag{5.18}$$

This replacement accounts for the $\lambda_s^2 s^d$ in (5.16). The sum $b^d \sum_x$ can be thought of as $\int d^d x$, which is $s^d \int d^d x'$ (or $s^d b^d \sum_{x'}$, taken over blocks in the volume L'^d).

Equations (5.14) and (5.16) are simply the statement that the average of the transformed variables over the transformed probability distribution is the same as the average of original variables over the original probability distribution. This statement can be made for all symmetry transformations. The usefulness of symmetry transformations will be diminished if a transformation is so complicated that it cannot be made simply. This point will become clearer when we discuss some specific applications of the RG.

3. ALTERNATIVES IN DEFINING THE RG

Let us mention some alternative definitions of the RG.

The definition given above involves the combination of s^d blocks to form a new block in the Kadanoff transformation. Instead of using discrete blocks, we can use the Fourier components σ_k and carry out Step (i), i.e., the Kadanoff transformation, by integrating out σ_q for $\Lambda/s < q < \Lambda$ [see Eq. (2.24)]. In Step (ii) we replace the remaining σ_k by $\lambda_s s^{d/2} \sigma_{sk}$. Putting two steps together, we have

$$e^{-\mathcal{H}'} = \left[\int e^{-\mathcal{H}} \prod_{\Lambda > q > \Lambda/s} d\sigma_q \right]_{\sigma_k \to \lambda_s s^{d/2} \sigma_{sk}} \tag{5.19}$$

It is possible, and in practice often desirable or even necessary, to modify the definition of the Kadanoff transformation at a technical level. This point will be brought up in subsequent chapters.

The value of s can be any real number ≥ 1 in using the Fourier components for carrying out the Kadanoff transformation, while it must be a positive integer when combining blocks. We shall see that, for the purpose of application to critical phenomena, the useful property is $R_s R_{s'} = R_{ss'}$ and it matters very little whether the values of s are discrete or continuous. In fact, it is convenient to restrict s to $s = 2^\ell$, $\ell = 0, 1, 2, 3, \ldots$ and define

$$R_s = (R_2)^\ell . \tag{5.19'}$$

This definition guarantees $R_s R_{s'} = R_{ss'}$. We can regard R_2 as the "generator" of the RG. Once R_2 is worked out, R_s can be obtained by repeating R_2. Of course, setting $s = 3^\ell$, $R_s = (R_3)^\ell$ would be equally acceptable. If $R_{1.5}$ is available, setting $s = (1.5)^\ell$, $R_s = (R_{1.5})^\ell$ will qualify also. If $R_{1+\delta}$ for an infinitesimal δ can be conveniently defined, then

$$R_s = e^{(\ln s)\Gamma}$$

$$\Gamma = \lim_{\delta \to 0} \frac{1}{\delta} (R_{1+\delta} - 1) \tag{5.20}$$

is again a good definition of the RG. The quantity Γ is often referred to as the "infinitesimal generator" of the RG.

There is additional room for flexibility in Step (ii) of the definition. In this step we replace σ_x by $\lambda_s \sigma_{x'}$. The factor λ_s must be of the form s^a in order that $R_s R_{s'} = R_{ss'}$ is maintained [see Eqs. (5.11) and (5.6)]. This way of performing Step (ii) will be referred to as linear and the RG so defined is a linear RG. Now if we generate the RG by a generator R_2 via (5.19′), we can generalize Step (ii) for R_2 by the replacement

$$\sigma_x \to f(\sigma_{x'}) \tag{5.21}$$

where $f(\sigma_{x'})$ is a monotonically increasing function of $\sigma_{x'}$. The function f can be chosen to simplify practical calculations. If $f(\sigma) = A\sigma$, with A = constant, then the RG is linear as previously defined with $\lambda_s = s^a$, $2^a = A$. Otherwise σ_x is replaced by a nonlinear function of $\sigma_{x'}$. The RG so defined will be called a <u>nonlinear RG</u>. Note that for a nonlinear RG, a relationship such as (5.14) will be modified and in general more complicated. For example, instead of (5.14), we have

$$\langle \sigma_x \rangle_P = \langle f^{\ell}(\sigma_{x/s}) \rangle_{P'}$$

$$\langle \sigma_x \sigma_{x+r} \rangle_P = \langle f^{\ell}(\sigma_{x/s}) f^{\ell}(\sigma_{(x+r)/s}) \rangle_{P'} , \tag{5.22}$$

where $f^{\ell}(\sigma)$ is defined by

$$f^1(\sigma) = f(\sigma) \ ,$$

$$f^2(\sigma) = f(f(\sigma)) \ ,$$

$$f^3(\sigma) = f(f(f(\sigma))) \ ,$$

$$\text{and so on.} \tag{5.23}$$

Here $s = 2^\ell$. In subsequent discussions, the RG will always be assumed to be linear unless otherwise stated.

4. CONCLUDING REMARK

It should not be forgotten that carrying out a symmetry transformation is not the same as solving a particular model. We simply obtain an equivalent model with different parameters. The RG is a set of symmetry transformations operating on a space of parameters. The concepts of temperature, averages, critical points, etc., play no part in R_s. We have not yet mentioned how we can use the RG to help us in solving physical problems. In the next chapter we explore the connection between the RG and critical phenomena.

VI. FIXED POINTS AND EXPONENTS

SUMMARY

We examine the connection between the mathematical structure of the RG and features of critical phenomena. Critical exponents are related to the transformation properties under R_s of points near a fixed point. If a certain simplicity in these properties is assumed, the correlation length ξ and assertions of the scaling hypothesis follow. The critical behavior of the correlation functions and the magnetization is discussed. The free energy is also examined. The concept of the critical region is introduced.

1. THE FIXED POINT AND ITS NEIGHBORHOOD

We mentioned in the beginning of Chapter V that a particular set of symmetry transformations is useful if the physical system of interest is at least approximately invariant under that set of transformations. Therefore, we need first to examine those points in the parameter space which are at least approximately invariant under R_s, before we use the RG to analyze critical phenomena.

A point μ^* which is invariant under R_s, i.e.,

$$R_s \mu^* = \mu^* , \tag{6.1}$$

will be called a <u>fixed point</u>. It plays the role of the completely symmetric H_0 of (5.1). Evidently if (6.1) is true for any finite $s > 1$, it will be true for $s \to \infty$, since we can repeatedly apply R_s.

Equation (6.1) may be viewed as an equation to be solved for μ^*. It is not expected to have a solution unless the value of a in $\lambda_s = s^a$ [see Eq. (5.11)] is properly chosen. It is quite plausible that, when $s \to \infty$, all factors of s must delicately balance to achieve (6.1). Such balance may not be possible for an arbitrary value of a. We have no general theorem so far to tell us whether there is a discrete, or a continuous, set of fixed points, or any fixed point at all. Let us assume now that there is at least one fixed point. We shall concentrate on a particular one with a definite value for a.

We define the <u>critical surface</u> of the fixed point μ^* as a particular subspace of the parameter space. All points μ in this subspace have the property

$$\lim_{s \to \infty} R_s \mu = \mu^* . \tag{6.2}$$

We can imagine that R_s drives μ to a different point. As s increases, all points on the critical surface are eventually driven to μ^*. For sufficiently large s, $R_s \mu$ will be in the immediate neighborhood of μ^*.

So far we have assumed the existence of a μ^* and that of a critical surface. Let us make further plausible assumptions in order to get a more concrete picture of the neighborhood of μ^*.

For a point μ near μ^* we write formally

$$\mu = \mu^* + \delta\mu , \tag{6.3}$$

where $\delta\mu$ is small in some sense. The equation $\mu' = R_s\mu$ can be written as

$$\delta\mu' = R_s^L\,\delta\mu + O((\delta\mu)^2) \quad , \tag{6.4}$$

where $\delta\mu' + \mu^* = \mu'$. We ignore terms of $O((\delta\mu)^2)$ in computing $\delta\mu'$, and R_s^L is thus a linear operator. If we denote the αth entry of μ by μ_α and the αth entry of $\mu' = R_s\mu$ by μ'_α, then R_s^L can be represented by a matrix

$$(R_s^L)_{\alpha\beta} = \left(\frac{\partial\mu'_\alpha}{\partial\mu_\beta}\right)_{\mu=\mu^*} \quad , \tag{6.5}$$

and (6.4) is a shorthand notation for the equations

$$\delta\mu'_\alpha = \sum_\beta \delta\mu_\beta \left(\frac{\partial\mu'_\alpha}{\partial\mu_\beta}\right)_{\mu=\mu^*} \quad . \tag{6.6}$$

Now we have a linear symmetry transformation R_s^L. Our experience in elementary quantum mechanics tells us what we should do. Recall that, in atomic physics, where the rotations are useful symmetry transformations, we determine eigenvalues and eigenvectors of the rotation operators. The eigenvectors (spherical harmonics) are then used as a set of basis vectors. The importance of these eigenvectors and eigenvalues in atomic physics is well established. Here the analogous and natural thing to do next is to determine the eigenvalues and eigenvectors of R_s^L.

Suppose that the eigenvalues of R_s^L are found to be $\rho_j(s)$ and the corresponding eigenvectors to be e_j. The subscript j runs through a set of labels. If there are ∞ entries in μ, the space of $\delta\mu$ will be infinite dimensional and there are then infinitely many eigenvalues and eigenvectors of R_s^L. Since $R_s R_{s'} e_j = R_{ss'} e_j$, we must have

$$\rho_j(s)\,\rho_j(s') = \rho_j(ss') \; . \tag{6.7}$$

It follows that

$$\rho_j(s) = s^{y_j} \; , \tag{6.8}$$

where y_j is independent of s.

We now use the eigenvectors as a set of basis vectors and write $\delta\mu$ as a linear combination

$$\delta\mu = \sum_j t_j e_j \; . \tag{6.9}$$

By (6.8), (6.4) is thus

$$\delta\mu' = \sum_j t'_j e_j \quad , \tag{6.10}$$

$$t'_j = t_j s^{y_j} \quad . \tag{6.11}$$

Clearly, if $y_j > 0$, t'_j will grow as s increases, and if $y_j < 0$, t'_j will diminish. If $y_j = 0$, t'_j will not change. The existence of one or more zero y_j implies that there is a continuous set of fixed points. In such a case we shall simply concentrate on a specific set of values for those t_j with $y_j = 0$, i.e., on one specific fixed point.

The vectors e_j span the linear vector space which is the neighborhood of μ^*. The subspace spanned by those e_j with $y_j < 0$ is the part near μ^* of the critical surface defined by (6.2).

2. LARGE s BEHAVIOR OF R_s AND CRITICAL EXPONENTS

So far no physical concept has entered into our discussion of the RG. We simply have defined R_s as representing some transformation of probability distributions. Now we want to connect the RG to the real world of critical phenomena.

Consider a ferromagnet at temperature T. The probability distribution of spin fluctuations is given by $P(T) \propto \exp[-H[\sigma]/T]$, where $H[\sigma]$ is the block Hamiltonian. This probability distribution is represented by a particular point which we denote by $\mu(T, h)$ in the parameter space. The entries of $\mu(T, h)$ are parameters which depend on the temperature T and the applied field h. They are smooth functions of T and h because they are local properties of a block. This point was discussed in Chapter II (see Sec. II. 3). The critical point of the ferromagnet is given by $T = T_c$ and $h = 0$.

The fundamental hypothesis linking the RG to critical phenomena is that $\mu(T_c, 0)$ is a point on the critical surface of a fixed point μ^*, i.e.,

$$\lim_{s \to \infty} R_s \mu(T_c, 0) = \mu^* \tag{6.12}$$

and $\mu(T, h)$ is not on the critical surface if $T \neq T_c$ or $h \neq 0$. Here μ^* is a mathematical object invariant under the RG, and $\mu(T_c, 0)$ represents all the relevant physics of spin fluctuations of a real material at its critical point.

Qualitatively (6.12) means the following. Suppose that we look at a sample of ferromagnet through a microscope, and that our eyes can see spin variations down to a size b. Thus R_s represents the operation of decreasing the magnification factor by a factor s, i.e., the sample seen appears to shrink by a factor s. Of course, we assume that the sample is sufficiently large so that the edges of the sample will not appear in the view through the microscope. The hypothesis (6.12) states that if we decrease the magnification by a sufficiently large amount, we shall not see any change if it is decreased further.

For simplicity let us subsequently restrict our discussion to cases with $h = 0$. Thus there will be no odd power of σ_x in $H[\sigma]/T$ and we can concentrate on the subspace of the parameter space representing $\exp(-\mathcal{K})$ without an odd power of σ_x in $\mathcal{K}$. $R_s \mu$ will stay in this subspace. Let $\mu(T)$ denote $\mu(T, 0)$.

Since $\mu(T)$ is a smooth function of T, and $\mu(T_c)$ is on the critical surface, then for sufficiently small $T - T_c$, $\mu(T)$ must be very close to the critical surface, and $R_s \mu(T)$ will move toward μ^* as s increases. For large enough s,

$R_s \mu(T)$ will be in the neighborhood of μ^*. But as $s \to \infty$, $R_s \mu(T)$ will go away from μ^* since $\mu(T)$ is not quite on the critical surface. The manner in which $R_s \mu(T)$ goes away from μ^* depends on the positive y_j's in (6.10) and (6.11). Since we do not know much about the y_j's at this moment, we need to make more assumptions. Suppose that just one of the y_j's, call it y_1, is positive. Then simplicity results. For very large s, by (6.11) μ' will be away from μ^* like $t_1 s^{y_1} e_1$. If $\mu(T)$ is close to μ^*, we can write

$$\delta \mu(T) = \mu(T) - \mu^* = \sum_j t_j(T) e_j \ . \tag{6.13}$$

For very large s, we have

$$R_s \mu(T) \approx \mu^* + R_s^L \, \delta \mu(T)$$

$$= \mu^* + t_1(T) s^{y_1} e_1 + O(s^{y_2}) \tag{6.14}$$

where $y_2 < 0$ is the greatest of all other y_j's. Since $t_1(T)$ is a smooth function of T and vanishes at $T = T_c$, we expand it

$$t_1(T) = A(T - T_c) + B(T - T_c)^2 + \cdots \tag{6.15}$$

and assume $A \neq 0$. Note that eigenvectors are defined up to a multiplicative constant. We can adopt the convention that $A > 0$, i.e., if A happens to be negative, we shall absorb the minus sign in the definition of e_1. Then we have, for very small $T - T_c$,

$$R_s \mu(T) \approx \mu^* + A(T - T_c)\, s^{y_1} e_1 + O(s^{y_2})$$

$$= \mu^* \pm (s/\xi)^{1/\nu} e_1 + O(s^{y_2}) \qquad (6.16)$$

where we have <u>defined</u>

$$1/\nu = y_1 \qquad (6.17)$$

$$\xi = |A(T - T_c)|^{-\nu}$$

and the $\pm$ in (6.16) is the sign of $(T - T_c)$. If $\mu(T)$ is not close to μ^*, the conclusion (6.16) still holds as long as $\mu(T)$ is very close to the critical surface and $R_s \mu(T)$ is close to μ^* for some values of s. This is because there is just one way ($\propto s^{1/\nu} e_1$) to go away from μ^* and the coefficient of $s^{1/\nu} e_1$ must be proportional to $T - T_c$ by the smoothness argument.

Now we apply (6.16) to examine the temperature and k dependence of $G(k, \mu(T))$ using the relation (5.16) and (5.11):

$$G(k, \mu(T)) = s^{2a+d}\, G(sk, R_s\mu(T))$$

$$= s^{2a+d}\, G(sk, \mu^* \pm (s/\xi)^{1/\nu} e_1 + O(s^{y_2})) \,. \tag{6.18}$$

Since this is true for any value of s provided s is very large, we can set $s = \xi$ to obtain

$$G(k, \mu(T)) = \xi^{2a+d}\, G(\xi k, \mu^* \pm e_1 + O(\xi^{y_2}))$$

$$\approx \xi^{2a+d}\, g(\xi k) \quad . \tag{6.19}$$

That is, when $T - T_c$ is so small that $O(\xi^{y_2})$ can be ignored, $G(k, \mu(T))$ is ξ^{2a+d} times a function of ξk. This is just what the scaling hypothesis tells us if we identify ξ as the correlation length and ν as the same ν defined in previous chapters.

In cases where $T = T_c$ ($\xi \to \infty$), and $k \neq 0$, we set $s = 1/k$ in (6.18) to obtain

$$G(k, \mu(T_c)) = k^{-(2a+d)} G(1, \mu^* + O(k^{-y_2})) \ . \qquad (6.20)$$

For sufficiently small k the $O(k^{-y_2})$ term can be ignored and we have

$$G(k, \mu(T_c)) \propto k^{-(2a+d)} \ . \qquad (6.21)$$

By the definition of η, we see that

$$\begin{aligned} 2 - \eta &= 2a + d \ , \\ a &= (2 - \eta - d)/2 \ . \end{aligned} \qquad (6.22)$$

Setting $k = 0$ in (6.19), we obtain, for sufficiently small $|T - T_c|$,

$$G(0, \mu(T)) \propto \xi^{2-\eta} G(0, \mu^* \pm e_1 + O(\xi^{y_2})) \propto |T - T_c|^{-\nu(2-\eta)} \ . \qquad (6.23)$$

From the definition of γ we get a previously obtained scaling law

$$\gamma = \nu(2 - \eta) \ . \qquad (6.24)$$

Now we generalize the above arguments to $h \neq 0$. The field h appears as an entry in μ representing the term

$$h b^d \sum_x \sigma_x \tag{6.25}$$

of the block Hamiltonian. The transformation of h under R_s can be easily worked out since (6.25), the total spin of the system times a constant, is not affected by the coarse graining procedure of Step (i), the Kadanoff transformation. Step (ii) changes σ_x into $\lambda_s \sigma_{x'}$, $b^d \sum_x$ into $(sb)^d \sum_{x'}$, and thus (6.25) into

$$h(sb)^d \sum_{x'} \lambda_s \sigma_{x'} = h s^{d+a} b^d \sum_{x'} \sigma_{x'} \quad , \tag{6.26}$$

$$\therefore \quad h' = h s^{\frac{1}{2}(d-\eta)+1} \quad , \tag{6.27}$$

where we have used the fact that $\lambda_s = s^a$, and $a = \frac{1}{2}(2 - \eta - d)$ from (6.22).

Assume that η is small; (6.27) shows that h' grows as s increases. We can regard (6.27) as an instance of (6.11) with a positive y. Let us use the notation

$$y_h = \frac{1}{2}(d - \eta) + 1 \tag{6.28}$$

and let e_h be the corresponding eigenvector. Equation (6.16) is now generalized to

$$R_s \mu(T,h) \approx \mu^* \pm (s/\xi)^{1/\nu} e_1 + hs^{y_h} e_h + O(s^{y_2}) . \quad (6.29)$$

We can use (6.29) to examine the T and h dependence of the magnetization $m = \langle \sigma_x \rangle$, which we write as $m(\mu(T,h))$. By (5.14), we have

$$m(\mu(T,h)) = s^a m(R_s \mu(T,h))$$

$$= s^a m(\mu^* \pm (s/\xi)^{1/\nu} e_1 + hs^{y_h} e_h + O(s^{y_2})) . \quad (6.30)$$

For $T = T_c$ and $h \neq 0$, we set $s = h^{-1/y_h}$ and obtain

$$m(\mu(T_c, h)) = h^{-a/y_h} m(\mu^* + e_h + O(h^{-y_2/y_h})) . \quad (6.31)$$

For sufficiently small h such that $O(h^{-y_2/y_h})$ can be neglected, we have

$$m(\mu(T_c, h) \propto h^{1/\delta}$$

$$\delta = -y_h/a = \frac{d + 2 - \eta}{d - 2 + \eta} \quad (6.32)$$

since $a = \frac{1}{2}(2 - \eta - d)$ [see Eq. (6.22)]. Finally, for $h = 0$ and $T \neq T_c$, we set $s = \xi$ in (6.30) to obtain

$$m(\mu(T)) = \xi^a\, m(\mu^* \pm e_1 + O(\xi^{y_2})) \; . \tag{6.33}$$

For very small $|T - T_c|$ such that $O(\xi^{y_2})$ is negligible,

$$m \propto |T - T_c|^\beta \; ,$$

$$\beta = -a\nu = \frac{1}{2}\,\nu\,(d - 2 + \eta) \; . \tag{6.34}$$

For small h and small $T - T_c$, we set $s = \xi$ in (6.30) and find

$$m(\mu(T,h)) = \xi^a\, w_\pm(h\xi^{y_h}) \; ,$$

$$w_\pm(\lambda) \equiv m(\mu^* \pm e_1 + \lambda e_h) \; . \tag{6.35}$$

This result is consistent with the conclusion (4.20), since $a = -\frac{1}{2}(d - 2 + \eta)$ is the same as $-d_\sigma$ of (4.11), and d_h of (4.17) is the same as y_h of (6.28).

Before proceeding further, we should again like to remind the reader that R_s only transforms the parameters, playing a role analogous to that of the rotation operations in atomic physics. The above conclusions on the dependence of various quantities on $T - T_c$ and on h are consequences of the assumed properties of R_s for large s near the fixed point. Note that these conclusions are drawn from (6.18) and (6.30) with further assumptions which have been implicit. Some of the implicit assumptions are incorrect and therefore some of the conclusions are meaningless or trivial. For example, (6.34) is obviously meaningless for $T > T_c$ since we know that the proportionality constant must be zero. The RG does not tell what the averages should be. It simply relates one average to another calculated with different parameters. It does not tell us whether $m(\mu^* \pm e_1)$ of (6.33) is zero or not. The conclusion (6.34) is based on the assumption that $m(\mu^* - e_1) \neq 0$, and the $O(\xi^{y_2})$ term can be thrown away. Likewise in (6.23) the RG says nothing about whether $G(0, \mu^* \pm e_1 + O(\xi^{y_2}))$ is zero, finite, or infinite. Additional information is needed.

3. THE FREE ENERGY

The application of the RG to the study of the free energy is less straightforward. In defining our parameter space, we did not keep track of any possible additive constant to $\mathcal{H}$. In defining R_s, we have also ignored the additive constant to $\mathcal{H}'$ produced by the Kadanoff transformation. In the calculation of the free energy, such constants do make a difference and must be kept track of. For definiteness, let us follow the convention that $\mathcal{H}[\sigma]$ is defined to be zero for $\sigma = 0$, i.e., to have no constant term. Any constant added to $\mathcal{H}[\sigma]$ will be written out explicitly. With this convention, (5.6c) must be written as

$$e^{-\mathcal{H}'[\sigma] - AL^d} = \int \prod_y d\sigma''_y \prod_{x'} \delta\left(\lambda_s \sigma_{x'} - s^{-d} \sum_y^x \sigma''_y\right) e^{-\mathcal{H}[\sigma'']} \tag{6.36}$$

where e^{-AL^d} is simply the value of the integral with all $\sigma_{x'}$ set to zero.

Let us define $\mathfrak{F}(\mu)$ and $\mathfrak{F}(\mu')$ by

$$e^{-\mathfrak{F}(\mu)L^d} = \int \prod_y d\sigma''_y \, e^{-\mathcal{H}[\sigma'']} , \tag{6.37}$$

$$e^{-\mathfrak{F}(\mu')L'^d} = \int \prod_{x'} d\sigma_{x'} e^{-\mathcal{H}'[\sigma]} . \tag{6.38}$$

To relate $\mathfrak{F}(\mu')$ to $\mathfrak{F}(\mu)$ we insert

$$1 = \int \prod_{x'} \lambda_s \, d\sigma_{x'} \, \delta\left(\lambda_s \sigma_{x'} - s^{-d} \sum_y^x \sigma_y''\right) \qquad (6.39)$$

in (6.37), and then use (6.36) and (6.38) to obtain

$$e^{-\mathfrak{F}(\mu)L^d} = e^{-\mathfrak{F}(\mu')L^d - AL^d - BL^d} \qquad (6.40)$$

where

$$e^{-BL^d} = (\lambda_s)^{L^d/(sb)^d} = e^{-(-s^{-d} b^{-d} a \ln s)L^d} \qquad (6.41)$$

since $\lambda_s = s^a$. Equation (6.40) relates $\mathfrak{F}(\mu)$ and $\mathfrak{F}(\mu')$. Note that A does depend on μ and s, according to (6.36).

Now consider the physical probability distribution represented by $\mu(T,h)$. Then the free energy per unit volume is

$$F(\mu(T,h)) = \mathfrak{F}(\mu(T,h))T \; . \qquad (6.42)$$

From (6.40) we obtain

$$F(\mu(T,h)) = s^{-d} F(R_s \mu(T,h)) + A(\mu(T,h), s)T + B(s)T \, .$$

Note that AT is the free energy per unit volume directly contributed by fluctuations over scales shorter than sb. Thus, <u>for a fixed s</u>, A must be a smooth function of $T - T_c$.

It is in fact independent of h since h couples only to the total spin, not to short scale variations. Thus if $F(\mu(T,h))$ is a singular function of T and/or h, the singularity is contained in the first term of the right-hand side of (6.42).

Now we apply (6.29) to (6.42), and obtain

$$F(\mu(T,h)) = s^{-d}\, F(\mu^* \pm (s/\xi)^{1/\nu} e_1 + hs^{y_h} e_h + O(s^{y_2}))$$

$$+ A(T,s)T + B(s)T \quad . \qquad (6.43)$$

For very small $T - T_c$, we again set $s = \xi$ and neglect $O(\xi^{y_2})$ to obtain

$$F(\mu(T,h)) = \xi^{-d} f_\pm(h\xi^{y_h}) + A(T,\xi)T + B(\xi)T , \qquad (6.44)$$

where the functions $f_\pm$ are defined by

$$f_\pm(\lambda) = F(\mu^* \pm e_1 + \lambda e_h) \quad . \qquad (6.45)$$

It would be aesthetically more pleasing if the terms A + B were absent in (6.43) and (6.44). In fact, ignoring A + B would give the "homogeneity property" of the free energy that is taken for granted in most of the literature on critical phenomena. However, the RG analysis leading to (6.43) and (6.44) shows that there is no reason to drop $A+B$. We argued above that $A(T, s) + B(s)$ is a smooth function of $T - T_c$ <u>at a fixed</u> s. Now we have set $s = \xi \propto |T - T_c|^{-\nu}$, it is not clear whether $A(T, \xi) + B(\xi)$ is a smooth function of $T - T_c$. The physical meaning of $A(T, \xi)T$ is the free energy of the modes with $k > \xi^{-1}$ in the absence of the modes with $k < \xi^{-1}$, while $\xi^{-d} f_{\pm}$ is the free energy of modes with $k < \xi^{-1}$. Thus we expect that they are equally singular for large ξ. Also,

$$B(\xi) = -a\xi^{-d} b^{-d} \ln \xi \propto |T - T_c|^{-d\nu} \ln |T - T_c| , \quad (6.46)$$

according to (6.41), is a singular function of $T - T_c$. Therefore, it is necessary to examine more closely the critical behavior of $A+B$. The singularity of the $A+B$ term was emphasized and explicitly demonstrated for the $n \to \infty$ case by Ma (1973, 1974a). The reader should be warned that in many authoritative papers this term is simply ignored.

Let us go back to (6.40) and write, setting $s = 2$,

$$\mathfrak{F}(\mu) = 2^{-d}\, \mathfrak{F}(R_2\mu) + \tilde{A}(\mu) \ , \tag{6.47}$$

$$\tilde{A}(\mu) \equiv A(\mu, 2) + B(2) \ . \tag{6.48}$$

Apply (6.47) to $\mathfrak{F}(R_2\mu)$ and substitute it to the left hand side. We obtain

$$\mathfrak{F}(\mu) = 2^{-d}\,(2^{-d}\, \mathfrak{F}(R_2^2\,\mu) + \tilde{A}(R_2\mu)) + \tilde{A}(\mu) \ . \tag{6.49}$$

Repeating this ℓ times, we obtain

$$\mathfrak{F}(\mu) = s^{-d}\, \mathfrak{F}(R_s\mu) + \sum_{m=0}^{\ell-1} 2^{-md}\, \tilde{A}(R_2^m\,\mu) \ , \tag{6.50}$$

for $s = 2^\ell$.* Thus we have found

$$A(\mu, s) + B(s) = \sum_{m=0}^{\ell-1} 2^{-md}\, \tilde{A}(R_2^m\,\mu) \ . \tag{6.51a}$$

If we have an RG with continuous s, (6.51a) can be expressed as

$$A(\mu, s) + B(s) = \int_1^s s'^{\,-d}\, \tilde{A}(R_{s'}\,\mu)\, \frac{ds'}{s'} \ , \tag{6.51b}$$

---

*This formula was studied and used in numerical work by Nauenberg and Nieuhuis (1974). See also Nelson (1975).

$$\tilde{A}(\mu) \equiv \frac{A(\mu, 1+\Delta\ell)+B(1+\Delta\ell)}{\Delta\ell}, \qquad (6.52)$$

for small $\Delta\ell$. We can regard (6.48) as (6.52) with $\Delta\ell = 1$. Note that $A(\mu, 1) = B(1) = 0$. Equation (6.51) expresses $A+B$ in terms of $\tilde{A}$ which has no explicit dependence on s. Its s′ dependence comes in only through $R_{s'}\mu$.

What we need is $A(\mu(T), \xi)+B(\xi)$. The easiest way to see its critical behavior is to differentiate it with respect to ξ. From (6.51b), we obtain

$$\xi \frac{\partial}{\partial\xi}(A+B) = \xi^{-d}\,\tilde{A}(R_\xi\mu)$$

$$= \xi^{-d}\,\tilde{A}(\mu^* \pm e_1 + O(\xi^{y_2})). \qquad (6.53)$$

Since $\xi = |t_1|^{-\nu}$, we have

$$\frac{\partial}{\partial t_1}(A+B) = -\frac{1}{\nu}\,|t_1|^{d\nu-1}\,\tilde{A}(\mu^* \pm e_1 + O(\xi^{y_2})). \quad (6.54)$$

Clearly, $A+B$ is just as singular as $\xi^{-d} f_\pm$ in (6.44). If we ignore $O(\xi^{y_2})$ and integrate (6.54) with respect to t_1, we obtain

$$T(A(\mu(T), \xi) + B(\xi)) = \tilde{f}_\pm |t_1|^{d\nu} + f_o \qquad (6.55)$$

where $\tilde{f}_{\pm}$, f_o are effectively constants. Substituting (6.55) in (6.44), we obtain

$$F(\mu(T,h)) = |t_1|^{\nu d}\,(f_{\pm}(h\xi^{y_h}) + \tilde{f}_{\pm}) + f_o \qquad (6.56)$$

Since $|t_1| \propto |T - T_c|$ and the specific heat C is $-T\partial^2 F/\partial T^2$, we obtain, for $|T - T_c| \to 0$ and $h = 0$,

$$C \propto |T - T_c|^{-\alpha} + \text{const.} ,$$
$$\alpha = -2 + \nu d . \qquad (6.57)$$

This result agrees with that obtained from the scaling hypothesis.

The reader might wonder why there are these unpleasant terms $A(T,\xi)T$ and $B(\xi)T$ in (6.44) for the free energy while there are no such terms in (6.19) and (6.35) for the correlation function $G(k)$ and the magnetization m. The answer is that $G(k) = \langle |\sigma_k|^2 \rangle$ and $m = \langle \sigma_x \rangle = L^{-d/2} \langle \sigma_k \rangle_{k=0}$ are average values of long wavelength (small k) Fourier components of the spin configuration. The Kadanoff transformation, Step (i) in defining R_s, does not affect these Fourier components, and Step (ii) is just a scale transformation. On the other hand the free energy involves all Fourier components directly. Complications like the A and B terms in (6.44) are to be expected. In fact, if we had used a nonlinear RG, there would have been extra terms in (6.19) and (6.35) as well. This point will come up again. There are quantities besides the free energy which involve σ_k with large k as well as small k. They will also come up in later chapters.

4. CRITICAL REGION

In deriving the leading singular T, h, k dependence of various quantities of (6.19), (6.20), (6.23), (6.32), (6.34), (6.35) and (6.44), we dropped terms of $O(s^{y_2})$ with $s = \xi$, k^{-1} or h^{-1/y_h}. Dropping these terms allowed us to set $R_s\mu(T,h)$ on the plane through μ^* spanned by e_1 and e_h [this is clear in view of Eq. (6.29)] and the results (6.19), ... (6.44) follow. The critical region in $T - T_c$ is defined as the range of $T - T_c$ over which $O(\xi^{y_2})$ is negligible. Likewise the critical region in k is the range of k in which $O(k^{-y_2})$ is negligible, and similarly for the critical region in h.

The relevant question to answer for determining the size of the critical region is how large the minimum s must be so that $R_s \mu(T,h)$ is well approximated by $\mu^* \pm (s/\xi)^{1/\nu} e_1 + hs^{y_h} e_h$. The answer depends on two factors: (a) How far $\mu(T,h)$ is away from the fixed point. The farther away it is, the larger the s needed for R_s to bring it in. This of course depends on the details of $\mu(T,h)$, i.e., on the details of the block Hamiltonian for the particular system of interest. (b) How fast R_s can bring $\mu(T,h)$ to the $e_1 e_h$ plane, i.e., how fast s^{y_j} vanishes as s increases for $y_j < 0$. The term s^{y_2} is the slowest term since by definition y_2 is the least negative among $y_j < 0$. Clearly the more negative y_2 is (i.e., the larger $|y_2|$ is) the faster s^{y_2} vanishes, and the smaller is the minimum required s.

Without making reference to any specific system, we cannot say much about factor (a). Let us simply estimate the minimum required s from factor (b) by saying that

$$s^{y_2} < 10\% \,,$$

i.e.,

$$s > 10^{1/|y_2|} \,. \tag{6.47}$$

Thus the critical region in $T - T_c$ is given by $\xi >$ const. $10^{1/|y_2|}$, i.e.,

$$|T - T_c| < 10^{-1/\nu |y_2|} \cdot \text{const.} \tag{6.48}$$

The critical region in k is given by

$$k < 10^{-1/|y_2|} \cdot \text{const.} \tag{6.49}$$

and that in h is given by

$$h < 10^{-y_h/|y_2|} \cdot \text{const.} \tag{6.50}$$

The "const." in (6.48) - (6.50) depends on factor (a), i.e., details of the specific physical system of interest.

5. SUMMARY AND REMARKS

In this and the previous chapter we have set up the scheme of the RG for analyzing critical phenomena. The main motivation behind the use of the RG is the same as that behind the use of symmetry transformations in various fields of physics.

We have the picture of a parameter space in which

we locate a fixed point μ^*, which is invariant under the RG. The fixed point sits on a critical surface. The RG transformation R_s drives any point on the critical surface toward μ^*. For a point not on the critical surface but very close to it, R_s will first drive it toward μ^* but eventually will drive it away from μ^* as $s \to \infty$. This mathematical scheme makes contact with the physics of critical phenomena through the hypothesis (6.12) that the point in the parameter space representing a system at a critical point is on the critical surface. By smoothness arguments, this point will be very close to the critical surface if the system is near the critical point. Critical phenomena are thus related to the properties of R_s near the fixed point. In particular, critical exponents are related to eigenvalues of the linearized R_s near the fixed point.

The linearized RG transformation R_s^L operating in the neighborhood of μ^* has eigenvalues s^{y_j}. If only two of the y_j are positive, called y_1 and y_h, then (6.29) and various assertions of the scaling hypothesis follow. The exponent ν is identified as $1/y_1$, and y_h is related to η by (6.28). Of course we have not exhausted all the consequences of (6.29).

The universality of critical exponents appears in a very natural way in this scheme. The critical surface is expected to be a large subspace of the parameter space. We expect that many different materials at their critical points can be represented by different points on the same critical surface. Since critical exponents are properties of R_s in the neighborhood of the fixed point, all these materials will share the same critical exponents.

Thus the important task is to classify and analyze all possible fixed points and associated critical surfaces and exponents. Then given any material near a critical point, all we have to do is to work out its block Hamiltonian, determine the representative point in the parameter space, and see which critical surface it is on or close to.

Rigorous mathematical work on the RG has been lacking. Precise numerical work has been attempted only quite recently even though the first RG analysis of Wilson (1971) was done numerically. We shall not discuss numerical techniques here nor attempt any rigorous proof of properties of the RG. To substantiate and clarify the mathematical picture of the fixed point, critical surface, etc. we shall discuss a few simple approximate RG analyses in the next two chapters. These analyses will show that this mathematical picture is not fictional, and the various assumptions we made above seem to be correct. There are fixed points with more complex structure than the kind we discussed above. For example, for a "tricritical" fixed point, there is one more positive y_j other than y_1 and y_h.

VII. THE GAUSSIAN FIXED POINT AND FIXED POINTS IN 4-ϵ DIMENSIONS

SUMMARY

In this chapter we first discuss the Gaussian fixed point and the linearized RG in its neighborhood as a simple illustration of the ideas explored in the last two chapters. The mathematics is quite simple and allows us to clarify some definitions as well as point out some ambiguities. The Gaussian fixed point provides a description of critical behavior for $d > 4$. Then the case $d = 4 - \epsilon$ with small $\epsilon > 0$ is discussed. In addition to the Gaussian fixed point, there will be a new fixed point, which is determined to $O(\epsilon)$. The calculation involved will be presented in detail in order that the reader may see the various undesirable complications as well as the nice simple results.

1. THE GAUSSIAN FIXED POINT

We begin with the Ginzburg-Landau model for n-component spins in d-dimensional space. For many subsequent discussions and calculations, it is more convenient to use the wave vector representation. Let us write down a few formulas for easy reference and to fix the notation:

$$\mathcal{H} = \frac{1}{2} \int d^d x \left(r_o \sigma^2 + \frac{1}{4} u \sigma^4 + c(\nabla\sigma)^2 \right)$$

$$= \frac{1}{2} \sum_{k,k} (r_o + ck^2) \, |\sigma_{ik}|^2$$

$$+ \frac{u}{8} L^{-d} \sum_{k,k',k'',ij} \sigma_{ik} \sigma_{ik'} \sigma_{jk''} \sigma_{j-k-k'-k''} \qquad (7.1)$$

where all wave vectors are restricted to less than Λ. We define our parameter space as the space of μ,

$$\mu = (r_o, u, c) \qquad (7.2)$$

and use the definition of $\mu' = R_s \mu$ with a sharp cutoff given by (5.19), namely

$$e^{-\mathcal{H}' - AL^d} = \left[\int \delta\phi \, e^{-\mathcal{H}} \right]_{\sigma_k \to s^{1-\eta/2} \sigma_{sk}} ,$$

$$\delta\phi = \prod_{\Lambda/s < q < \Lambda} \prod_{i=1} d\sigma_{iq} . \tag{7.3}$$

Here $\mathcal{H}'$ is defined to include no additive constant. The additive constant generated by the integral over $\int d\sigma_{iq}$ is AL^d.

Now we must face the task of performing the integral in (7.3). If $u = 0$, then the integral is very simple because each σ_{iq} has an independent Gaussian distribution $\exp\left(-\frac{1}{2}(r_o + cq^2)\,|\sigma_{iq}|^2\right)$. The integration over $d\sigma_{iq}$ will contribute to AL^d the amount of the free energy of the mode i, q. If $u \neq 0$, the integral immediately becomes very difficult. To avoid mathematical complications at this early stage, let us for the moment look for fixed points with $u = 0$. It turns out that there are such fixed points and we can illustrate a few important concepts with them. Assuming $u = 0$, we perform the integral in (7.3) and make the replacement $\sigma_k \to s^{1-\eta/2} \sigma_{sk}$ to obtain

$$\mathcal{H}' = \frac{1}{2} \sum_{i, k < \Lambda/s} s^{2-\eta} (ck^2 + r_o) |\sigma_{isk}|^2$$

$$= \frac{1}{2} \sum_{i, k' < \Lambda} (r_o s^{2-\eta} + s^{-\eta} ck'^2) |\sigma_{ik'}|^2, \quad (7.4a)$$

$$AL^d = -\frac{1}{2} \sum_{\Lambda/s < q < \Lambda} n \ln\left(\frac{2\pi}{r_o + cq^2}\right) . \quad (7.4b)$$

Clearly, Eq. (7.4) has the same form as (7.1), with

$$r_o s^{2-\eta} = r_o' , \quad s^{-\eta} c = c' \quad (7.5a)$$

replacing r_o and c, respectively. Therefore

$$\mu' = R_s \mu = (r_o s^{2-\eta}, 0, cs^{-\eta}) . \quad (7.5b)$$

When we choose $\eta = 0$ we find the fixed point

$$\mu_o^* = (0, 0, c) ,$$
$$\eta = 0 . \quad (7.6)$$

This is called a Gaussian fixed point. The value of c is arbitrary. We can always choose the unit of σ so that c assumes any positive value. This fixed point represents the probability distribution $\exp(-\mathcal{H}_o^*)$,

$$\mathcal{H}_o^* = \frac{c}{2} \int d^d x \, (\nabla \sigma)^2 \, ,$$

$$e^{-\mathcal{H}_o^*} = \prod_{k,i} e^{-\frac{c}{2} k^2 |\sigma_{ki}|^2} \, . \tag{7.7}$$

Clearly, (7.5b) shows that by setting $\eta = 2$, we obtain another fixed point

$$\mu_\infty^* = (r_o, 0, 0) \, ,$$

$$\eta = 2 \, . \tag{7.8}$$

Here the value of $r_o > 0$ is arbitrary and we can choose the appropriate unit of σ so that $r_o = 1$. In analogy with Eq. (7.7), we have

$$\mathcal{H}_\infty^* = \frac{r_o}{2} \int d^d x \, \sigma^2 \, ,$$

$$e^{-\mathcal{H}_\infty^*} = \prod_{x,i} e^{-r_o \frac{b^d}{2} \sigma_i^2(x)} \, . \tag{7.9}$$

There is still another fixed point like (7.9) except that $r_o < 0$:

$$\mu^*_{-\infty} = (-|r_o|, 0^+, 0) \quad , \tag{7.10a}$$

where 0^+ means a very small u to keep the probability normalizable. In analogy with Eq. (7.9) we have

$$e^{-\mathcal{H}^*_{-\infty}} = \prod_{x,i} e^{+|r_o| \frac{b^d}{2} \sigma_i^2(x) - u\sigma_i^4(x)} \tag{7.10b}$$

where $u = 0^+$.

The physical meanings of the above three fixed points are very different. Equation (7.10b) describes a situation where each block spin is independent of all the other block spins. Each block spin tends to have a large value since u is small. If $c = 0^+$, a uniform magnetization results. This describes a system far below T_c. Equation (7.9) describes a situation in which each block spin has a Gaussian distribution and is independent of all other blocks. This is the situation at very high temperatures. Recall that $(\nabla\sigma)^2$ measures the coupling between neighboring block spins. As (7.7) shows, μ_o^* describes the limit of strong coupling. That it reflects the properties of a critical point for $d > 4$ will become clear later. Let us proceed to study R_s operating in the neighborhood of μ_o^*.

2. THE LINEARIZED RG NEAR THE GAUSSIAN FIXED POINT

As soon as u becomes nonzero, the quartic term in $\mathcal{H}$ makes the integral of (7.3) very complicated. Since we are interested in the linearized R_s, the algebra is still not too bad. The following procedure for obtaining R_s^L is instructive and can be generalized and applied to other kinds of fixed points. We go through it in detail as an exercise. We write

$$\mathcal{H} = \mathcal{H}^* + \Delta\mathcal{H} ,$$

$$\mathcal{H}' + AL^d = \mathcal{H}^* + A^* L^d + \Delta\mathcal{H}' + \Delta A L^d , \qquad (7.11)$$

where $\Delta\mathcal{H}$, $\Delta\mathcal{H}'$, and ΔA are regarded as small quantities. Substituting (7.11) in (7.3) and expanding these small quantities, we obtain the first-order terms

$$\Delta\mathcal{H}' + \Delta A L^d$$

$$= \langle \Delta\mathcal{H} \rangle_{\sigma_k \to s^{1-\eta/2} \sigma_{sk}} , \tag{7.12a}$$

where

$$\langle \Delta\mathcal{H} \rangle = \frac{\int \delta\phi \, e^{-\mathcal{H}^*} \Delta\mathcal{H}}{\int \delta\phi \, e^{-\mathcal{H}^*}} , \tag{7.12b}$$

which means averaging over ϕ, i.e., over σ_q, $\Lambda/s < q < \Lambda$, with fixed σ_k, $k < \Lambda/s$. Equation (7.12b) is the linearized Step (i), i.e., the linearized Kadanoff transformation.

Before proceeding, we remark that in general one can break $\mathcal{H}$ into two pieces in any manner

$$\mathcal{H} = \mathcal{H}_0 + \mathcal{H}_1 \tag{7.13a}$$

and expand (7.3) to obtain

$$\mathcal{H}' + AL^d = -\ln\left(\int \delta\phi \; e^{-\mathcal{H}}\right)_{\sigma_k \to s^{1-\eta/2}\sigma_{sk}}$$

$$= \left[-\ln \int \delta\phi \; e^{-\mathcal{H}_o} - \sum_{m=1}^{\infty} \frac{(-)^m}{m!} \langle \mathcal{H}_1^m \rangle_c \right]_{\sigma_k \to s^{1-\eta/2}\sigma_{sk}}$$

(7. 13b)

where

$$\langle \mathcal{H}_1^m \rangle = \frac{\int \delta\phi \; e^{-\mathcal{H}_o} \mathcal{H}_1^m}{\int \delta\phi \; e^{-\mathcal{H}_o}} \qquad (7.13c)$$

and the subscript c means taking the cumulant. The first two cumulants are

$$\langle \mathcal{H}_1 \rangle_c = \langle \mathcal{H}_1 \rangle \; ,$$

$$\langle \mathcal{H}_1^2 \rangle_c = \langle \mathcal{H}_1^2 \rangle - \langle \mathcal{H}_1 \rangle^2 = \langle (\mathcal{H}_1 - \langle \mathcal{H}_1 \rangle)^2 \rangle \; . \qquad (7.14)$$

For (7. 12), we have taken $\mathcal{H}_o = \mathcal{H}^*$.

Equation (7. 12) gives a convenient formula for finding R_s^L, if $\mathcal{H}^*$ is already known. For our Gaussian fixed point, $\mathcal{H}^*$ is simply $\frac{c}{2} \int d^d x \, (\nabla\sigma)^2$ and

$$\langle \Delta\mathcal{H} \rangle = \frac{1}{2}\int d^d x \left(r_o \langle \sigma^2 \rangle + \frac{1}{4} u \langle \sigma^4 \rangle \right) . \qquad (7.15)$$

To evaluate the averages over ϕ, let us separate ϕ from σ by defining

$$\sigma_i = \sigma_i' + \phi_i , \qquad (7.16a)$$

$$\sigma_i' = L^{-d/2} \sum_{k < \Lambda/s} \sigma_{ik} \, e^{ik \cdot x} , \qquad (7.16b)$$

$$\phi_i = L^{-d/2} \sum_{\Lambda/s < q < \Lambda} \sigma_{iq} \, e^{iq \cdot x} . \qquad (7.16c)$$

We need to evaluate $\langle \sigma^2 \rangle$ and $\langle \sigma^4 \rangle$ over ϕ keeping σ' fixed. We obtain

$$\langle \sigma^2 \rangle = \langle \sigma'^2 + 2\sigma' \cdot \phi + \phi^2 \rangle$$

$$= \sigma'^2 + \langle \phi^2 \rangle \qquad (7.16d)$$

since $\langle \phi \rangle = 0$. The average $\langle \phi^2 \rangle$ is over the independent Gaussian distributions for σ_{iq} and is easily done. Note that $\langle \sigma_{iq} \sigma_{jq'} \rangle = \delta_{ij} \delta_{-q'q} \langle \sigma_{iq} \sigma_{i-q} \rangle$. We have, squaring (7.16c), and then applying (7.7),

$$\langle \phi^2 \rangle = \sum_i \langle \phi_i^2 \rangle$$

$$= L^{-d} \sum_{\Lambda/s < q < \Lambda,\, i} \langle \sigma_{iq}\, \sigma_{i-q} \rangle$$

$$= n(2\pi)^{-d} \int d^d q\, (1/cq^2)$$

$$= nK_d \int_{\Lambda/s}^{\Lambda} dq\, q^{d-3}/c$$

$$= n_c(1 - s^{2-d}) , \tag{7.17}$$

where

$$n_c \equiv (n/c)\, K_d\, \Lambda^{d-2}/(d-2) , \tag{7.18}$$

$$K_d \equiv 2^{-d+1}\, \pi^{-d/2}/\Gamma\left(\tfrac{1}{2}\, d\right) . \tag{7.19}$$

Here K_d is the surface area of a unit sphere in d-dimensional space divided by $(2\pi)^d$.

Now $\langle \sigma^4 \rangle$ is given by

$$\langle (\sigma'^2 + 2\sigma' \cdot \phi + \phi^2)^2 \rangle$$

$$= \sigma'^4 + 2\sigma'^2 \langle \phi^2 \rangle + 4\langle (\phi \cdot \sigma')^2 \rangle + \langle \phi^4 \rangle . \tag{7.20}$$

The average in the third term is

$$\langle (\phi \cdot \sigma')^2 \rangle = \sum_{i,j} \sigma_i' \sigma_j' \langle \phi_i \phi_j \rangle$$

$$= \sum_i \sigma_i'^2 \langle \phi_i^2 \rangle = \sigma'^2 \langle \phi^2 \rangle / n$$

$$= \sigma'^2 (n_c/n)(1 - s^{2-d}) \qquad (7.21)$$

The last term of (7.20) is left as an exercise for the reader:

$$\langle \phi^4 \rangle = (n^2 + 2n)(n_c/n)^2 (1 - s^{2-d})^2 \; . \qquad (7.22)$$

Substituting (7.16d) and (7.20) in (7.15), then using the results (7.17), (7.21) and (7.22), we obtain

$$\langle \Delta \mathcal{H} \rangle = \frac{1}{2} \int d^d x \left[\left(r_o + u\left(\frac{n}{2} + 1\right) \frac{n_c}{n} (1 - s^{2-d}) \right) \sigma'^2 + \frac{1}{4} u \sigma'^4 \right] + \Delta A L^d \; , \qquad (7.23)$$

$$\Delta A = \frac{1}{2}\left[r_o n_c(1 - s^{2-d}) + \frac{1}{4}\, u(n^2 + 2n)\left(\frac{n_c}{n}\right)^2 (1 - s^{2-d})^2\right] . \tag{7.24}$$

Now we have to make the substitution $\sigma_k \to s\sigma_{sk}$ in (7.23). As we noted much earlier [see (5.6) and note $\lambda_s = s^{1-\eta/2-d/2}$ with $\eta = 0$], this substitution is the same as

$$\sigma'(x) \to \sigma(x')\, s^{1-d/2} \quad ,$$

$$x' = x/s \quad , \tag{7.25}$$

$$\int d^d x = s^d \int d^d x' \quad .$$

Substituting (7.25) in (7.23), we get

$$\langle \Delta \mathcal{H} \rangle_{\sigma_k \to s\sigma_{sk}} = \frac{1}{2}\int d^d x' \left(r_o' \sigma^2 + \frac{1}{4}\, u'\sigma^4\right) + \Delta A L^d$$

$$= \Delta\mathcal{H}' + \Delta A L^d \quad ,$$

$$r_o' = s^2 \left(r_o + u\left(\frac{n}{2} + 1\right) \frac{n_c}{n} (1 - s^{2-d})\right) ,$$

$$u' = s^{4-d} u \quad . \tag{7.26}$$

This completes the formula for R_s^L within the parameter space of $\mu = (r_o, u, c)$ near $\mu_o^* = (0, 0, c)$.

We can represent (7.26) by the matrix equation

$$\begin{pmatrix} r_o' \\ \\ u' \end{pmatrix} = R_s^L \begin{pmatrix} r_o \\ \\ u \end{pmatrix}$$

$$R_s^L = \begin{pmatrix} s^2 & (s^2 - s^{4-d})\,B \\ \\ 0 & s^{4-d} \end{pmatrix} \tag{7.27}$$

where $B = \left(\frac{n}{2} + 1\right) n_c/n$. The eigenvalues of R_s^L are obviously s^2 and s^{4-d}. The eigenvectors are

$$e_1 = \begin{pmatrix} 1 \\ 0 \end{pmatrix} , \quad e_2 = \begin{pmatrix} -B \\ 1 \end{pmatrix} \tag{7.28}$$

In the language of Chapter VI, (7.26) gives the exponents

$$\begin{aligned} y_1 &= 1/\nu = 2 , \\ y_2 &= 4 - d . \end{aligned} \tag{7.29}$$

The conclusions concerning critical behavior are valid provided $y_2 < 0$. However, Eq. (7.29) shows that this is possible only for $d > 4$. Otherwise, $y_2 > 0$ and the picture given in Chapter VI is not valid. We say that

$$\begin{aligned} &\mu_o^* \text{ is stable for } d > 4 \ , \\ &\mu_o^* \text{ is unstable for } d < 4 \ . \end{aligned} \tag{7.30}$$

In Fig. 7.1, we show the fixed point μ_o^* and the orientations of e_1 and e_2. The arrows show the direction of $R_s\mu$ as s increases. For $d < 4$, there is no critical surface for μ_o^*. For $d > 4$, the critical surface is the line along e_2. Any μ on this line is driven toward μ_o^* by R_s, as shown in Fig. 7.1a. The equation for the critical surface is given by $t_1 = 0$, with

$$t_1 = r_o + uB \ . \tag{7.31}$$

In the neighborhood of μ_o^*, we get

$$\mu = \mu_o^* + t_1 e_1 + u e_2 + h e_h \tag{7.32}$$

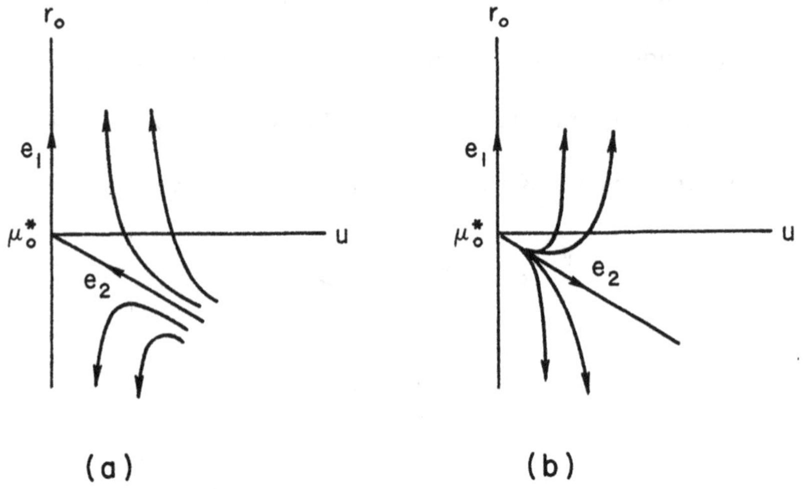

Figure 7.1. The Gaussian fixed point $r_o^* = 0$, $u^* = 0$, and the eigenvectors e_1 and e_2 of R_s^L. The flow lines and arrows show how $R_s\mu$ moves as s increases. (a) $d < 4$. (b) $2 < d < 4$.

Here we have added a new direction e_h in the parameter space to account for the presence of a uniform magnetic field h. The transformation of h under R_s is trivial and was given before by (6.27):

$$h' = h\, s^{1/h} , \tag{7.33}$$

$$y_h = \frac{1}{2} d + 1 .$$

3. RELEVANT, IRRELEVANT, AND MARGINAL PARAMETERS, SCALING FIELDS, AND CROSSOVER

Let us digress to introduce some terminology which is now standard in the literature.

The parameters $t'_1 = t_1 s^{y_1}$ and $h' = h s^{y_h}$ discussed above grow as s increases (since $y_1, y_h > 0$). They are called <u>relevant</u> parameters with respect to the Gaussian fixed point. The parameter $u' = u s^{4-d}$ diminishes (for $d > 4$) and is termed <u>irrelevant</u>. The parameter $c' = c$ does not change and is called <u>marginal</u>.

In general, relevance is defined with respect to a fixed point. If R_s drives a parameter toward its fixed point value, that parameter is irrelevant. If R_s drives it away, it is relevant. If R_s does not change it, it is marginal.

According to the discussions of Chapter VI, only relevant parameters are relevant to critical behavior. There are, however, cases where irrelevant parameters play very important roles in critical exponents. We shall see an example shortly.

The concept of the scaling field is very useful. Scaling fields $g_i(\mu)$ are special functions of the parameters having the following transformation property

$$g_i(R_s \mu) = g_i(\mu)\, s^{y_i} ,$$
$$g_i(\mu^*) = 0 . \tag{7.34}$$

For μ very close to μ^*, g_i can be determined by determining the eigenvectors of R_s^L. In fact

$$g_i(\mu^* + \delta\mu) = t_i + O(\delta\mu^2) . \tag{7.35}$$

See (6.9) and (6.11) for t_i. The scaling fields are natural extensions of t_i. We call y_i the exponent of g_i. Clearly the scaling fields form a set of convenient parameters. Also, g_i is relevant, marginal, or irrelevant according to whether y_i is greater, equal to, or less than zero.

We can regard $\mathcal{H}$ as a function of the scaling fields. Near the fixed point, we can expand $\mathcal{H}$:

$$\mathcal{H} = \mathcal{H}^* + \sum_i t_i \mathcal{O}_i + O(t_i^2) . \tag{7.36}$$

The variable $\mathcal{O}_i$ is again termed relevant, marginal, or irrelevant according to the sign of y_i. We can use the scaling fields to obtain a general definition of $\mathcal{O}_i$:

$$\mathcal{O}_i = \partial \mathcal{H} / \partial g_i \quad , \tag{7.37}$$

which is meaningful for μ away from μ^* as well as in the neighborhood of μ^*. The $\mathcal{O}_i$'s are called <u>scaling variables</u> or scaling operators. They are analogous to the tensor operators of the rotation group. The scaling variable associated with the field h is just the total spin $\mathcal{O}_h = -\int d^d x \, \sigma$. That associated with t_1 is, for the Gaussian fixed point, $\mathcal{O}_1 = \int d^d x \, \sigma^2$ since t_1 is linear in r_o for our model. Note that $\mathcal{O}_i$ must be defined with respect to a fixed point since g_i is.

More generally, $\mathcal{H}$ may have, in addition to the terms considered in the Ginzburg-Landau model, other terms of various powers of σ and gradients of σ. For example,

$$\int d^d x [u_6 \sigma^6 + v_1 (\nabla \sigma^2)^2 + v_2 \sigma^2 (\nabla \sigma)^2] \ . \tag{7.38}$$

Here we need to extend the parameter space and write

$$\mu = (r_o, u, c, u_6, v_1, v_2) . \tag{7.39}$$

The Gaussian fixed point remains, in the sense that $\mathcal{K}^*$ stays the same and $\mu_o^* = (0, 0, c, 0, 0, 0)$. We obtain the linearized formulas for R_s^L

$$r_o' = s^2 [r_o + uB(1 - s^{2-d}) + u_6 B_6 (1 - s^{2-d})^2 + v_2 C_2 (1 - s^{-d})] ,$$

$$u' = s^{4-d} [u + u_6 D_6 (1 - s^{2-d})] ,$$

$$c' = c + (v_1 E_1 + v_2 E_2)(1 - s^{2-d}) ,$$

$$u_6' = s^{6-2d} u_6 ,$$

$$v_1' = s^{2-d} v_1 ,$$

$$v_2' = s^{2-d} v_2 , \tag{7.40a}$$

where B_6, C_2, D_2, E_1, E_2 are constants depending on n, d and Λ. The derivation of (7.40) is left as an exercise for the reader. The eigenvalues of R_s^L are s^{y_i}, with the values of y_i:

$$2,\ 4-d,\ 0,\ 6-2d,\ 2-d,\ 2-d\ . \tag{7.40b}$$

These values are simply those powers of s produced by the replacement (7.25), and are easily obtained by inspection. Namely, for a given $\int t_i D_i \, d^d x$ in $\mathcal{H}$, we have the addition formula for y_i

$$\begin{aligned} &\left(1 - \frac{d}{2}\right) \times (\text{power of } \sigma \text{ in } D_i) \\ &+ (-1) \times (\text{power of } \nabla \text{ in } D_i) \\ &+ d \\ &= y_i \end{aligned} \tag{7.41}$$

Thus, the more powers of σ and ∇, the more negative y_i, and therefore the more irrelevant the corresponding parameter. Clearly only r_o is relevant for $d > 4$. For $4 > d > 3$, u becomes relevant also. For $3 > d > 2$, u_6 turns relevant. For $d < 2$, all are relevant except that c is always marginal.

The terms included in (7.38) are all invariant under rotations in spin space and rotations in coordinate space. One can of course add terms which are not invariant under these rotations. The term $-h\int d^d x\, \sigma$ is the simplest example. Here h transforms as a vector under spin rotation. We can have parameters that transform as tensors and other types. Such non-invariant parameters represent anisotropic perturbations which will be discussed in Chapter IX.

Conventionally, a fixed point is called unstable if there are one or more relevant parameters other than $t_1 \propto (T - T_c)$ and h. If there is just one more such parameter, e.g., the u for the Gaussian fixed point for $3 < d < 4$ (ignoring anisotropic perturbation), the fixed point is often referred to as a tricritical fixed point. If there are two, it is called tetracritical.

For an unstable fixed point, there is one or more crossover exponent ϕ_i associated with the one or more relevant parameters, with $y_i > 0$, defined by

$$\phi_i = y_i \nu \quad . \tag{7.42}$$

The formula (6.16), which determines the critical behavior implied by μ^*, becomes

$$R_\xi \mu = \mu^* \pm e_1 + h\,|t_1|^{-\nu y_h} e_h + t_i\,|t_1|^{-\phi_i} e_i + \cdots , \tag{7.43}$$

where we have set $s = \xi = |t_1|^{-\nu}$. One can define ϕ_h by $\nu y_h = \phi_h$ if one desires. The crossover exponent simply tells how important t_i is for a given t_1. The smaller the t_1, the more important the e_i term.

4. CRITICAL EXPONENTS FOR d > 4

Now we have $\nu = 1/2$, $\eta = 0$, and $y_2 < 0$ for $d > 4$. The RG argument of Chapter VI gives

$$\alpha = 2 - \nu d = 2 - d/2 ,$$

$$\beta = \frac{\nu}{2}(d - 2 + \eta) = \frac{1}{4}(d - 2) ,$$

$$\gamma = \nu(2 - \eta) = 1$$

$$\delta = \frac{d + 2 - \eta}{d - 2 + \eta} = \frac{d + 2}{d - 2} . \tag{7.44}$$

The value of β and that of δ here do not agree with those given by (3.58) obtained by Gaussian approximation,

$$\beta = 1/2 \ ,$$
$$\delta = 3 \ , \tag{7.45}$$

although the rest of the exponents agree. We expect that the Gaussian approximation, which is based on the assumption of small u, will be valid, since, near μ_o^*, u is small. In fact, (7.45) is correct and consistent with the RG. The β and δ given by (7.44) are wrong. The reasons are as follows.

The RG arguments lead to Eq. (6.30), which can be written as

$$m(t_1, h, u) = s^{1-d/2}\, m(t_1 s^2, hs^{1+d/2}, us^{4-d}) \tag{7.46a}$$

since $y_1 = 2$, $y_2 = 4-d$ and $y_h = 1 + d/2$. If we set $h = 0$, $s = \xi = |t_1|^{-1/2}$, then we get (6.33), which is

$$m = |t_1|^{\frac{1}{4}(d-2)}\, m\left(\pm 1, 0, u\,|t_1|^{\frac{1}{2}(d-4)}\right). \tag{7.46b}$$

If we neglect the t_1 dependence of the last entry, and assume $m(\pm 1, 0, 0)$ = constant, we get $m \propto |t_1|^{1/4(d-2)}$ and thus the β given by (7.44). However, it turns out that the last entry is important and $m(\pm 1, 0, 0)$ is meaningless. In fact, Gaussian approximation gives m proportional to $u^{-1/2}$ for $t_1 < 0$ and u cannot be neglected. Equation (7.46b) and the fact that $m(t_1, 0, u) \propto u^{-1/2}$ imply that

$$m \propto |t_1|^{\frac{1}{4}(d-2)} \left(u\,|t_1|^{\frac{1}{2}(d-4)}\right)^{-1/2}$$

$$\propto |t_1|^{1/2} \tag{7.47}$$

i.e., $\beta = 1/2$, as (7.45) says.

If we set $t_1 = 0$, $s = h^{-(d/2+1)^{-1}}$ in (7.46), and again notice that $m \propto u^{-1/3}$ for $t_1 = 0$, we find $\delta = 3$. The Gaussian approximation is consistent with the RG for $d > 4$. One just has to be very careful. We emphasize again that the RG provides no explicit solution of the model. We can only deduce transformation properties. The information that $m \propto u^{-1/2}$ for $t_1 < 0$ and $m \propto u^{-1/3}$ for $t_1 = 0$ is not provided by the RG. It is provided by explicitly solving the model. The solution by Gaussian approximation happens to be adequate. Note that u is an irrelevant variable, yet it plays an important role in determining critical exponents. The fact that m must be proportional to some inverse power of u for $T \le T_c$ is a characteristic of the Ginzburg-Landau form of $\mathcal{H}$. The quartic term, which is proportional to u, is needed to keep m finite. If u is zero, then $m \to \infty$.

5. THE RG FOR $d = 4 - \epsilon$ AND FIXED POINTS TO $O(\epsilon)$

The Gaussian fixed point becomes unstable for $d < 4$. In order to search for a stable fixed point for $d = 3$, we can no longer regard u as small, and the task of performing (7.3) becomes difficult. Fortunately, we can still learn a great deal about the stable fixed point by studying the RG for d very close to 4. It turns out that the stable fixed point lies very close to μ_o^* if d is less than but very close to 4. One might think that this should be obvious by continuity. Since the stable fixed point is at $r_o^* = u^* = 0$ for $d \ge 4$, it must be close to (0, 0) for $d = 4 - \epsilon$, with small $\epsilon > 0$. Such a view is incorrect since the Gaussian fixed point μ_o^* remains a fixed point, although becoming unstable, for $d < 4$. There are now two, not one, fixed points for $d < 4$. The stable one is new, and not a continuation of the Gaussian fixed point from $d \geq 4$.

It turns out that for $d > 4$ there is another, although unphysical, fixed point with $u^* < 0$. The new fixed point for $d < 4$ can be viewed as the continuation of this unphysical fixed point. As d is decreased from above 4, u^* moves toward 0. This unphysical fixed point and the Gaussian fixed point merge as $d \to 4$. As d is decreased below 4, u^* becomes positive and the fixed point emerges as a physical fixed point. At $d = 4$, the two fixed points coincide. Some strange critical behaviors appear as consequences of such "degeneracy" of fixed points. This is why $d = 4$ is not a very simple case for studying critical behavior.

The valid reason for expecting that the stable fixed point, if any, should be close to μ_o^* for very small ϵ is the following. The growth of $u' = s^{4-d} u = s^{\epsilon} u$ as s increases has to be held back by nonlinear terms so far not included in this linearized equation. For small ϵ, the growth rate is very small. It would take only a small non-linear term to hold it back. Therefore u^* must be small. Perhaps this idea can be seen more easily if we write a differential equation

$$\frac{du'}{d\ell} = \epsilon u' , \tag{7.48}$$

$$\ell \equiv \ln s , \tag{7.49}$$

which is equivalent to $u' = s^{\epsilon} u$. To hold back the growth rate $\epsilon u'$, an additional term is needed in (7.48). The lowest nonlinear term should be proportional to u'^2:

$$\frac{du'}{d\ell} = \epsilon u' - g u'^2 \tag{7.50}$$

where g is some constant. To get a fixed point, we set $\epsilon u^* - g u^{*2} = 0$. If $g < 0$, we still have just $u^* = 0$ ($u^* < 0$ excluded). If $g > 0$, we have

$$u^* = \epsilon / g \tag{7.51}$$

in addition to $u^* = 0$. Thus u^* is small for small ϵ.

Of course, we don't know g yet. If $g = O(\epsilon)$, the above argument won't work. We have to calculate g explicitly.

The study of the RG for $d = 4 - \epsilon$ initiated by Wilson and Fisher (1972), and Wilson's subsequent method of ϵ-expansion (1972) have stimulated a great deal of progress in the field of critical phenomena.

If there is a fixed point close to μ_o^*, we should be able to locate it using the RG formula for μ close to μ_o^*. Linearized formulas are not good enough. We need to include the next order in $\mu - \mu_o^*$. To find the new formulas, we utilize the expansion formulas (7.13) choosing

$$\mathcal{H}_o = \frac{1}{2} \int d^d x \, (r_o \sigma^2 + c(\nabla\sigma)^2)$$

$$\mathcal{H}_1 = \frac{1}{2} \int d^d x \, \frac{u}{4} \sigma^4 \; . \tag{7.52}$$

We could also choose $\mathcal{H}_o = \mathcal{H}_o^*$, the Gaussian fixed point Hamiltonian, and include $r_o \sigma^2$ in $\mathcal{H}_1$. It makes no real difference. The choice of (7.52) makes the intermediate steps easier to follow. $\mathcal{H}_1$ is proportional to u. Keeping terms up to second order in $\mathcal{H}_1$ in (7.13b), we obtain

$$\mathcal{H}' + AL^d = \mathcal{H}_o' + A_o' L^d$$

$$+ \left[\langle \mathcal{H}_1 \rangle - \tfrac{1}{2}\langle (\mathcal{H}_1 - \langle \mathcal{H}_1 \rangle)^2 \rangle\right]_{\sigma_k \to s^{1-\eta/2}\sigma_{sk}} \tag{7.53}$$

where $\mathcal{H}_o'$ and A_o' are just those given by (7.4a) and (7.4b), respectively, and

$$\langle \mathcal{H}_1 \rangle = \frac{u}{8} \int d^d x \, \langle \sigma^4 \rangle \;, \tag{7.54}$$

$$-\frac{1}{2}\langle (\mathcal{H}_1 - \langle \mathcal{H}_1 \rangle)^2 \rangle = -\frac{u^2}{128} \int d^d x \, d^d y \, \langle (\sigma^4(x) - \langle \sigma^4(x) \rangle)(\sigma^4(y) - \langle \sigma^4(y) \rangle) \rangle \;. \tag{7.55}$$

In these formulas, we have to write $\sigma' + \phi$ for σ as given by (7.16) and then average over ϕ keeping σ' fixed. We shall restrict our attention to the case where u and r_o are of $O(\epsilon)$ and evaluate (7.54) and (7.55) to an accuracy of $O(\epsilon^2)$. Then we shall be able to use the results to determine the fixed point values r_o^* and u^*, if they are of $O(\epsilon)$. Since there is already $u^2 = O(\epsilon^2)$ in front of (7.55), it is sufficient to evaluate the integral with $d = 4$,

$r_o = 0$. For (7.54) we need to evaluate the integral to first order in ϵ.

Let us go through the evaluation of (7.54) and (7.55) carefully and in some detail because all RG calculations of various models to $O(\epsilon)$ must go through evaluations similar to this one.

The evaluation of $\langle \sigma^4 \rangle$ has essentially been done. It follows (7.20) and (7.21):

$$\langle \sigma^4 \rangle = \sigma'^4 + 2\sigma'^2 (1 + 2/n) \langle \phi^2 \rangle + \langle \phi^4 \rangle \qquad (7.56)$$

where $\langle \phi^2 \rangle$ is given by (7.17) with $(r_o + cq^2)^{-1}$ replacing $(cq^2)^{-1}$ for $\langle \sigma_{iq} \sigma_{i-q} \rangle$:

$$\begin{aligned} \langle \phi^2 \rangle &= nK_d \int_{\Lambda/s}^{\Lambda} dq\, q^{d-1} (r_o + cq^2)^{-1} \\ &= n_c (1 - s^{-2+\epsilon}) \\ &\quad - nK_4 c^{-2} r_o \ln s \\ &\quad + O(\epsilon^2) , \end{aligned} \qquad (7.57a)$$

where n_c is given by (7.18) and K_d by (7.19).

When we substitute (7.57a) in (7.56) and then (7.56) in (7.54), we obtain

$$\langle \mathcal{H}_1 \rangle = \frac{1}{2} \int d^d x \left(r_o^{(1)} \sigma'^2 + \frac{u}{4} \sigma'^4 \right) , \tag{7.57b}$$

$$r_o^{(1)} = (u/c) \left(\frac{n}{2} + 1 \right) \frac{1}{2} \Lambda^2 K_4 (1 - s^{-2}) + uC\epsilon$$

$$- r_o (u/c^2) \left(\frac{n}{2} + 1 \right) K_4 \ln s , \tag{7.57c}$$

where C is independent of r_o and u. Note that there is no term in $\langle \mathcal{H}_1 \rangle$ proportional to $(\nabla \sigma')^2$. Therefore c is unchanged up to $O(\epsilon)$. This means that

$$\eta = 0 , \tag{7.58}$$

at least to $O(\epsilon)$. We shall find that $(\nabla \sigma')^2$ will appear in (7.55) and therefore $\eta = O(\epsilon^2)$. The formula for r_o is obtained after carrying out the replacement (7.25) in (7.57b)

$$r_o' = s^2 (r_o + r_o^{(1)} + u^2 D)$$

$$= s^2 \Big[r_o + (u/c) \left(\frac{n}{2} + 1\right) \frac{\Lambda^2}{2} K_d (1 - s^{-2}) + uC\epsilon$$

$$- r_o (u/c^2) \left(\frac{n}{2} + 1\right) K_4 \ln s + u^2 D \Big] + O(\epsilon^3) \qquad (7.59)$$

where $u^2 D$ comes from (7.55), to be discussed later. The first three terms in the square bracket are just (7.26) with $d = 4 - \epsilon$. The $r_o u \ln s$ and $u^2 D$ terms are new nonlinear terms. As it turns out, C and D terms play no role in determining the fixed point and exponents to $O(\epsilon)$.

The u term in (7.57b) is just that in the old $\mathcal{H}_1$ and gives $u' = s^{\epsilon} u$. The nonlinear terms will come from (7.55).

Now we proceed to evaluate (7.55), which is of course much more complicated. First, we express $\sigma^4 - \langle \sigma^4 \rangle$ in terms of σ' and ϕ:

$$\sigma^4 - \langle \sigma^4 \rangle = \phi^4 - \langle \phi^4 \rangle$$

$$+ 4 ((\sigma' \cdot \phi)^2 - \langle (\sigma' \cdot \phi)^2 \rangle)$$

$$+ 2\sigma'^2(\phi^2 - \langle\phi^2\rangle)$$

$$+ 4\sigma'\cdot\phi(\sigma'^2 + \phi^2) \quad . \qquad (7.60)$$

After substituting this in (7.55), we have an extremely large number of terms, the collection of which becomes a very tedious task. The graph technique explained in Chapter IX can help us in sorting out terms and writing down the integrals that we need to evaluate. Of course, we do not need graphs to calculate (7.55). We can simply do it by brute force. It is tedious, but not difficult. The result is

$$-\frac{1}{2}\langle(\mathcal{H}_1 - \langle\mathcal{H}_1\rangle)^2\rangle$$

$$= -u^2\int d^dx\, d^dy \left[\frac{1}{128}\langle(\phi^4(x) - \langle\phi^4\rangle)(\phi^4(y) - \langle\phi^4\rangle)\rangle \right. \qquad (a)$$

$$+ \frac{1}{8}\sigma'^2(x)(n+2)^2\, G^2(x-y)\, G(0) \qquad (b)$$

$$+ \frac{1}{4}\sigma'(x)\cdot\sigma'(y)(n+2)\, G^3(x-y) \qquad (c)$$

$$+ \frac{1}{16}\sigma'^2(x)\,\sigma'^2(y)(n+4)\, G^2(x-y) \qquad (d)$$

$$+\frac{1}{4}(\sigma'(x)\cdot\sigma'(y))^2\, G^2(x-y) \qquad \text{(e)}$$

$$+\frac{1}{2}\sigma'^2(x)\,\sigma'(x)\cdot\sigma'(y)(n+2)\, G(x-y)\, G(0) \qquad \text{(f)}$$

$$+\frac{1}{4}\sigma'^2(x)\,\sigma'(x)\cdot\sigma'(y)\,\sigma'^2(y)\, G(x-y)\Big] \qquad \text{(g)}$$

(7.61)

where

$$\delta_{ij}\, G(x-y) = \langle \phi_i(x)\, \phi_j(y)\rangle \quad ,$$

$$G(0) = \langle \phi_i(x)\, \phi_i(x)\rangle = \langle \phi^2\rangle /n \ . \qquad (7.62)$$

The function $G(r)$ is given by

$$G(r) = L^{-d} \sum_{\Lambda/s < q < \Lambda} e^{iq\cdot r}\, q^{-2}\, c^{-1} \qquad (7.63a)$$

$$= (2\pi)^{-2}\, r^{-2}\, c^{-1}\, (J_o(\Lambda r/s) - J_o(\Lambda r)),\ d = 4\ ,$$

(7.63b)

where J_o is a Bessel function. Intuitively, we expect $G(r)$ to fall off to zero very fast for $r > s\Lambda^{-1}$ because $G(r)$ contains Fourier components for $q > \Lambda/s$ only, and we have the qualitative picture of making blocks of size $s\Lambda^{-1}$. However, G(r) does not turn out as expected. Instead, it has a long oscillating tail arising from the Bessel function in (7.63b). This oscillating tail is a consequence of the sharp cutoff in the q integral. It will be washed out if we integrate r over an odd power of $\mathcal{G}(r)$, but if we integrate over an even power of $\mathcal{G}(r)$, we may run into trouble, as we shall see later.

Term (a) in (7.61) is a constant independent of σ' and is a contribution to A'. Term (b) has the form $\sigma'^2(x) \cdot$ constant and can be identified as a contribution to r_o', i.e., the $u^2 D$ term of (7.59). The rest of the terms in (7.61) involve $\sigma'(x)$ and $\sigma'(y)$, and apparently cannot be identified as contributions to r_o' or u'. However, we note that only Fourier components with $k < \Lambda/s$ are left in σ'. Thus $\sigma'(x)$ is not much different from $\sigma'(y)$ if $|x - y| < s\Lambda^{-1}$. If the function $\mathcal{G}(x-y)$ falls off very rapidly for $|x - y| > s\Lambda^{-1}$, as we expected intuitively, then the integrand of (7.61) is small unless x and y are within a distance $s\Lambda^{-1}$. Let us write $\sigma'(y) = \sigma'(x + r)$, $r = y - x$, and expand in powers of r:

$$\sigma'(y) = \sigma'(x) + r \cdot \nabla\sigma(x) + \frac{1}{2}(r \cdot \nabla)^2 \sigma(x) + \cdots \quad . \tag{7.64}$$

Substituting this expansion in (7.61) and integrating over r, we get the so called gradient expansion for (7.61). The reason for this gradient expansion is not that it converges rapidly (in fact it does not), but that with its help we can easily separate the part of (7.61) which has the Ginzburg-Landau form as the lowest-order terms of the expansion. We shall study such lowest-order terms first. The first term of the expansion, i.e. $\sigma'(y) = \sigma'(x)$ of (7.64), contributes to (7.61) the terms

$$\frac{1}{2}\int d^d x \left(\sigma'^2(x)\, u^2 D + \sigma'^4(x)\, \frac{\Delta u}{4}\right) , \tag{7.65}$$

where

$$u^2 D = -u^2 \int d^d r \left[\frac{1}{4}(n+2)^2\, \mathcal{G}^2(r)\, \mathcal{G}(0) + \frac{1}{2}(n+2)\, \mathcal{G}^3(r)\right] , \tag{7.66}$$

$$\Delta u = -u^2 \int d^d r \left(\frac{n}{2} + 4\right) \mathcal{G}^2(r) \quad . \tag{7.67}$$

Note that (f) and (g) of (7.61) do not contribute because $\int d^d r\, \mathcal{G}(r) = 0$. The quantity $u^2 D$ contributes to r_o', and Δu to u'. The integral of (7.67) is easily evaluated. Setting $d = 4$ and using (7.63) for $G(r)$, we obtain

$$\int d^4 r\, \mathcal{G}^2(r) = \frac{K_4}{c^2} \int_0^\infty \frac{dr}{r} \left(J_o(\Lambda r/s) - J_o(\Lambda r)\right)^2$$

$$= \frac{K_4}{c^2} \ln s \quad . \tag{7.68}$$

Note that $K_4 = (8\pi^2)^{-1}$, and

$$\int_0^\infty \frac{dx}{x} J_o(x)\left(1 - J_o(sx)\right) = \theta(s-1) \ln s \quad .$$

The integral is more obvious in wave vector space:

$$\int d^4 r\, \mathcal{G}^2(r) = (K_4/c^2) \int_{\Lambda/s}^{\Lambda} q^3\, dq\, (cq^2)^{-2}$$

$$= (K_4/c^2) \ln s \quad .$$

If we ignore the oscillating part of $J_o(\Lambda r)$ and regard $J_o(\Lambda r)$ as a step function $\theta(\Lambda^{-1} - r)$, then $J_o(\Lambda r/s) - J(\Lambda r) \sim \theta(r - \Lambda^{-1})\,\theta(s\Lambda^{-1} - r)$. We would still get the same result as in (7.68). Apparently the result is insensitive to the oscillating tail of $J_o(\Lambda r)$.

Substituting (7.68) in (7.67), we readily obtain the formula for u':

$$u' = s^{\epsilon}(u + \Delta u)$$

$$= s^{\epsilon}(u - (u^2/2c^2)(n+8)\,K_4 \ln s) , \qquad (7.69)$$

where the factor s^{ϵ} is the result of the replacement (7.25). This is an important result. It shows that u' is determined to $O(\epsilon^2)$ by u alone. Before justifying this result [it should be recalled that we have not examined all the terms in (7.61) and have only kept the first term in the gradient expansion (7.64)], let us examine the consequences of (7.69). We write $u' = u = u^*$, and $1 + \epsilon \ln s$ for s^{ϵ} in (7.69). We get

$$\ln s(\epsilon u^* - (u^{*2}/2c^2)(n+8)K_4) = 0 .$$

Besides the solution $u^* = 0$, which is for the Gaussian fixed point μ_o^*, there is the solution

$$u^*/c^2 = 2\epsilon\,(n+8)^{-1}\,K_4^{-1} \; , \qquad (7.70)$$

as was expected from (7.51). We can then obtain r_o^* from (7.59) to $O(\epsilon)$. Only the first two terms on the right-hand side are needed. We obtain

$$\begin{aligned} r_o^* &= -(u^*/c)\left(\frac{n}{2}+1\right)\frac{\Lambda^2}{2}K_4 \\ &= -\left(\frac{n+2}{n+8}\right)\epsilon\,\frac{\Lambda^2 c}{2} \end{aligned} \qquad (7.71)$$

with u^* given by (7.70). Let $\delta r_o = r_o - r_o^*$ and $\delta u = u - u^*$. The linearlized formula

$$\begin{pmatrix} \delta r_o' \\ \\ \delta u' \end{pmatrix} = R_s^L \begin{pmatrix} \delta r_o \\ \\ \delta u \end{pmatrix} \qquad (7.72)$$

is easily obtained by linearizing (7.59) and (7.69) in the neighborhood of (r_o^*, u^*). One finds

$$R_s^L = \begin{pmatrix} s^{y_1} & B(s^{y_1} - s^{y_2}) \\ 0 & s^{y_2} \end{pmatrix} \tag{7.73a}$$

$$y_1 = 2 - \frac{n+2}{n+8}\epsilon = 1/\nu \quad , \tag{7.73b}$$

$$y_2 = -\epsilon \; , \tag{7.73c}$$

$$B = \left(\frac{n}{2} + 1\right) \frac{\Lambda^2}{2} K_4 \; . \tag{7.73d}$$

The intermediate steps are recommended as a simple exercise. It should be noted that the C and D terms of (7.59) are not needed in determining y_1, but the $r_o u \ln s$ term is needed. Without evaluating C and D, we know the $B(s^{y_1} - s^{y_2})$ element of the matrix (7.73a) only as

$$B(s^2 - 1) + O(\epsilon) \; . \tag{7.74}$$

Our lack of knowledge of the $O(\epsilon)$ term does not affect the determination of the eigenvalues s^{y_1}, s^{y_2} of R_s^L because the lower left element turns out to be zero. This happy situation reflects the fact that (7.69) is independent of r_o.

The fact that $y_2 = -\epsilon$ is now negative means that the fixed point is stable. The stabilization is owing to the nonlinear term of (7.69). The Gaussian fixed point and the new fixed point are shown in Fig. 7.2. The critical surface is given by

$$t_1 \equiv r_o + u\left(\frac{1}{2}n + 1\right)\frac{1}{2c}\Lambda^2 K_4 = 0 \qquad (7.75)$$

which is accurate to $O(\epsilon)$. In Fig. 7.2 we also display the "flow pattern" or "stream lines" which indicate how R_s pushes various points in the parameter space.

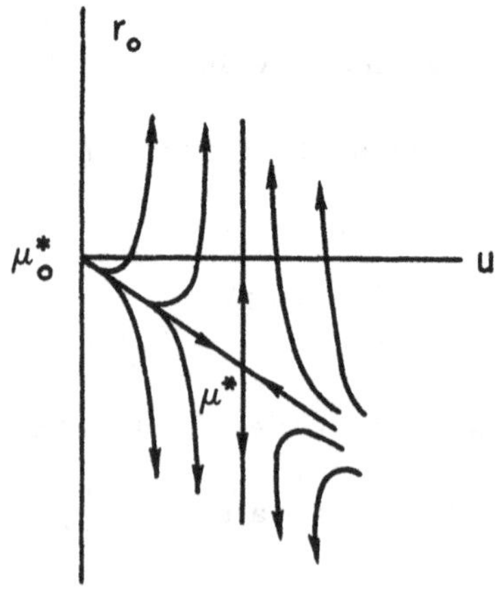

Figure 7.2. The Gaussian fixed point μ_o^* and the stable fixed point μ^* for $d = 4 - \epsilon$, $\epsilon > 0$.

Now we have a complete picture to $O(\epsilon)$. Using ν given by (7.73), and that $\eta = 0$, other critical exponents can be obtained via the scaling laws discussed in Chapter VI. The results are given in Table 7.1.

The important role played by the parameter $u - u^*$ in the case $d > 4$ discussed earlier is not expected here. For the Gaussian fixed point $(d > 4)$ u is driven to zero by R_s. Here u goes to u^*, a constant. The magnetization

Table 7.1

Critical exponents to $O(\epsilon)$

$$\alpha = \frac{(4-n)\epsilon}{2(n+8)}$$

$$\beta = \frac{1}{2} - \frac{3\epsilon}{2(n+8)}$$

$$\gamma = 1 + \frac{(n+2)\epsilon}{2(n+8)}$$

$$\delta = 3 + \epsilon$$

$$\eta = 0$$

$$\nu = \frac{1}{2} + \frac{(n+2)\epsilon}{4(n+8)}$$

m for $T \leq T_c$ is still proportional to some inverse power of u, but not $u - u^*$. Ignoring $(u - u^*)s^{-\epsilon}$ is reasonably safe.

The important concept of critical region was discussed near the end of Chapter VI. The size of the critical region depends crucially on y_2, which measures how fast $R_s\mu$ moves toward μ^* as s increases. Since $y_2 = -\epsilon$, the motion toward μ^* is very slow for small ϵ. This is what is often referred to as the "slow transient" behavior of $R_s\mu$ for small ϵ. Substituting $y_2 = -\epsilon$ in Eqs. (6.48)-(6.50), we obtain the size of the critical region. For example, (6.48)

$$|T - T_c| < 10^{-2/\epsilon} \times \text{constant} \qquad (7.76)$$

which is exponentially small for small ϵ. This means that T must be extremely close to T_c in order for us to observe the behavior described by the critical exponents of Table 7.1.

In the above discussion we have freely exponentiated the logarithm of s, for example replacing $1 - \epsilon \ln s$ by $s^{-\epsilon}$, to obtain exponents. How is this justified? Obviously it is not justified if s is very large. Large s is just what we need to discuss critical behavior.

Such exponentiations are extrapolations. They are based on the form s^{y_i} of the eigenvalues of R_s^L. By construction, $R_s^L R_{s'}^L = R_{ss'}^L$. Of course, we have not constructed R_s to all orders in ϵ, and therefore we cannot observe $s^{-\epsilon}$, but only $1 - \epsilon \ln s$. We are assuming that if we constructed R_s^L exactly, obtained the eigenvalues, and then expanded in powers of ϵ, we would get, to $O(\epsilon)$, $1 - \epsilon \ln s$ from s^{y_2}. We are making an approximation on the exponent. A small error on y_i would of course produce a large error on s^{y_i} if s is very large.

6. EFFECT OF OTHER $O(\epsilon^2)$ TERMS IN $R_s\mu$

So far we have taken into account only the first term in the gradient expansion (7.64). This amounts to ignoring the variation of $\sigma(x)$ over a distance $s\Lambda^{-1}$. We must now examine the higher-order terms in the gradient expansion. We shall see from some of the unpleasant complications that our formulation of the RG is still primitive.

The linear term in r will not contribute to (7.61), because $G(r) = G(-r)$ and r will be integrated over. In fact only even powers of r contribute. Let us collect the contribution of r^2 terms to (7.61),

$$\text{(c)} \quad \int d^d x \, \frac{1}{2} (\nabla \sigma')^2 \, \frac{u^2}{4} (n+2) \int d^d r \, \frac{r^2}{d} \, \not{G}^3(r)$$

$$\text{(d)} \quad \int d^d x \, \frac{1}{2} (\nabla \sigma'^2)^2 \, \frac{u^2}{16} (n+4) \int d^d r \, \frac{r^2}{d} \, \not{G}^2(r)$$

$$\text{(e)} \quad \int d^d x \left[\frac{1}{4} (\nabla \sigma'^2)^2 + \sigma^2 (\nabla \sigma)^2 \right] \frac{u^2}{4} \int d^d r \, \frac{r^2}{d} \, G^2(r) \tag{7.77}$$

where (c), (d), (e) refer to those in (7.61). We have used integration by parts to write $f(x) \nabla^2 g(x)$ as $-\nabla f(x) \cdot \nabla g(x)$. Also we have replaced $r_\alpha r_\beta$ by $\delta_{\alpha\beta} r^2/d$ on account of the spherical symmetry of $\not{G}(r)$. Note that there is still no contribution from (f) and (g) of (7.61) because $\int d^d r \, r^2 G(r) = 0$. In fact, $\int d^d r \, r^m G(r) = 0$ for any finite m because $G(r)$ contains no Fourier component with $q < \Lambda/s$.

The (c) term of (7.77) has the form $\int d^d x \, \frac{1}{2} (\nabla\sigma')^2 \Delta c$, which contributes to c'. Setting $d = 4$ in (7.77c) and using (7.63b) for $G(r)$, we obtain

$$\Delta c = \frac{u^2}{2(8\pi^2)^2 c^3} (n+2) \int_0^\infty \frac{dr}{r} (J_0(r\Lambda/s) - J_0(r\Lambda))^3$$

$$= \frac{u^2}{2(8\pi^2)^2 c^3} (n+2)(\ln s + Q) \ , \qquad (7.78)$$

where

$$Q = -3 \int_0^\infty \frac{dr}{r} J_0(r)(1 - J_0(r/s))(J_0(r/s) - J_0(r)) . \qquad (7.79)$$

Note that the r integral of (7.78) is independent of Λ. The results of (7.78) and (7.79) can be roughly understood as follows. If we ignore its oscillating tail, $J_0(r)$ would look roughly like a step function, i.e., $J_0(r) \sim \theta(1 - r)$. Likewise $J_0(r/s) \sim \theta(s - r)$ and $J_0(r/s) - J_0(r) \sim \theta(r - 1)\,\theta(s - r)$, which is nonzero only over $1 < r < s$. Thus the r integral of (7.78) is roughly $\int_1^s dr/r = \ln s$. The integrand of (7.79) would be $\theta(1 - r)\,\theta(r - s) \times \theta(r - 1)\,\theta(s - r) = 0$ in this rough picture. Therefore (7.79) is due entirely to the oscillating part of the J_o's and is a result of the sharp cutoff in wave vector q. If we formulate the RG with a different kind of cutoff, (7.79) will be different, but the $\ln s$ term would remain the same.

The transformation formula for c is

$$c' = s^{-\eta}(c + \Delta c + \Delta \tilde{c})$$

$$= s^{-\eta}\left[c + \frac{u^2}{2(8\pi^2)^2 c^3}(n+8)(\ln s + Q) + \Delta\tilde{c}\right] , \qquad (7.80)$$

which is the same as (7.5) except for the Δc, $\Delta\tilde{c}$ terms. At the fixed point, u^* is given by (7.70), and $8\pi^2$ is just K_4^{-1}. We adjust η to remove the logarithmic term:

$$\eta = \frac{u^{*2}}{2(8\pi^2)^2 c^4}(n+2)$$

$$= \frac{\epsilon^2 (n+2)}{2(n+8)^2} . \qquad (7.81)$$

The $\Delta\tilde{c}$ in (7.80) is the contribution from <u>extra terms</u> which we have not included in the Ginzburg-Landau form of $\mathcal{H}$. At the fixed point $\Delta\tilde{c}$ must cancel the Q term so that $c = c'$. We now turn our attention to these extra terms.

So far we have studied those terms in (7.61) which are of the same form as those in the Ginzburg-Landau Hamiltonian. The (d) and (e) terms of (7.77) and all terms from higher orders of the gradient expansion are not of the same form as the Ginzburg-Landau $\mathcal{H}$; $(\nabla\sigma'^2)^2$ and $\sigma'^2(\nabla\sigma')^2$ of (7.77) are the simplest among them. In other words, R_s does not preserve the Ginzburg-Landau form. It generates additional terms of different form. We need more entries in μ in addition to (r_o, u, c) in order to define completely the action of R_s. Let us write

$$\mu = (r_o, u, c; \mu_2)$$

$$\mathcal{H} = \mathcal{H}_{GL} + \mathcal{H}_2$$

$$= \mathcal{H}_o + \mathcal{H}_1 + \mathcal{H}_2 \qquad (7.82)$$

where μ_2 includes all additional entries needed, and $\mathcal{H}_2$ gives the extra terms specified by μ_2. The kind of terms that are generated from $\mathcal{H}_{GL}$ by R_s to $O(\epsilon^2)$ must be included in $\mathcal{H}$ to start with in order to have a consistent set of formulas for $\mu' = R_s \mu$ to $O(\epsilon^2)$. In other words,

$$\mathcal{H}_2 = \int d^d x \left[v_1 (\nabla \sigma^2)^2 + v_2 \sigma^2 (\nabla \sigma)^2 \right.$$

$$+ \text{ higher gradients}$$

$$\left. + \int d^d r \, \sigma^2(x) \, \sigma(x) \cdot \sigma(x+r) \, \sigma^2(x+r) \, u_6(r) \right] \tag{7.83}$$

where the "higher gradients" include $\sigma^2(\nabla^2\sigma)^2$, $(\nabla^2\sigma^2)^2$, etc., with more ∇'s but always with four powers of σ. The last term has the form of (7.61g). The function $u_6(r)$ has only Fourier components with wave vectors greater than Λ. This term cannot be approximated by a finite number of terms in the gradient expansion. The μ_2 in (7.82) must include v_1, v_2, and all parameters specifying the higher gradients and $u_6(r)$. Note that only the last term of (7.83) has six powers of σ. The rest have only four. It is consistent to take $v_1, v_2, \ldots, u_6$ as of $O(\epsilon^2)$. To this order, we need not include terms with more than six powers of σ. The reason is that to $O(\epsilon^2)$, the contribution of $\mathcal{H}_2$ to the transformed Hamiltonian $\mathcal{H}'$ is just

$$\langle \mathcal{H}_2 \rangle_{\sigma_k \to s\sigma_{sk}} \tag{7.84}$$

which does not give any term with more than six powers of σ. To $O(\epsilon^2)$, the fixed point must include μ_2^* in addition to r_o^* and u^*. It is consistent to keep only terms in $\mathcal{H}^*$ with six or fewer powers of σ.

Our concern now is not the explicit values of $v_1^*, v_2^*, \ldots, u_6^*(r)$, but the question of whether $\mathcal{H}_2$ generates an extra contribution to Δu of (7.67), which is of $O(\epsilon^2)$. If it does, then our u^* previously obtained to $O(\epsilon)$ will be changed and, if so, we need to know whether the values of exponents to $O(\epsilon)$ will alter. [Do not forget that to determine u^* to $O(\epsilon)$, one needs some information about $R_s\mu$ to $O(\epsilon^2)$.] The contribution to Δu from $\mathcal{H}_2$ is obtained from (7.84) as the coefficient of $(1/8)\int d^dx\, \sigma'^4$ in $\langle \mathcal{H}_2 \rangle$. In view of (7.83), the gradient terms cannot give such a contribution. The only possibility is from the $u_6(r)$ term, which gives

$$\int d^d x \, d^d r \, \sigma'^2(x) \, \sigma'^2(x+r) \, u_6(r) \langle \phi(x) \, \phi(x+r) \rangle \, . \tag{7.85}$$

To get σ'^4, we ignore the r in $\sigma'^2(x+r)$ and obtain

$$\int d^d x \, \sigma'^4 \int d^d r \; u_6(r) \, \mathcal{G}(r) \; . \tag{7.86}$$

Since $u_6(r)$ contains Fourier components with wave vectors greater than Λ, while $\mathcal{G}(r)$ contains only those smaller than Λ, the r-integral of (7.86) must vanish. We thus conclude that $\langle \mathcal{H}_2 \rangle$ does not contribute to Δu and our previous results to $O(\epsilon)$ are not modified. There is a contribution from $\langle \mathcal{H}_2 \rangle$ to r_o, but, like the $uC\epsilon$ and $u^2 D$ terms in (7.59), it makes no difference in our results to $O(\epsilon)$. The reason is again that the equation for u' alone determines u^*, which in turn determines v_1^* and v_2^* without involving r_o. This completes our discussion of results to $O(\epsilon)$.

$\langle \mathcal{H}_2 \rangle$ will contribute to $\Delta \tilde{c}$ of (7.80), but we shall not compute it. There are more serious problems with the (d) and (e) of (7.77). The integral

$$\int d^4r\; r^2\, \mathcal{G}^2(r) \propto \int_0^\infty dr\; r\,(J_o(\Lambda r/s) - J_o(\Lambda r))^2 \qquad (7.87)$$

actually diverges owing to the oscillating tail of J_o. The square of the oscillating tail is always positive, which makes the integrand go like $rr^{-1} \sim$ constant for $r \to \infty$, since $J_o \sim r^{-1/2} \cos(\Lambda r - \pi/4)$. As a result, we get a long-range interaction in $\mathcal{H}'$. This long-range interaction is a purely mathematical artifact due to the sharp cutoff in q-integration and has no physical consequence. To remove it, we need to use a smooth cutoff, that given by (A.19), for example. The calculation of the RG with a smooth cutoff is much more complicated, and has been done only so far as verifying (7.81) for the $O(\epsilon^2)$ term of η.

It should be noted that even if we use a smooth cutoff and thereby remove the infinities like (7.87), the gradient expansion is still not useful for studying the RG except for very special cases. The reason is that we need to keep track of spin variations down to a scale of a block size in order to define the RG completely. For the fixed points and exponents to $O(\epsilon)$, we barely got by without getting into these troubles.

The reader very probably feels uneasy and dubious because of these various complications which already appear in $O(\epsilon^2)$. It seems quite hopeless to carry out R_s explicitly to $O(\epsilon^3)$. This feeling is entirely justified. In fact, it is one of the purposes of the above lengthy discussion to emphasize that our understanding and command of the RG is still at a very primitive stage. A useful approximate method of carrying out R_s in general is still not available in spite of the rapid development of this field. The RG carried out to $O(\epsilon)$ for various other models, some of which will be discussed in later chapters, accounts for much of the success of the RG in studying critical phenomena. In every case, it is carried through the way discussed above. The exponents have been calculated to higher orders in ϵ, without working the RG out to higher orders in ϵ, by the ϵ-expansion method of Wilson (1972), which we shall discuss in Chapter IX.

The reader undoubtedly wonders whether, as a tool for computation, the whole business of the RG is worthwhile if it is so complicated. Might not one do just as well by solving the model with a computer in the old-fashioned way? Computer calculation has been necessary in carrying out some approximate RG. However, let us emphasize again that the difficulty of carrying out the RG is still much less than that of solving the model because R_s is a transformation of nonsingular parameters and has much more room for approximations. Some of the numerical work will be discussed briefly in the next chapter.

We conclude this chapter with a remark on the implication of the simplicity of fixed points for $d > 4$ and for $d = 4 - \epsilon$ with very small ϵ. The mathematical complication stems from the large and strongly interacting spin fluctuations. This point was stressed in Chapter III, where we also argued that the smaller d is, the more important these fluctuations are in determining critical behavior. For the models studied above, the interaction among such fluctuations is unimportant for $d > 4$. The Gaussian fixed point is stable. Points near this fixed point describe weakly interacting spin fluctuations.

For d below 4, the interaction becomes important and the behavior of long wavelength modes is strongly affected. One cannot use u as a small parameter in solving the Ginzburg-Landau model for $d < 4$. The advantage of the RG approach is apparent. We have been able to use u as a small parameter in studying the RG for small ϵ, and get the exponents. The distinction between the model of interest and the fixed point must be borne in mind. The strength of interaction of a realistic model is usually strong and depends on the details of the physical system of interest. The fixed point is a mathematical object defined by $R_s \mu^* = \mu^*$. The weakness of the interaction at the fixed points discussed above certainly does not imply that in general interactions are weak for d above or near 4. The hypothesis is that even for a strongly interacting model near its critical temperature, the RG will transform the model into that described by a point in the neighborhood of the fixed point.

VIII. RENORMALIZATION GROUPS IN SELECTED MODELS

SUMMARY

We continue our study of the basic structure of the RG. First we shall briefly touch upon the case of $n \to \infty$, i.e., an infinite number of spin components. The RG can be worked out exactly in this case. Then we introduce the famous approximate recursion formula of Wilson (1971). We shall derive it, state the approximations involved, and then apply it to the $n \to \infty$ case as an illustration. The RG defined for discrete spins will be introduced next. Numerical results of some two-dimensional calculations are sketched. Questions involved in the definitions and the truncation necessary for carrying out the RG transformations are discussed. The main purpose of this chapter is to point out some general features of the kind of approximations or truncations involved in all the RG analyses so far. We want to make clear the fact that much remains to be learned about some very basic definitions and some fundamental aspects of the RG before we discuss its successful applications in the next few chapters.

1. THE RG IN THE LARGE-n LIMIT

There is another case, besides the small ϵ case discussed in the previous chapter, in which fixed points can be determined easily. That is the case of $n \to \infty$, i.e., an infinite number of spin components. The reason for the simplicity in the large-n limit is qualitatively explained as follows.

The terms in $\mathcal{H}$ which cause mathematical complications are the σ^4 and possibly σ^6, etc., i.e., terms which are not quadratic, in an isotropic spin model. For large n, $\sigma^2 = \sum^n \sigma_i^2$ is the sum of a large number of terms. We thus expect σ^2 to have very small fractional fluctuation $\sim 1/\sqrt{n}$. This turns out to be true for $d > 2$. When such small fractional fluctuation is ignored, $\sigma^2 \approx \langle \sigma^2 \rangle$

becomes a constant. The spin fluctuations become independently coupled to an average field. This simplification allows one to obtain fixed points and exponents without much mathematical labor.

The values of n of physical interest are of course not large, $n = 1$ for Ising systems, $n = 2, 3$, for XY and Heisenberg models. There are certain crystal symmetry critical points with order parameters of $n = 4, 8$ (Mukamel, 1975). The RG analysis for $n \to \infty$ serves as the basis for the $1/n$ expansion of exponents. Such expansion will be discussed, together with the ϵ expansion, in the next chapter, where perturbation methods will be explored.

For the purposes of our discussion the $n \to \infty$ limit means that we have finite but large n and keep only the leading order in $1/n$ in calculations. Stanley (1968) showed that the $n \to \infty$ limit produces the same thermodynamics as the spherical model of Berlin and Kac (1952). The spherical model is essentially a Gaussian model (of any n):

$$\mathcal{H} = \frac{1}{2} \int d^d x \, (c(\nabla\sigma)^2 + r_o \sigma^2) \tag{8.1a}$$

with a constraint

$$\int d^d x \langle \sigma^2 \rangle = \text{fixed constant} . \qquad (8.1b)$$

This constraint is an attempt to simulate the effect of restricting the magnitude of σ, i.e., the effect of the σ^4 term in the Ginsburg-Landau Hamiltonian. It is much weaker. The σ^4 term restricts the value of σ^2 in every block while Eq. (8.1b) is only a constraint on the average over the whole system.

In the following we shall briefly summarize what we know about the RG in the large-n limit. An attractive feature of the large-n limit is that there is no restriction on d. Some properties of the RG which are difficult to see in the small-ϵ case are easily revealed by the $n \to \infty$ analysis. Many of the scaling fields and scaling variables are available in the $n \to \infty$ limit but not easy to work out for small ϵ. We shall skip most of the details of the analysis, which can be found in Ma (1973b, Sec. IV; 1974a, Sec. IV). The reader is cautioned, however, that these papers use a different notation convention.

The Ginzburg-Landau form of $\mathcal{H}$ must be extended

in order to describe the RG in the $n \to \infty$ limit. It is sufficient to use a parameter space defined as follows. Let

$$\mathcal{H} = \int d^d x \left(\frac{1}{2} c(\nabla\sigma)^2 + U(\sigma^2) \right) ,$$

$$U(\sigma^2) = \sum_{m=1}^{\infty} \frac{1}{m} u_{2m} (\sigma^2/2)^m . \tag{8.2}$$

The parameter space is defined by values of the infinite set

$$\mu = (u_2, u_4, \ldots ; c) . \tag{8.3}$$

Or, we can simply regard the parameter space as the space of

$$\mu = (U; c)$$

i.e., the direct product space of a positive real number and the space of real polynomials.

An important function is

$$t(\sigma^2) = 2\partial U(\sigma^2)/\partial\sigma^2$$

$$= \sum_{m=1}^{\infty} u_{2m} (\sigma^2/2)^{m-1} . \tag{8.4}$$

We can also write $\mu = (t; c)$, since t contains the same parameters as U does.

The meaning of $t(\sigma^2)$ is seen as follows. We mentioned earlier that the fractional fluctuation of σ^2 is small for large n. Thus $\sigma^2 - \langle \sigma^2 \rangle$ can be considered small compared to $\langle \sigma^2 \rangle$. Here $\langle \ldots \rangle$ is the usual statistical average over $\exp(-\mathcal{H})$. Expanding $U(\sigma^2)$, we have

$$U(\sigma^2) \approx U(\langle \sigma^2 \rangle) + \frac{1}{2} t(\langle \sigma^2 \rangle)(\sigma^2 - \langle \sigma^2 \rangle) + \cdots \tag{8.5}$$

since $\frac{1}{2} t(\sigma^2) = \partial U / \partial \sigma^2$. Thus (8.2) is, for large n,

$$\mathcal{H} \approx \frac{1}{2} \int d^d x \, (c(\nabla \sigma)^2 + t(\langle \sigma^2 \rangle) \sigma^2) + L^d \left(U(\langle \sigma^2 \rangle) - \frac{1}{2} t(\langle \sigma^2 \rangle) \langle \sigma^2 \rangle \right) . \tag{8.6}$$

The first term is quadratic in σ and the second term is a constant. Thus $\exp(-\mathcal{H})$ is approximately a Gaussian distribution for σ, and $t(\langle \sigma^2 \rangle)$ plays the role of r_o in (7.1). Namely, $t(\langle \sigma^2 \rangle)$ is the effective field seen by σ^2. It plays the role of the self-consistent field in the Hartree approximation in atomic and many-body theory.

The transformation $\mu' = R_s \mu$ can be expressed through the function

$$t'(\sigma^2) = \sum_{m=1}^{\infty} u'_{2m} (\sigma^2/2)^{m-1} , \tag{8.7}$$

which is determined by solving the equations

$$t'(\sigma^2) = s^2 \, t(\langle \phi^2 \rangle + s^{2-d} \sigma^2) , \tag{8.8a}$$

$$\langle \phi^2 \rangle = nK_d \int_{\Lambda/s}^{\Lambda} dq \, q^{d-1} [cq^2 + t'(\sigma^2)/s^2]^{-1} , \tag{8.8b}$$

$$c' = c , \tag{8.8c}$$

$$\eta = 0 . \tag{8.8d}$$

Here ϕ has the same meaning as in (7.16) and $\langle \phi^2 \rangle$ is the same as in (7.17) except for the extra t'/s^2 in the denominator. [Equation (8.8) is the same as (4.35) of Ma (1973b).]

Equation (8.8) is more complicated than those for the case of small ϵ. An infinite number of parameters is involved in t and t'. The determination of the fixed points and exponents is less straightforward. Let us summarize the results.

(i) For $d \geq 4$, there is only the stable Gaussian fixed point, $t^*(\sigma^2) = 0$.

(ii) For $2 < d < 4$, the Gaussian fixed point is still there but becomes unstable. The stable fixed point $t^*(\sigma^2)$ can be obtained by solving the equation

$$\sigma^2/n_c = 1 - (d-2)\,\Lambda^{2-d} \int_\Lambda^\infty dp\, p^{d-1}\,[(t^* + p^2)^{-1} - p^{-2}] \tag{8.9}$$

where n_c is given by (7.18). We can obtain

$$\mathcal{H}^* = \int d^d x \left(\frac{1}{2}\, c(\nabla\sigma)^2 + U^*(\sigma^2) \right) \quad \text{via}$$

$$U^*(\sigma^2) = \frac{1}{2} \int_0^{\sigma^2} d\lambda\, t^*(\lambda)\ . \tag{8.10}$$

Representative plots of U^* and t^* are given in Figs. 2 and 3 of Ma (1973b). Note that for every positive value of c there is a fixed point, just as in the cases in the previous chapter. Some other properties are given in Table 8.1.

Table 8.1

Some properties of the stable fixed point of $2 < d < 4$, $n \to \infty$

Critical Surface: $t(n_c) = 0$

Exponents:

$\eta = 0$	$\alpha = (d-4)/(d-2)$
$y_1 = 1/\nu = d-2$	$\beta = 1/2$
$y_2 = d-4$	$\gamma = (d/2 - 1)^{-1}$
	$\delta = (d+2)/(d-2)$

$$n_c = \frac{n}{c} K_d \Lambda^{d-2} (d-2)^{-1}$$

The scaling fields, defined by (7.34), namely, those functions g_i of μ with the simple property $g_i(R_s\mu) = g_i(\mu) s^{y_i}$, can be easily found using (8.8). First we solve the equation $t = t(\sigma^2)$ and get $\sigma^2 = f(t)$. The function f of course depends on μ. We also solve $t' = t'(\sigma^2)$ for σ^2 to obtain $\sigma^2 = f'(t')$ where $f'(t')$ is just $f(t)$ with μ', t' replacing μ, t, respectively. From (8.8a), we obtain

$$f(t'/s^2) = s^{2-d}\sigma^2 + \langle\phi^2\rangle ,$$

which implies

$$f'(t') = s^{d-2} f(t'/s^2) - s^{d-2}\langle\phi^2\rangle . \qquad (8.11)$$

Here $\langle\phi^2\rangle$ is a function of t'/s^2 as given by (8.8b). Equation (8.11) is just another way of writing (8.8). The transformation of f to f' gives all the information concerning $R_s\mu = \mu'$. Here t' can be viewed as just a dummy variable. The fixed point $f^*(t')$ is obtained by setting $f = f' = f^*$ in (8.11) and then solving for f^*. This is easily done by expanding everything in powers of t', and is left as an exercise for the reader. Subtracting the fixed point equation from (8.11), we obtain

$$f'(t') - f^*(t') = s^{d-2}(f(t'/s^2) - f^*(t'/s^2)) . \qquad (8.12)$$

We expand $f(t') - f^*(t')$ in powers of t':

$$f(t') - f^*(t') = \sum_{m=1}^{\infty} g_m t'^{m-1} . \qquad (8.13a)$$

Then (8.12) tells us that

$$g_m' = s^{y_m} g_m ,$$

$$y_m = d - 2m, \quad m = 1, 2, 3, \ldots \quad . \tag{8.13b}$$

The g_m's are of course scaling fields. Within the parameter space defined by (8.2) with a fixed c, (8.13b) gives all exponents for the stable fixed point. Only $y_1 = d - 2$ is positive for $2 < d < 4$.

2. WILSON'S RECURSION FORMULA

Wilson's recursion formula (1971) was the first explicit realization of the RG and launched the whole new theory of critical phenomena. Let us give a quick derivation of R_2 and then discuss its consequences.

We again take $\mathcal{H}$ to be of the form (8.2). The first step is to write $\sigma = \sigma' + \phi$ as before and then to integrate over ϕ. Again

$$\phi(x) = L^{-d/2} \sum_{\Lambda/2 < q < \Lambda} \sigma_q \, e^{iq \cdot x} \tag{8.14}$$

Here we take $s = 2$. In our previous discussions we used σ_q as integration variables in performing $\int \delta\phi$. The advantage of using σ_q is that the gradient term $\int (\nabla\phi)^2 d^d x$ becomes simply a sum $\sum_q q^2 |\sigma_q|^2$. Each term in the sum involves only one q. The disadvantage is that the quartic term and higher powers of $\sigma(x)$ become complicated products of σ_q with different q's. These terms involve $\phi(x)$ with the same x. Thus, while the plane wave expansion (8.14) is convenient for the gradient term, the other terms would be more simply handled if $\phi(x)$ were expanded in more localized functions. Wilson's approach is to expand $\phi(x)$ in a set of functions that are the most localized functions one can construct by superimposing plane waves with $\Lambda/2 < q < \Lambda$. These are the wave packets $W_z(x)$ (which are just the Wannier functions for the band of plane waves $\Lambda/2 < q < \Lambda$). $W_z(x)$ is very small except for x near z within a distance $\sim 2\Lambda^{-1}$. The set of z's forms a lattice. We expand $\phi(x)$ in these functions and write

$$\phi(x) = \sum_z \phi_z W_z(x) \tag{8.15}$$

$$\int d^d x \; W_z(x)\, W_{z'}(x) = \delta_{zz'} \quad , \tag{8.16}$$

$$\int \delta\phi = \prod_z \int d\phi_z \quad . \tag{8.17}$$

The set of wave packets is orthonormal. Again, we write $\mathcal{H}$ as

$$\mathcal{H} = \int d^d x \, \frac{c}{2} (\nabla\sigma')^2 + \int d^d x \left[\frac{c}{2} (\nabla\phi)^2 + U(\phi + \sigma')\right] . \tag{8.18}$$

The first integral does not enter in $\int \delta\phi \, e^{-\mathcal{H}}$.

Now come the bold approximations:

(a) Ignore the overlap between wave packets.

This means that at any point x there is just one $W_z(x)$ which is taken as nonzero. We then have a picture of blocks. Each block has a $W_z(x)$. The second integral of (8.18) becomes a sum over blocks

$$\sum_z \left(\frac{c}{2} \overline{q^2} \, \phi_z^2 + \int d^d x \, U\big(W_z(x)\,\phi_z + \sigma'(x)\big) \right) \tag{8.19}$$

where

$$\overline{q^2} \equiv \int d^d x \, (\nabla W_z(x))^2 \tag{8.20}$$

is the mean square wave number of the packet W_z, and the integral $\int d^d x$ in (8.19) is taken over the block z. The multiple $\int \delta\phi$ is now a product of independent integrals $\prod_z \int d\phi_z$.

(b) <u>Ignore the variation of $\sigma'(x)$ within a block.</u>

This means, for a given z, $\sigma'(x)$ is considered as a constant.

(c) <u>Ignore the variation of $|W_z(x)|$.</u>

This means that $W_z(x)$ is either $|W_z(x)|$ or $-|W_z(x)|$. Note that $W_z(x)$ is orthogonal to $\sigma'(x)$, since it contains only Fourier components $> \Lambda/2$ while $\sigma'(x)$ contains only those $< \Lambda/2$. Thus, within a block,

$$\int d^d x \; W_z(x) = 0 \quad , \tag{8.21}$$

i.e., $W_z(x)$ must be positive ($= |W_z|$) over a half of the block and negative ($= -|W_z|$) over the other half.

Under these approximations, the integral of (8.19) is simply

$$\frac{\Omega}{2}\left[U(\Omega^{-1/2}\phi_z + \sigma') + U(-\Omega^{-1/2}\phi_z + \sigma')\right]$$

$$\equiv \Omega\,\overline{U}(\phi_z, \sigma') \tag{8.22}$$

where Ω is the volume of a block. Note that $|W_z(x)| \approx \Omega^{-1/2}$ since $\int W_z^2(x)\, d^d x = 1$. The integral over $d\phi_z$ can now be written as

$$I(\sigma') \equiv \int d\phi_z\; e^{-\frac{c}{2}\overline{q^2}\phi_z^2 - \Omega\overline{U}}$$

$$= \Omega^{-1/2}\int dy\; e^{-\frac{c}{2}\overline{q^2}\,\Omega y^2 - \frac{1}{2}\Omega\,[U(\sigma'+y) + U(\sigma'-y)]}$$

$$\equiv e^{-\Omega\overline{U}'(\sigma') - \Omega A} \tag{8.23}$$

where A is a constant so that $\overline{U}'(0) = 0$.

Taking the product over all the blocks, we get

$$\int \delta\phi\; e^{-\mathcal{H}} = e^{-\int d^d x \frac{c}{2}(\nabla\sigma')^2}\prod_z I(\sigma')$$

$$= e^{-\int d^d x\left[\frac{c}{2}(\nabla\sigma')^2 + \overline{U}'(\sigma')\right] - AL^d} \tag{8.24}$$

where $\sum_z \Omega$ has been written as $\int d^d x$. This completes Step (i) of the RG, i.e., the Kadanoff transformation.

Step (ii) is just the simple replacement,

$$\sigma'(x) \rightarrow \sigma(x')\, 2^{1-\frac{d}{2}-\frac{\eta}{2}} , \quad \int d^d x \rightarrow 2^d \int d^d x' \quad \text{with } x' = x/2.$$

We obtain

$$\mathcal{H}' = \int d^d x' \left[\frac{c'}{2} (\nabla\sigma)^2 + U'(\sigma) \right]$$

$$c' = c\, 2^{-\eta} ,$$

$$U'(\sigma) = 2^d\, \bar{U}'(\sigma 2^{1-d/2-\eta/2}) ,$$

$$= 2^d\, \Omega^{-1} \ln \left[I(\sigma 2^{1-d/2-\eta/2})/I(0) \right] \qquad (8.25)$$

$$A = -\Omega \ln I(0) ,$$

with $I(\sigma)$ given by (8.23). This completes an approximate formula for R_2. Repeated application of (8.25) generates $R_s = (R_2)^\ell$, $s = 2^\ell$.

Clearly, to keep $c' = c \neq 0$, we need to choose $\eta = 0$. Formula (8.25) can be simplified somewhat by defining

$$\Omega U(\sigma) = Q(\sigma) , \qquad (8.26)$$

and choosing the value of c such that

$$\frac{c}{2}\,\overline{q^2}\,\Omega = 1 \ . \qquad (8.27)$$

The parameter space is now simply the space of Q. Then we have the transformation formula for Q under R_2 :

$$Q'(\sigma) = 2^d \ln \left[I(\sigma 2^{1-d/2})/I(0) \right] ,$$

$$I(\sigma) = \int dy \ e^{-y^2 - \frac{1}{2}[Q(\sigma+y)+Q(\sigma-y)]} \ . \qquad (8.28)$$

This is the recursion formula of Wilson (1971).

In the above derivation we have taken $n = 1$. The generalization to n-component spins with $n > 1$ is straightforward. The σ and y in (8.28) are simply regarded as denoting vectors. The dy in the integral is replaced by $d^n y = \prod_{i=1}^{n} dy_i$.

Now one needs to look for fixed points and exponents. Numerical studies of (8.28) for $d = 3$ were carried out by Wilson (1971), and subsequently by Grover (1972), Grover, Kadanoff, and Wegner (1972), and other authors.

Fixed points and exponents were found. Again, besides the Gaussian fixed point $Q_o^* = 0$, which is unstable for $d < 4$, there is a stable fixed point Q^*. We shall not go into any detail concerning the numerical program except to mention briefly the following procedure for determining Q^* [see Grover (1972)]. This should give a rough idea as to how numerical work can be done.

Clearly, any help from other sources of information concerning Q^* will help to get the search procedure started. Suppose that one gets a good guess for Q^* by extrapolating from the small-ϵ results. Call such a guess Q_o, which is off by δQ:

$$Q_o = Q^* + \delta Q \quad . \tag{8.29}$$

Let us expand δQ in the eigenvectors of R_s^L, assuming that δQ is small,

$$\delta Q = t_1 e_1 + q_o \ ,$$

$$q_o = t_2 e_2 + \cdots \ , \tag{8.30}$$

where $R_s q_o \to 0$, for large s, assuming $y_2, y_3, \ldots < 0$.

Now apply R_2 to Q_o to obtain Q_1

$$Q_1 = Q^* + 2^{y_1} t_1 e_1 + q_1 \; . \tag{8.31}$$

Clearly, if we are lucky enough that t_1 happens to be zero, then we will get Q^* by applying R_2 a few times. If we are not that lucky, we need to remove the $t_1 e_1$ term. From (8.29) and (8.31) we obtain

$$t_1 e_1 = (Q_1 - Q_o - q_1 + q_o)/(2^{y_1} - 1) \; , \tag{8.32}$$

$$Q^* = Q_o - t_1 e_1 - q_o$$

$$= (2^{y_1} Q_o - Q_1)/(2^{y_1} - 1) - 2^{y_1} q_o/(2^{y_1} - 1). \tag{8.33}$$

If we knew y_1, then

$$Q_o^{(1)} = (2^{y_1} Q_o - Q_1)/(2^{y_1} - 1) \tag{8.34}$$

would be Q^* except for the last term of (8.33), which contains no e_1. One can make an estimate of y_1 by applying R_2 to (8.32) and dividing to obtain

$$2^{y_1} = \frac{Q_2 - Q_1 - q_2 + q_1}{Q_1 - Q_o - q_1 + q_o} \; . \qquad (8.35)$$

A rough estimate would be obtained by $(Q_2 - Q_1)/(Q_1 - Q_o)$ evaluated at some arbitrary value of σ. Now we can use $Q_o^{(1)}$ with this estimate of 2^{y_1} as our new guess of Q^*, and repeat the above steps to improve $Q_o^{(1)}$ to $Q_o^{(2)}$. The result of a few iterations of these steps converges to Q^*.

Wilson (1971) first obtained by numerical means the fixed point and exponents (not by the procedure mentioned above). His results agree well with those calculated by numerical work using the old fashioned series expansion techniques. The important question is how good are the approximations (a), (b) and (c), which made the simplicity of the recursion formula possible. Clearly, the approximations are very crude. We know that the wave packets must overlap in the coordinate space if they do not overlap in the wave vector space. The variation of $\sigma'(x)$ over a wave-packet size must be accounted for, otherwise η would always be zero. Also $W_z(x)$ surely is more complicated than either $+\Omega^{-1/2}$ or $-\Omega^{-1/2}$. Wilson (1971) discussed in detail the implications of these approximations. The reader is referred to his paper for the discussion. Golner (1972) improved the approximations in many respects. His results do not agree as well as those of the unimproved recursion formula with the series expansion results (which are regarded as very accurate). This implies that the improvements needed are more subtle than expected and a deeper understanding of the original recursion formula and approximations is still lacking.

Although Wilson's recursion formula has a dubious accuracy, it has been the single most important and successful formula. Not only did it start the whole new theory of critical phenomena with numerical success, it also produced the first small-ϵ results (Wilson and Fisher, 1972), i.e., it gave exact results for the exponents to $O(\epsilon)$. The expansion of the recursion formula to $O(\epsilon)$ is left as an exercise for the reader. As an illustration, we shall in the next section obtain exact results for the case $n \to \infty$ using Wilson's recursion formula. Thus, in both the case of small ϵ and the case of $n \to \infty$, the recursion formula gives exact results. This is indeed remarkable, especially since very little has been known exactly about the RG besides these.

3. APPLICATION TO THE $n \to \infty$ CASE

Let us see how the Wilson formula (8.28) works for the case $n \to \infty$. We write $Q(\sigma^2)$ instead of $Q(\sigma)$ for convenience. The dy in (8.28) is now replaced by $d^n y$. Now we have

$$(\sigma \pm y)^2 = \sigma^2 + y^2 \pm 2\sigma \cdot y$$

$$\approx \sigma^2 + y^2 . \tag{8.36}$$

Here σ^2 and

$$y^2 = \sum_{i=1}^{n} y_i^2$$

are of $O(n)$, while $\sigma \cdot y = O(1)$ since only one component of y, i.e., that parallel to σ, contributes to $\sigma \cdot y$. Therefore (8.28) gives

$$I(\sigma) \approx \int d^n y \; e^{-y^2 - Q(\sigma^2 + y^2)}$$

$$\propto \int_0^\infty dy \; y^{n-1} \; e^{-y^2 - Q(\sigma^2 + y^2)}$$

$$\approx \int_0^\infty dy \; e^{n \ln y - y^2 - Q(\sigma^2 + y^2)} \quad . \tag{8.37}$$

For large n, the integrand is the exponential of a large quantity, and thus peaks strongly at the maximum of the exponent, which is determined by setting the derivative of the exponent to zero:

$$\frac{n}{y} - 2y - 2y\, t(\sigma^2 + y^2) = 0 \tag{8.38}$$

where $t(\sigma^2) \equiv \partial Q(\sigma^2)/\partial \sigma^2$. To the leading order in n, we have

$$\ln I(\sigma) = n \ln y - y^2 - Q(\sigma^2 + y^2) + \text{constant} \tag{8.39}$$

with y satisfying (8.38), i.e.,

$$y^2 = \frac{n}{2} \frac{1}{1 + t(\sigma^2 + y^2)} \quad . \tag{8.40}$$

Solving (8.40) for y^2 and substituting the result in (8.39) and then in (8.28), we obtain $Q'(\sigma^2)$. It is convenient to express the recursion formula in terms of $t'(\sigma^2) = \partial Q'(\sigma^2)/\partial\sigma^2$:

$$t'(\sigma^2) = 2^2\, t(\sigma^2\, 2^{2-d} + y^2) \;, \tag{8.41a}$$

$$y^2 = \frac{n}{2}\,(1 + t'(\sigma^2)/2^2)^{-1} \;, \tag{8.41b}$$

where (8.41b) follows from (8.40) and (8.41a). Equation (8.41) is an approximate form for the exact equation (8.8).

To find the fixed points and exponents, we proceed as follows: we solve σ^2 from $t = t(\sigma^2)$. We get σ^2 as a function of t:

$$\sigma^2 = f(t) \;. \tag{8.42}$$

Likewise the transformed function $t'(\sigma^2)$ can also be solved for σ^2:

$$\sigma^2 = f'(t') \;. \tag{8.43}$$

From (8.41) we obtain

$$f(t'/4) = \sigma^2\, 2^{2-d} + \frac{n}{2}(1+t'/4)^{-1} \quad , \tag{8.44}$$

which means, by the definition (8.43),

$$f'(t') = 2^{d-2} f(t'/4) - \frac{n}{2} 2^{d-2} (1+t'/4)^{-1} \quad . \tag{8.45}$$

This is another way of writing the recursion relation (8.41). The fixed point is obtained by setting $f' = f = f^*$ in (8.45). One easily obtains

$$f^*(t') = \sum_{m=1}^{\infty} b_m^* \, t'^m$$

$$b_m^* = \frac{n}{2}(1 - 2^{-d+2m})^{-1} \quad . \tag{8.46}$$

Subtracting the fixed point equation from (8.45), we obtain

$$f'(t') - f^*(t') = 2^{d-2} \left(f(t'/4) - f^*(t'/4) \right) \tag{8.47}$$

and write

$$f(t') - f^*(t') = \sum_{m=1}^{\infty} g_m \, t'^m \quad . \tag{8.48}$$

Then it is obvious from (8.47) that

$$g'_m = g_m 2^{y_m} , \tag{8.49}$$

$$y_m = d - 2m .$$

The exponents y_m thus agree with the exact results given by (8.13), although the detailed form of the fixed point is different. The Wilson recursion formula thus gives very much the same structure of RG for the $n \to \infty$ as the exact formula except that the algebra is much simpler.

Evidently Wilson's recursion formula has in a very crude way captured the essence of the RG. Attempts to improve upon it have not been all that successful. Likewise, attempts to carry out the RG to the next order in ϵ or in $1/n$ have met with a great deal of mathematical complexity.

4. DEFINITIONS OF THE RG FOR DISCRETE SPINS

So far we have studied only models with continuous spin variables. In Sec. II.3, where we first discussed the coarse graining procedure, we pointed out that discrete spins will become essentially continuous after coarse graining. The discreteness is washed out if we consider interactions over a scale a few times larger than the unit cell.

However, discrete spins are far easier to study numerically. It is possible to build into the RG a truncation procedure such that the discreteness of the spins is formally not washed out. There are many ways to truncate. They simplify the RG greatly and have made possible many recent numerical advances by many authors, e.g., Niemeijer and van Leeuwen (1973, 1974, 1975); Nauenberg and Nienhuis (1974, 1975); Kadanoff and Haughton (1974); Wilson (1975); and others. These numerical studies have given us important insights into RG procedures in general as well as specific numbers. Here we shall not discuss any details of the numerical calculations, but only comment on the basic setup and various unanswered questions.

A basic ingredient of the RG for discrete spins is the truncation scheme illustrated by the following examples.

Consider a two-dimensional triangular lattice of Ising ($n = 1$) spins as shown in Fig. 8.1a. Each spin can be

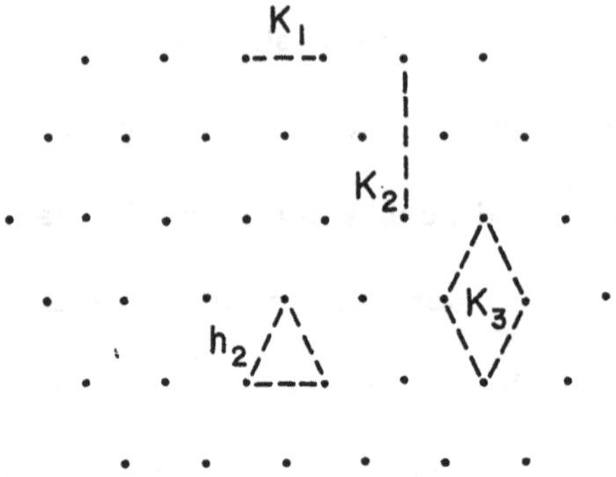

(a)

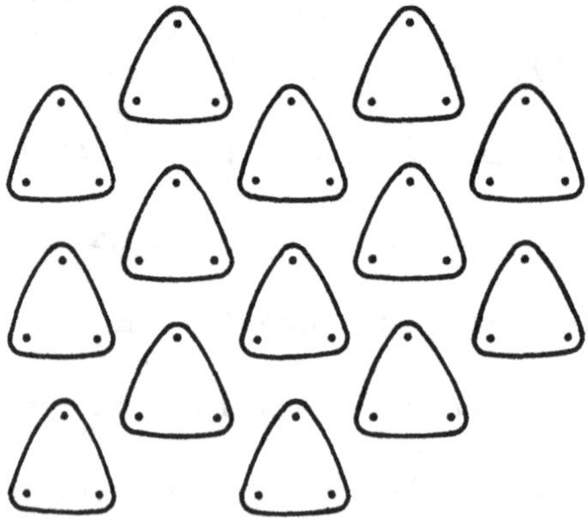

(b)

Figure 8.1. (a) Triangular lattice of Ising spins. K_1, K_2, K_3, h_2 are interaction parameters of (8.53). (b) Block construction used by Niemeijer and van Leeuwen (1973, 1974).

+1 or -1. A Kadanoff transformation can be defined by forming blocks each of which contain three spins as shown in Fig. 8.1b and then defining a new block spin for each block. The mean of the three spins $\sigma^1_{r'}$, $\sigma^2_{r'}$, $\sigma^3_{r'}$ in the block r' can take four values, namely ± 1, $\pm 1/3$, instead of just ± 1. Clearly, it is not convenient to use the mean as the new block spin as we did before. One can force the four values into two values by defining the new block spin as

$$\sigma_{r'} = \mathrm{sgn}(\sigma^1_{r'} + \sigma^2_{r'} + \sigma^3_{r'})$$

$$= \frac{1}{2}(\sigma^1_{r'} + \sigma^2_{r'} + \sigma^3_{r'} - \sigma^1_{r'}\sigma^2_{r'}\sigma^3_{r'}) , \qquad (8.51)$$

which of course is either +1 or -1. This is not quite the same as the coarse graining we are used to. It can be viewed as having an additional coarse graining in the spin space. The RG of Niemeijer and van Leeuwen (1973, 1974) is based on Eq. (8.51).

For discrete spins, it is convenient to use a parameter space of

$$\mu = (K_1, K_2, K_3, \ldots) \tag{8.52}$$

with K_i defined by

$$-\mathcal{H} = \sum_{\mathbf{r}} \Bigg(K_1 \sum_{\delta = \mathrm{nn}} \sigma_{\mathbf{r}} \sigma_{\mathbf{r}+\delta} + K_2 \sum_{\delta = \mathrm{nnn}} \sigma_{\mathbf{r}} \sigma_{\mathbf{r}+\delta}$$

$$+ K_3 \sigma_{\mathbf{r}} \sigma_{\mathbf{r}+\delta} \sigma_{\mathbf{r}+\delta'} \sigma_{\mathbf{r}+\delta''} + \cdots \tag{8.53}$$

$$+ h\sigma_{\mathbf{r}} + h_2 \sigma_{\mathbf{r}} \sigma_{\mathbf{r}+\delta} \sigma_{\mathbf{r}+\delta'} + \cdots \Bigg),$$

where K_1 is the nearest-neighbor (nn) coupling parameter, K_2 is the next-nearest-neighbor (nnn) coupling parameter, and K_3 is the "4 spin" coupling parameter, h is the magnetic field, and $h_2, \ldots$ are higher odd spin coupling parameters. Figure 8.1a shows which spins are coupled in these terms. There are of course more terms with other coupling parameters. The RG transformation $R_{\sqrt{3}}\, \mu = \mu'$ is then defined by

$$e^{-\mathcal{H}[\sigma'] - AL^d} = \sum_{\sigma} \rho[\sigma', \sigma]\, e^{-\mathcal{H}[\sigma]} \tag{8.54a}$$

where $\rho[\sigma', \sigma]$ is a product taken over the new blocks

$$\rho[\sigma',\sigma] = \prod_{r'} p(\sigma'_{r'};\ \sigma^1_{r'},\ \sigma^2_{r'},\ \sigma^3_{r'}) \ . \tag{8.54b}$$

The function p is a projector. It is one if the arguments satisfy (8.51) and is zero otherwise. It plays the role of the δ-function in (2.23) for the continuous spin case. Explicitly

$$p = \frac{1}{2}\left[1 + \frac{1}{2}\sigma'_{r'}(\sigma^1_{r'} + \sigma^2_{r'} + \sigma^3_{r'} - \sigma^1_{r'}\sigma^2_{r'}\sigma^3_{r'})\right] . \tag{8.55}$$

Obviously,

$$p(+1;\sigma^1,\ \sigma^2,\ \sigma^3) + p(-1;\ \sigma^1,\ \sigma^2,\ \sigma^3) = 1 \ , \tag{8.56a}$$

since one of the two terms is one and the other zero. It follows that

$$\sum_{\sigma'} \rho[\sigma',\sigma] = 1 \ . \tag{8.56b}$$

The definition (8.54a) is of course a generalization of (2.23) or (A.17) to the discrete spins. In (8.54) the constant term AL^d independent of σ' is explicitly written. Note that we need to shrink the new lattice by a factor $\sqrt{3}$

to obtain the same lattice spacing as the old one. That is why (8.54) defines $R_{\sqrt{3}}$. The whole RG is defined by

$$R_s = (R_{\sqrt{3}})^{\ell} \quad , \quad s = 3^{\ell/2} \quad , \tag{8.57}$$

$$\ell = 0, 1, 2, \ldots$$

Note that there is no rescaling of the spin variable σ in contrast to the replacement $\sigma \rightarrow s^{1-\eta/2-d/2}\sigma$ in the earlier definition of the linear RG. Here the value of σ is always ± 1. No rescaling can be defined, nor is it necessary.

When we sum (8.54a) over σ' on both sides and apply (8.56), we obtain, after taking the logarithm,

$$\mathfrak{F}(\mu')3^{-1} + A = \mathfrak{F}(\mu) \tag{8.58}$$

where $\mu' = R_{\sqrt{3}}\mu$ and $\mathfrak{F}$ is the free energy per volume. Note that $d = 2$ here. The result (6.50) applies here with trivial modifications:

$$\mathfrak{F}(\mu) = s^{-2}\mathfrak{F}(R_s\mu) + \sum_{m=0}^{\ell-1} 3^{-m} A(R_{3^{m/2}}\mu) \tag{8.59}$$

where $s = 3^{\ell/2}$. The first term can be dropped in calculating $\mathfrak{F}$ if ℓ is chosen to be large enough.

The transformation formulas for various average values are more complicated. One easily shows via (8.54) that

$$\left\langle e^{\sum_{r'} \lambda_{r'} \sigma'_{r'}} \right\rangle_{\mu'} = \left\langle e^{\sum_{r'} \lambda_{r'} t_{r'}} \right\rangle_{\mu} ,$$

$$t_{r'} = \frac{1}{2}(\sigma^1_{r'} + \sigma^2_{r'} + \sigma^3_{r'} - \sigma^1_{r'} \sigma^2_{r'} \sigma^3_{r'}) , \qquad (8.60)$$

which looks obvious enough in view of (8.51). From (8.60) we obtain

$$m(\mu') = \langle \sigma'_{r'} \rangle_{\mu'} = \langle t_{r'} \rangle_{\mu} ,$$

$$G(r/\sqrt{3}, \mu') = \langle \sigma'_{r'} \sigma'_o \rangle_{\mu'} = \langle t_{r'} t_o \rangle_{\mu} \qquad (8.61)$$

etc.

where the $1/\sqrt{3}$ in G is again due to the shrinking of the new lattice of blocks. We need to know how $m(\mu')$ and $G(r/\sqrt{3}, \mu')$, etc., are related to the corresponding quantities calculated with the original Hamiltonian, $m(\mu) = \langle \sigma_r \rangle_\mu$, $G(r, \mu) = \langle \sigma_r \sigma_o \rangle_\mu$, etc. Because of the nonlinear definition of the new block spin $\sigma'_{r'}$ in terms of the old spins, such relationships are very complicated; for example,

$$m(\mu') = \frac{3}{2} m(\mu) - \frac{1}{2} \langle \sigma^1_{r'} \sigma^2_{r'} \sigma^3_{r'} \rangle_\mu \quad . \tag{8.62}$$

Neither can one simply relate $G(r, \mu)$ to $G(r, \mu')$. This difficulty will be discussed again shortly.

Similar constructions have been applied to discrete two-dimensional Ising spins on a square lattice. Figure 8.2a shows the formation of new blocks, each of which includes four spins. New block spins can be defined as in (8.51)

$$\sigma'_{r'} = \mathrm{sgn}\,(\sigma^1_{r'} + \sigma^2_{r'} + \sigma^3_{r'} + \sigma^4_{r'}) \tag{8.63}$$

except that special attention has to be paid to the cases where the four spins sum to zero. One can arbitrarily assign $\sigma'_{r'} = 1$ for the configurations shown in Fig. 8.2b and $\sigma'_{r'} = -1$ for the inverse of these configurations, as

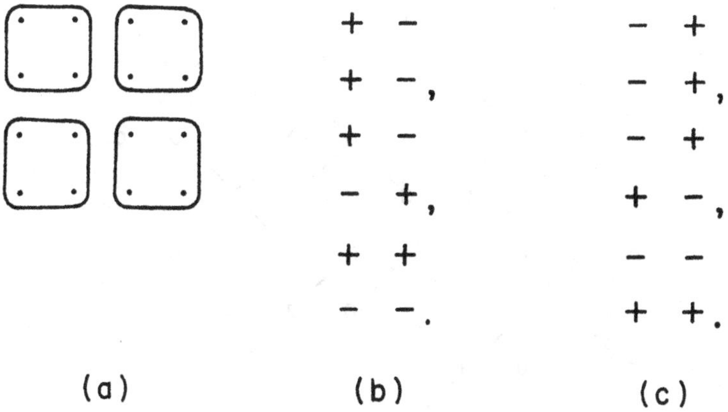

Figure 8.2. (a) Block construction on a square lattice. (b) Configurations in a block which are counted has having $\sigma' = +1$, and (c) Those counted as having $\sigma' = -1$.

shown in Fig. 8.2c. Nauenberg and Nienhuis (1974) used this block construction to define their RG.

Another construction (van Leeuwen, 1975) shown in Fig. 8.3 is more interesting. It takes five spins to make a new block. New block spins are defined by

$$\sigma'_{r'} = \mathrm{sgn}(\sigma^1_{r'} + \cdots + \sigma^5_{r'}) \quad . \tag{8.64}$$

This block construction has the advantage that the spins on one of the sublattices (labeled by a in Fig. 8.3) and those on the other (labeled by b) are not mixed, in contrast to

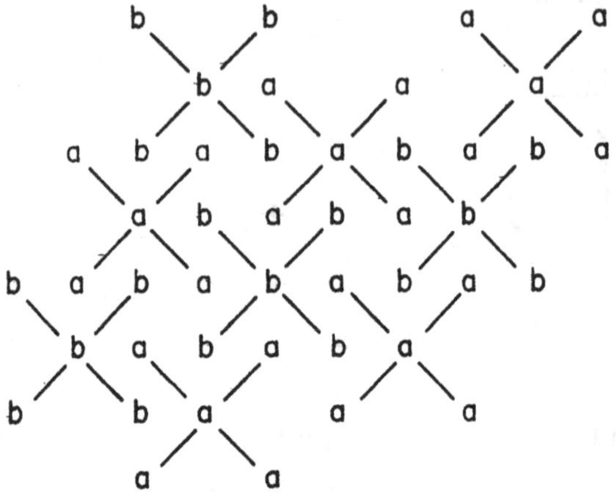

Figure 8.3. The block construction of van Leeuwen (1975). The sublattices a and b are not mixed.

the above constructions. This is very handy because both ferromagnetic and antiferromagnetic critical points can be studied by the same RG.

Relations similar to (8.60) - (8.62) can be derived for RGs constructed with (8.63) and (8.64).

The renormalization groups defined above have achieved a great deal of success in getting good values of critical exponents. In addition, Nauenberg and Nienhuis (1974b) have used their RG to study the free energy and the equation of state. Some of their results will be briefly sketched later. On the other hand, these RG's suffer from the difficulty that some average values of interest, the magnetization and the correlation functions, do not have simple transformation laws. That is to say, the average calculated with the original Hamiltonian cannot be simply related to that calculated with the transformed Hamiltonian. This is the price one pays for keeping the spins discrete by a highly nonlinear procedure in defining the new block spins, i. e., by taking the sign of the mean spin in a block as the new block spin of that block. The new block spin bears only a faint resemblance to the mean of the old spins in the block. Consequently such RGs are not very good representations of the coarse graining procedure which is a basic ingredient desired in an RG. Since the physical meaning of the new block spin is quite different from the mean spin over a block, it becomes difficult to analyze spin fluctuations in the original model with the transformed Hamiltonian.

There are alternative definitions of the RG which are formally free from such a difficulty while still keeping spins discrete. They have been studied by Kadanoff and Houghton (1975), and by Wilson (1975). These RGs are generally defined by (8.54) with a p of the form

$$p = \frac{1}{2}(1 + \sigma'_{r'} t_{r'}[\sigma]) \tag{8.65}$$

where the new spin $\sigma'_{r'}$ is either +1 or -1, and $t_{r'}[\sigma]$ is some linear function of the old spins in the block located at r'. Note that (8.65) is general and includes (8.55), (8.63), (8.64) or any transformation satisfying (8.56) if we also allow $t_{r'}[\sigma]$ to be nonlinear in the old spin variables. To make p a projector like (8.55), we have to make $t_{r'}[\sigma]$ nonlinear, i.e., include the $\sigma^1_{r'} \sigma^2_{r'} \sigma^3_{r'}$ term. In allowing p not to be a projector, we can no longer have a precise relationship like (8.55) or (8.64) defining the new block spin in terms of the old spins. The RG is directly defined in terms of p. At this moment, giving up a precise definition of $\sigma'_{r'}$ sounds like a step backwards in the effort to make $\sigma'_{r'}$ a better representation of the mean of old spins in a block. It may actually be the case. However, mathematically there is an improvement, as will be seen shortly.

Let us give a couple of examples. The simplest example is (Kadanoff, in Gunton and Green, 1973),

$$t_{r'}[\sigma] = \rho\, \sigma_{r'} \tag{8.66}$$

where r' are those points not slashed in Fig. 8.4. Here each block has just two spins. The constant ρ is an adjustable parameter. It bears some resemblance to the parameter λ_s [see (5.6)], in defining the linear RG. One needs to choose ρ properly in order that the right fixed point can be found. The special choice $\rho = 1$ happens to be useful for the study of the one-dimensional Ising model (Nelson and Fisher, 1974). Note that for the special case $\rho = 1$, $p = \frac{1}{2}(1 + \sigma'_{r'}\sigma_{r'})$ is a projector, i.e., it is one if $\sigma'_{r'} = \sigma_{r'}$ and zero otherwise.

Equation (8.66) defines $\mu' = R_{\sqrt{2}}\,\mu$. The $\sqrt{2}$

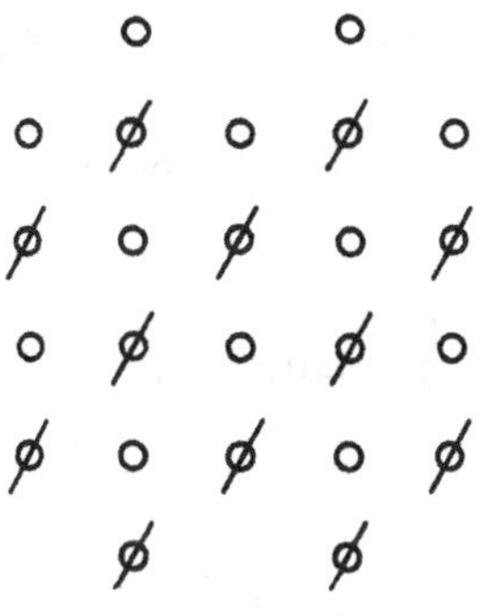

Figure 8.4. Block construction by eliminating every other spin.

reflects the fact that the new lattice must be shrunk by a factor of $\sqrt{2}$ to get back the old lattice spacing. Another example is, for the triangular lattice, as shown in Fig. 8.1,

$$t_{r'}[\sigma] = \rho \bar{\sigma}_{r'} ,$$

$$\bar{\sigma}_{r'} \equiv \frac{1}{3}(\sigma^1_{r'} + \sigma^2_{r'} + \sigma^3_{r'}) , \qquad (8.67)$$

where $\bar{\sigma}_{r'}$ is the mean spin in the block r'. This implies that

$$p = \frac{1}{2}(1 + \rho\, \sigma'_{r'} \bar{\sigma}_{r'}) . \qquad (8.68)$$

Equation (8.61) still holds even when $p = \frac{1}{2}(1 + \sigma'_{r'} t_{r'})$ is not a projector, as long as $r' \neq r''$, i.e., as long as we restrict our attention to correlations of spins in different blocks. For the case (8.66), we have, using (8.61),

$$m(\mu') = \frac{1}{2} \rho\, m(\mu) ,$$

$$G(r/\sqrt{2}, \mu') = \left(\frac{\rho}{2}\right)^2 G(r, \mu) \qquad (8.69)$$

where $\mu' = R_{\sqrt{2}}\mu$. For the case (8.68), we have

$$m(\mu') = \frac{1}{2}\rho\, m(\mu) \ ,$$

$$G(r/\sqrt{3}, \mu') = \left(\frac{\rho}{2}\right)^2 \langle \bar{\sigma}_r \bar{\sigma}_o \rangle_\mu \qquad (8.70)$$

etc.,

where $\mu' = R_{\sqrt{3}}\mu$. Of course $\langle \bar{\sigma}_r \bar{\sigma}_o \rangle_\mu$ is not quite $G(r, \mu) = \langle \sigma_r \sigma_o \rangle_\mu$. But if we do not care about short-range (over a block size) variations of σ_r, $\langle \bar{\sigma}_r \bar{\sigma}_o \rangle_\mu$ is effectively $G(r, \mu)$. Thus (8.69) and (8.70) show that physical quantities of interest calculated with μ' are simply related to those calculated with μ, as they were in our earlier study of the linear RG. These simple relations are what we would expect in view of scale transformations and coarse graining procedures. However, these appealingly simple relations are a bit deceptive and must be interpreted with caution. The reader must be reminded again that they are not the results of defining the new spin as the mean of the old spins in a block, but the results of the mathematical simplicity of the transformation. As we mentioned earlier, the meaning of the new spin is even more vague than that defined by (8.51) or (8.63).

5. NUMERICAL WORK ON THE RG FOR TWO-DIMENSIONAL ISING SYSTEMS

The fixed points and exponents of the RG for discrete spins defined above have been found numerically by the authors quoted above. The exponents and other data for the two-dimensional Ising model are known exactly from Onsager's solution and provide a check for the numerical RG results. So far the results reported all agree well with exact results. However, there are still questions which are unanswered. In this and in the next two sections we shall attempt to give some qualitative discussion of what is involved in numerical calculations and some of the results and unanswered questions.

In principle, there is an infinite number of parameters in $\mu = (K_1, K_2, K_3, \ldots)$. If one starts with a μ with only a small number of nonzero entries, the transformed μ' will have more entries. If the transformation is repeated, still more entries will become nonzero and interactions involving more and more spins must be included in the Hamiltonian. It is clear that one has to truncate in some way in order to carry out numerical calculations. The choice of a proper truncation scheme is a very difficult problem which is still not well understood. So far there have been a number of truncation schemes. We shall briefly discuss the scheme of Nauenberg and Nienhuis (1974) as an illustration. In their scheme, 16 spins as shown in Fig. 8.2a are analyzed. A periodic boundary condition is imposed. One then carries out the RG defined by Eq. (8.63) on this system, which is divided into four blocks. It is sufficient to consider the parameter space of

$$\mu = (K_1, K_2, K_3, h, h_2) \quad . \tag{8.69}$$

The meaning of the parameters is given by (8.53). Since there are only four new block spins, $\mathcal{K}'$ can only have the nearest neighbor K_1', next nearest neighbor K_2', 4-spin K_3', magnetic field h', and 3-spin h_2' coupling terms. Thus μ' will have the same set of entries as μ.

One then calculates $R_2\mu = \mu'$ exactly on the 16-spin system, searches for the fixed points, linearizes R_2

near the fixed points, and obtains exponents. Note that once the program for getting μ' from the input μ is written, one can forget about the lattice and the spins. R_s, $(s = 2^\ell)$, is obtained by repeating this program ℓ times. The free energy can be calculated using (8.59) (with 2 replacing $\sqrt{3}$). The first term of (8.59) can be dropped for a sufficiently large ℓ. The series is rapidly convergent. The specific heat can be deduced from the results for the free energy. Figure 8.5 shows some results of Nauenberg and Nienhuis (1974b). Table 8.2 lists their results as regards the fixed point and exponents.

Other authors have used different schemes of truncation, kept different numbers of coupling parameters, or applied different numerical techniques. In all cases, they all get about the same answers for the exponents, within $\sim 5\%$ of the exact values. The number of spins involved in numerically defining the RG is not many more than 10. The reason for keeping only ~ 10 spins is that, to account for all possible interactions among N spins, one needs to analyze $\sim 2^N$ configurations. The task becomes formidable when N gets large.

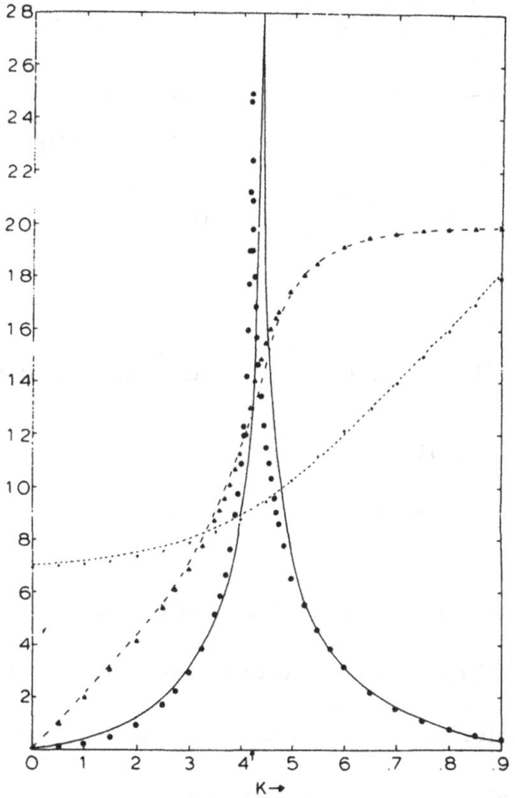

Figure 8.5. (Taken from Fig. 1 of Nauenberg and Nieuhuis (1974b).) Dashed curve, free energy from Onsager's exact solution; dash-dotted curve, exact energy; solid curve, exact specific heat. The points are numerical results of Nauenberg and Nieuhuis. K (the K_1 in the text) is the nearest neighbor coupling parameter. All other parameters are taken to be zero.

Table 8.2

The ferromagnetic fixed point and associated exponents of the RG of Nauenberg and Nienhuis (1974b)

$K_1^* = 0.307$		(nearest neighbor)
$K_2^* = 0.084$		(next nearest neighbor)
$K_3^* = -0.004$		(4 spin)
$y_1 = 1/\nu = 0.937$		(the exact value is 1)
$y_2 = -2.012$		

We have emphasized repeatedly that the whole purpose of the RG is to study a transformation at a local level, i.e., a transformation of interactions among only a small number of spins. Thus one should be able to do the RG analysis and obtain results of interest with a small number of spins and coupling parameters. Numerical results have qualitatively supported this conclusion. The nearest-neighbor interaction K_1 has always been found to be the most important one, and is the largest at the fixed point. All other parameters are smaller by a factor of 4 or more.

Besides exponents, other features of the RG have also been studied numerically. For example, plots of critical surfaces have been made by Nauenberg and Nienhuis (1974a). One can thereby get a qualitative picture of the global structure of the fixed points and critical surfaces. We now give a brief sketch of the gross features of this structure.

Let us keep track of only K_1 and K_2, i.e., the nearest-neighbor and the next-nearest-neighbor interaction parameters. Assume also that there is no magnetic field nor any interaction involving an odd number of spins. Imagine that we define the RG for a two-dimensional square-lattice Ising model with the block construction of van Leeuwen (1975) shown in Fig. 8.3. Note that the next-nearest-neighbor interaction (K_2) operates only within a sublattice, i.e., between either the spins in sublattice a, or those in sublattice b. On the other hand, the nearest-neighbor interaction (K_1) always involves a spin in sublattice a and another in sublattice b. If $K_1 = 0$, then the sublattices become independent. The block construction in Fig. 8.3 does not mix the sublattices and would maintain this independence.

In Fig. 8.6, we sketch some qualitative results, which are explained below.

The (K_1, K_2) plane is of course a subspace of the parameter space of $\mu = (K_1, K_2, \ldots)$. The point F is the ferromagnetic fixed point. The line $F'FK_cB$ is the critical surface associated with this fixed point. The point K_c on this line is the Ising model with only the nearest-neighbor interaction at its critical point. Points on this

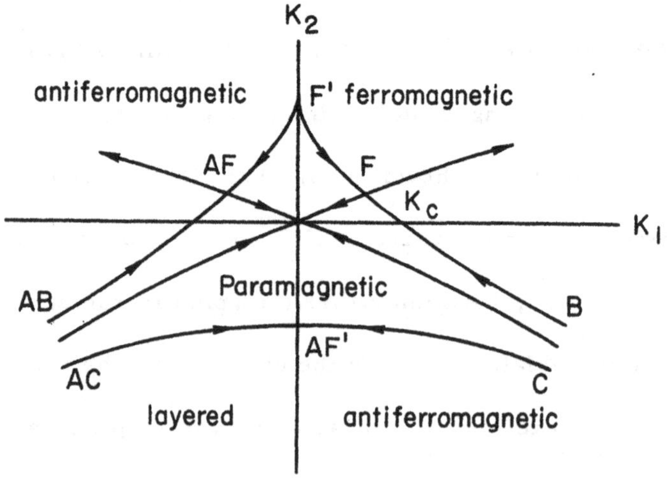

Figure 8.6. Sketch of fixed points and critical surfaces in the parameter space for two-dimensional Ising systems.

line are pushed toward F by R_s, as the arrows indicate.

The point AF is the antiferromagnetic fixed point. The nearest-neighbor interaction K_1 is negative, but the next-nearest-neighbor interaction K_2 is positive. The line F'-AF-AB is the critical surface for this fixed point.

Note that Fig. 8.6 is symmetric about the K_2 axis. This is easily understood. Under the transformation in which we change the sign of K_1 and at the same time change the sign of all spins on one of the two sublattices in Fig. 8.3, the Hamiltonian is unchanged. Thus we can obtain results for the case $K_1 < 0$ by applying this transformation to those for $K_1 > 0$. This transformation does not interfere with the RG defined by the block construction, Fig. 8.3. Note that the invariance of the Hamiltonian under this transformation will not hold if there is a magnetic field or any interaction which involves an odd number of spins in either sublattice, or in any way destroys the symmetry between the two sublattices.

As we mentioned earlier, if $K_1 = 0$, we have two independent and identical sublattices. The point F' is the ferromagnetic fixed point for these sublattices. The point

AF′ is the antiferromagnetic fixed point for these sublattices. The fixed point F′ is unstable against turning on a small K_1 while AF′ is stable.

Figure 8.6 can also be viewed as a phase diagram. Points with $K_1 > 0$ and above the line F′ - F - B represent ferromagnetic Ising systems. Points with $K_1 < 0$, and above the line F′ - AF - AB represent antiferromagnetic systems. Below these lines but above the line C - AF′ - AC, points represent paramagnetic systems. Below the line C - AF′ - AC, points represent layered antiferromagnetic systems. Figure 8.7 shows the ground state configuration of a layered antiferromagnetic system. Each of the two sublattices is antiferromagnetic owing to a large negative K_2.

+	−	+	−	+	−
+	−	+	−	+	−
+	−	+	−	+	−
+	−	+	−	+	−
+	−	+	−	+	−

Figure 8.7. A ground state configuration of a layered antiferromagnet.

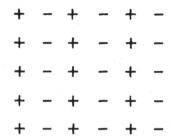

A system at a very low temperature will have large values of K_1 and K_2 since $\mathcal{K} \propto 1/T$. Thus the outer regions of the parameter space, i.e., where K_1, K_2 are large, represent systems at very low temperatures. The asymptotic behavior of the critical surfaces and other characteristics are therefore determined by low-temperature properties of Ising systems. For example, the asymptote of B or C is determined by equating the energy of the ground ferromagnetic configuration (all spins line up) to that of the layered antiferromagnetic configuration (Fig. 8.7), i.e.,

$$-2K_1 - 2K_2 = +2K_2 \tag{8.70}$$

which gives the asymptote $K_1 + 2K_2 = 0$.

Likewise for high temperatures we have $K_1, K_2 \rightarrow 0$. The origin of Fig. 8.6 thus describes the high-temperature limit. It is a stable fixed point.

We shall close this section by mentioning the 8-vertex model of Baxter (1971). The Baxter model starts with a square lattice Ising model with $K_2 > 0$ and $K_1 = 0$, i.e., on the K_2 axis of Fig. 8.6. This means two identical independent sublattices of Ising spins. Then a small K_3, i.e., a 4-spin interaction

$$\sum_r K_3 \sigma_r \sigma_{r+\delta} \sigma_{r+\delta'} \sigma_{r+\delta''} \tag{8.71}$$

for $-\mathcal{K}$ [see (8.53)] is turned on. Two of the four spins in (8.71) belong to sublattice a and the other two belong to sublattice b. As a result, the two sublattices are no longer independent. Baxter solved the model for small K_3 and found that critical exponents depend on K_3. What is happening can be described in the RG language as follows. Imagine what we have a K_3 axis pointing out of the paper from Fig. 8.6. Calculations show that there is a line of fixed points in the (K_1, K_2, K_3) space intersecting the (K_1, K_2) plane at F′. The critical exponents of the Baxter model depend on which fixed point on this line is involved. There is another line of fixed points going through AF′. On the other hand, the ferromagnetic and antiferromagnetic fixed points F and AF will be lifted off the K_1K_2 plane when K_3 is included but they remain isolated points.

It should be emphasized that the existence of a line of fixed points of an RG does not automatically imply different exponents for different fixed points on the line. In our study of the linear RG, for example, the parameter c [see (8.3), (8.8)] can assume different values and give different fixed points but with the same exponents.

The cusp formed by the critical surfaces at the point F′ is also of interest. This and other features will not be elaborated here. The reader is referred to the papers of van Leeuwen (1975) and Nauenberg and Nienhuis (1974).

6. DISCUSSION

In spite of the success of the numerical RG program for two-dimensional Ising systems so far, a great deal remains to be understood, in order to be able to apply the program to study three-dimensional systems.

The most pressing question is how to choose the optimal definition of the RG which will allow us to get good approximate answers using only a small number of spins. Before answering this question, one must note that one may get very bad answers using a small number of spins if the choice of RG is not optimum, even though good answers can be obtained using a larger number of spins. Bell and Wilson (1974) studied this question using a linear RG applied to a Gaussian model. Let us sketch some of their basic ideas.

If there are many fixed points other than the ones we are looking for, we may get a wrong fixed point because we cannot distinguish it from the right one with just a small number of spins. For example, there are fixed points which have an infinite range of interaction among spins, although such fixed points cannot be reached if we apply the RG to a Hamiltonian with short-range (i. e., finite) interactions. The fixed points of interest are those with short-range interactions. The exact range of a short-range fixed point depends on the detailed definition of the RG. Suppose that we happened to pick an RG whose short-range fixed point has an interaction range of three-lattice spacings. Then, if we carry out the RG program on 3×3 spins, we shall not be able to distinguish the short-range fixed point and longer-range fixed points. We would need more, say 6×6, spins to sort out the short-range fixed point.

There are often adjustable parameters in the definition of the RG, such as the ρ in (8.66). It may be possible that the right fixed point cannot be obtained unless such parameters are chosen to be certain values. Otherwise, one gets a wrong fixed point. It becomes crucial to devise practical criteria for choosing the right values. Kadanoff and Houghton (1974), Bell and Wilson (1974), have discussed various possibilities. More recently, Kadanoff and coworkers (1975) have studied variational approaches to optimize the choice of RG, and applied them to obtain exponents for three-dimensional Ising models.

Recent numerical efforts have not been limited to discrete spins. For example, Golner and Riedel (1975), Myerson (1975), and others have derived approximate RG equations from the exact differential equation of the RG of Wilson and Kogut (1974), and carried out numerical calculations for $d = 3$. These important developments will not be reviewed here and the reader is referred to the original papers.

In this and previous chapters we have reviewed several methods of constructing and analyzing the RG transformations. All of these examples have one thing in common, namely, that some form of truncation is necessary in order to get any result. In the small ϵ case the truncation is effected by dropping higher-order terms in ϵ. In the large-n case it is done by dropping higher orders in $1/n$. To obtain Wilson's recursion formula, the truncation is made by plausible approximations. In carrying out the RG for discrete spins one has to truncate by sticking to only a small number of spins. Furthermore, all these truncations are very difficult to improve. Carrying out calculations to one more order in ϵ or including more spins requires an order of magnitude greater effort. We simply have not understood enough about the RG to be able to systematically improve the approximation in practice, although we can formally define the RG exactly. We do not know how much more accurate the exponents will be if we include two more spins in the calculation, for example.

We notice that the simplicity of the fixed points and critical surfaces in the examples given above is a consequence of bold truncations. A fundamental question arises concerning the operational meaning of the exact RG free from any truncation, despite the fact that its formal meaning is easily defined. There has been no proof that the fixed points, critical surface, etc., found in a truncated RG exist for the exact RG (except in cases like the one-dimensional Ising model whose T_c is zero and which probably does not reflect many of the important features which are relevant for the RG in two or three dimensions).

We should note that there is a difference between the truncation in the small-ε (or the $n \to \infty$) case and that in Wilson's recursion formula or the numerical calculations described above. The former kind of truncation is formally exact to the order calculated in ε or $1/n$. This formal exactness cannot be taken as rigorously established since questions concerning the order of limits, the convergence of expansions, etc., are unresolved. The latter kind of truncation is based more on experience and plausible arguments, and is not well understood either.

At this moment it is fair to say that our understanding of the RG is still quite incomplete, although no inconsistency has been found in the basic ideas.

In the following chapters we shall assume the validity of these basic ideas and the existence of the RG, exact or truncated, and discuss various extended concepts and applications.

IX. PERTURBATION EXPANSIONS

SUMMARY

We introduce the perturbation expansion of the Ginzburg-Landau model by means of graphs. The use of the perturbation theory as a mathematical device for the ϵ expansion and the $1/n$ expansion is discussed, with emphasis upon the basic assumptions of these expansions. Elementary techniques are illustrated and critical behavior below T_c and the problem of anisotropy are briefly discussed. We also caution against certain misinterpretations which have been common in the literature. Tables of exponents in the ϵ and the $1/n$ expansions are provided.

1. USE OF PERTURBATION THEORY IN STUDYING CRITICAL PHENOMENA

The success of perturbation methods in various branches of physics has been impressive indeed. Not only has the perturbation expansion proven invaluable for practical calculations, but it also has contributed to our basic qualitative understanding of many phenomena. The expansion in powers of the electronic charge in electrodynamics, for example, is the basis for understanding physical processes such as emission and absorption of photons.

The study of critical phenomena is among the many cases where perturbation expansions do not work. As we have seen in Chapter III, if T is very close to T_c, the fluctuation of the order parameter is so large that the perturbation theory does not provide an adequate description of the phenomena involved. One needs to analyze critical phenomena with concepts and methods that are free from any perturbation expansion. The method of the RG is designed for this purpose.

If the perturbation expansion does not work for critical phenomena, why, then, are we going to study it?

The reason is that many important results have been derived from expansions in powers of $\epsilon = 4 - d$ or $1/n$. Perturbation theory is useful in the study of such expansions.

These small parameters, ϵ and $1/n$, are quite artificial and have no physical meaning, in contrast to the electron charge in electrodynamics. Thus the perturbation method which we shall discuss will be no more than a handy mathematical tool whose application to the study of critical phenomena will rely on various assumed properties of the RG. It should be borne in mind that this method is used essentially for the purpose of mathematical extrapolations and not for the physical interpretation of critical phenomena.

In the next few sections we shall introduce the general technique of using graphs and its application to the study of critical phenomena. Rather than review material which is already extensively discussed in the literature on the ϵ expansion and the $1/n$ expansion, we shall discuss those elementary aspects of the method which seem to have been left out or taken for granted in the literature.

2. PERTURBATION EXPANSION OF THE GINZBURG-LANDAU MODEL

For the moment, let us forget about critical phenomena and the RG. We begin with the Gaussian probability distribution $P_o \propto e^{-\mathcal{H}_o}$ with our old notation

$$\mathcal{H}_o = \frac{1}{2}\int d^dx \left[(\nabla\sigma)^2 + r_o\sigma^2\right]$$

$$= \sum_{k,i} \frac{1}{2} |\sigma_{ik}|^2 (r_o + k^2) \ . \tag{9.1}$$

Note that $\sigma_{ik} = \sigma^*_{i-k}$ since $\sigma_i(x)$ is real. Let α_{ik} and β_{ik} be, respectively, the real and the imaginary parts of σ_{ik}. Thus $\alpha_{ik} = \alpha_{i-k}$ and $\beta_{ik} = -\beta_{i-k}$. For every pair of wave vectors (k, -k) and for every component i we have an independent Gaussian probability distribution

$$e^{-\alpha^2_{ik}/G_o(k)} \ e^{-\beta^2_{ik}/G_o(k)} \tag{9.2}$$

where $G_o(k) = (r_o + k^2)^{-1}$. The averages over the probability distribution $P_o \propto e^{-\mathcal{H}_o}$ are therefore

$$\langle \sigma_{ik} \rangle_o = 0 \ ,$$

$$\langle \sigma_{ik}^2 \rangle_o = \langle \alpha_{ik}^2 \rangle_o - \langle \beta_{ik}^2 \rangle_o = 0 \ ,$$

$$\langle |\sigma_{ik}|^2 \rangle_o = G_o(k) \ ,$$

$$\langle \sigma_{ik} \sigma_{jk'} \rangle_o = \delta_{-kk'} \delta_{ij} G_o(k) \ . \tag{9.3}$$

It is a good exercise to show, using (9.2), that

$$\langle \sigma_{ik}^m \sigma_{i-k}^m \rangle_o = \langle |\sigma_{ik}|^{2m} \rangle_o$$

$$= \frac{(2m)!}{m!\, 2^m} G_o(k)^m \ . \tag{9.4}$$

The factor $(2m)!/(m!\, 2^m)$ happens to be just the number of ways of dividing up $2m$ objects into m pairs. More generally we can figure out the average

$$A = \left\langle \sigma_{i_1 k_1} \sigma_{i_2 k_2} \cdots \sigma_{i_\ell k_\ell} \right\rangle_o \tag{9.5}$$

for any $i_1, \ldots, i_\ell$, $k_1, \ldots, k_\ell$, by the following rules:

(a) if ℓ is odd, $A = 0$, and

(b) if ℓ is even, A is the sum of products of pairwise average, summing over all possible ways of pairing up the ℓ σ's. Each pair gives a factor shown in the last line of (9.3).

If we take the Fourier transform of (9.5), we get

$$\left\langle \sigma_{i_1}(x_1)\, \sigma_{i_2}(x_2) \ldots \sigma_{i_\ell}(x_\ell) \right\rangle_o \tag{9.6}$$

which is then the sum of products of pairwise averages; each pair gives

$$\begin{aligned} \langle \sigma_i(x)\, \sigma_j(x') \rangle_o &= \delta_{ij}\, G_o(x - x') \\ &\equiv \delta_{ij}\, L^{-d} \sum_k e^{ik\cdot(x-x')}\, G_o(k) \;. \end{aligned} \tag{9.7}$$

Now we turn to the Ginzburg-Landau model with

$$\mathcal{H} = \mathcal{H}_o + \mathcal{H}_1 \;,$$

$$\mathcal{H}_1 = \frac{1}{2} \int d^d x\, d^d x' \, \frac{1}{2}\, \sigma^2(x)\, u(x - x')\, \frac{1}{2}\, \sigma^2(x') \;, \tag{9.8}$$

which we have encountered many times earlier with $u(x - x') = u\,\delta(x - x')$. The reason for writing 1/2's in this special way will be clear shortly. We want to treat $\mathcal{H}_1$ as a perturbation and expand

$$e^{-\mathcal{H}} = e^{-\mathcal{H}_o} e^{-\mathcal{H}_1}$$

$$= e^{-\mathcal{H}_o} \sum_{m=0}^{\infty} \frac{(-)^m}{m!} \mathcal{H}_1^m \quad . \tag{9.9}$$

The average value of any quantity Q over the distribution $P \propto e^{-\mathcal{H}}$ is then expressed in terms of the average over the Gaussian distribution $P_o \propto e^{-\mathcal{H}_o}$:

$$Q = \int \delta\sigma\, e^{-\mathcal{H}} Q \Big/ \int \delta\sigma\, e^{-\mathcal{H}}$$

$$= \left\langle e^{-\mathcal{H}_1} Q \right\rangle_o \Big/ \left\langle e^{-\mathcal{H}_1} \right\rangle_o$$

$$= \sum_{m=0}^{\infty} \frac{(-)^m}{m!} \langle \mathcal{H}_1^m Q \rangle_o \Big/ \sum_{m=0}^{\infty} \frac{(-)^m}{m!} \langle \mathcal{H}_1^m \rangle_o \ . \tag{9.10}$$

The free energy per volume $\mathfrak{F}$ is obtained through

$$e^{-\mathfrak{F}L^d} = \int \delta\sigma\, e^{-\mathcal{H}}$$

$$= \left\langle e^{-\mathcal{H}_1} \right\rangle_o \int \delta\sigma\, e^{-\mathcal{H}_o} \quad ,$$

$$\mathfrak{F} = \mathfrak{F}_o - L^{-d} \ln \left\langle e^{-\mathcal{H}_1} \right\rangle_o$$

$$= \mathfrak{F}_o - L^{-d} \ln \sum_{m=0}^{\infty} \frac{(-)^m}{m!} \langle \mathcal{H}_1^m \rangle_o \qquad (9.11)$$

is an obvious notation.

In (9.10) and (9.11) the task is to calculate averages of powers of $\mathcal{H}_1$ and their products with Q over the Gaussian distribution. Since Q is in general a product of σ's and so is $\mathcal{H}_1$, each term in (9.10) and (9.11) is of the form (9.6) before integrating over space. Our task becomes one of combining pairwise averages. The use of graphs helps greatly in this task. Let us see how it helps in evaluating the first-order term in the sum of (9.11), namel ,

$$-\langle \mathcal{H}_1 \rangle_o$$

$$= -\frac{1}{2} \int d^d x \; d^d x' \left\langle \frac{1}{2} \sigma^2(x) \, u(x - x') \frac{1}{2} \sigma^2(x') \right\rangle_o \; . \tag{9.12}$$

In Fig. 9.1a we pick two points x and x′ and connect them with a dashed line to represent $u(x - x')$. The solid lines sticking out at x and x′ represent the σ's. The rule for averaging established earlier is to pair up the σ's in all possible ways. Graphically this is represented by joining the ends of lines in Fig. 9.1a pairwise in all possible ways. One way is shown in Fig. 9.1b representing

$$-\frac{1}{2} \cdot \frac{1}{2} \langle \sigma^2 \rangle_o \; u(x - x') \; \frac{1}{2} \langle \sigma^2 \rangle_o$$

$$= -\frac{1}{8} \left(\sum_i G_o(x - x) \right) u(x - x') \left(\sum_i G_o(x' - x') \right)$$

$$= -\frac{1}{8} n^2 \; G_o^2(0) \, u(x - x') \; . \tag{9.13}$$

Figure 9.1c shows another way giving

$$-\frac{1}{2} \cdot \frac{1}{2} \left(\sum_i \langle \sigma_i(x) \, \sigma_i(x') \rangle_o^2 \right) = -\frac{1}{4} n G_o(x - x')^2 \, u(x - x') \; . \tag{9.14}$$

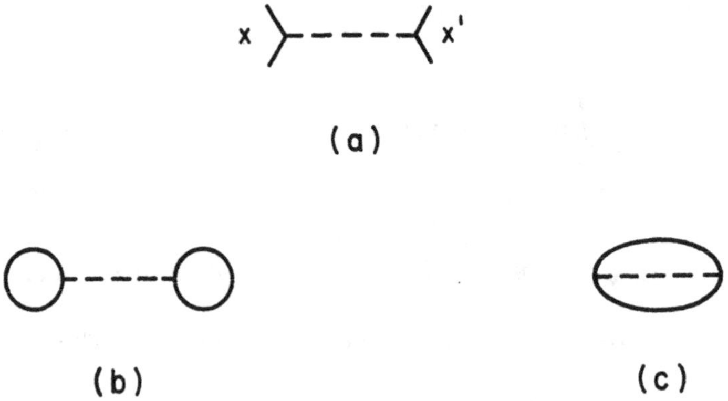

Figure 9.1. (a) The graph representation of the interaction $u(x-x')$. (b) and (c) First-order terms in the free energy.

One of the 1/2 factors is gone because there are two ways to join $\sigma_i(x)$ to one of the two $\sigma_i(x')$'s in $\sigma^2(x')$. Thus we have

$$-\langle \mathcal{H}_1 \rangle_o = -\int d^d x\, d^d x' \left(\frac{n^2}{8} G_o(0)^2\, u(x-x') \right.$$

$$\left. + \frac{n}{4} G_o(x-x')^2\, u(x-x') \right) . \qquad (9.15)$$

This example illustrates the way graphs can be constructed to represent terms in the perturbation expansion. However, in practice the use of graphs is the other way around. Namely, under a set of rules, one can draw graphs first and then write down the perturbation terms. One needs a set of efficient and easily memorized rules. How these rules are set up and used is flexible and to some extent a matter of personal preference. The rules developed below are our suggestions.

Let us stay with the sum of (9.11) for the moment.

For a given m there are m factors of the interaction, i.e.,

$$\frac{1}{m!}\left(\frac{1}{2}\right)^m \int (-u(x_1-x_1'))(-u(x_2-x_2')) \cdots (-u(x_m-x_m'))$$
$$\times \frac{1}{2}\sigma^2(x_1)\frac{1}{2}\sigma^2(x_1') \cdots \frac{1}{2}\sigma^2(x_m)\frac{1}{2}\sigma^2(x_m')\, d^dx_1 \ldots d^dx_m' \ , \tag{9.16}$$

and we need m dashed lines to represent them. The integrations over $x_1, x_1', \ldots x_m'$ sum over all the configurations of the dashed lines. The factor $(1/m!)(1/2)^m$ tells us that those configurations differing by only a permutation without a change of geometry are counted as only one configuration, i.e., every geometrically distinct configuration is counted once and once only. To carry out

$\langle \sigma^2(x_1) \ldots \sigma^2(x'_m) \rangle_o$ we join the ends of dashed lines by solid lines in all possible ways. Let us write down the rules for computing terms of mth order.

(a) Draw m dashed lines. Each of the two ends of a dashed line joins the ends of two solid lines. Solid lines are thus continuously joined into closed loops. Figures 9.1 and 9.2 show the first- and the second-order terms. One must draw all possible graphs consistent with this rule.

(b) For each graph, label the coordinates of the joints ("vertices"). Write $-u(x-x')$ if a dashed line joins x and x', and write $G_o(x-x')$ if a solid line joins x and x'.

(c) Integrate over all coordinates and give a factor n for each closed loop of solid lines. This takes care of the integrations and summations over components in (9.16). $\left(\text{Recall that } \sigma^2 = \sum_{i=1}^{n} \sigma_i^2 .\right)$

(d) Divide the result of (c) by the number of permutations of vertices under which the graph is unchanged. This accounts for the above emphasized fact that each geometrically distinct configuration must be counted only once.

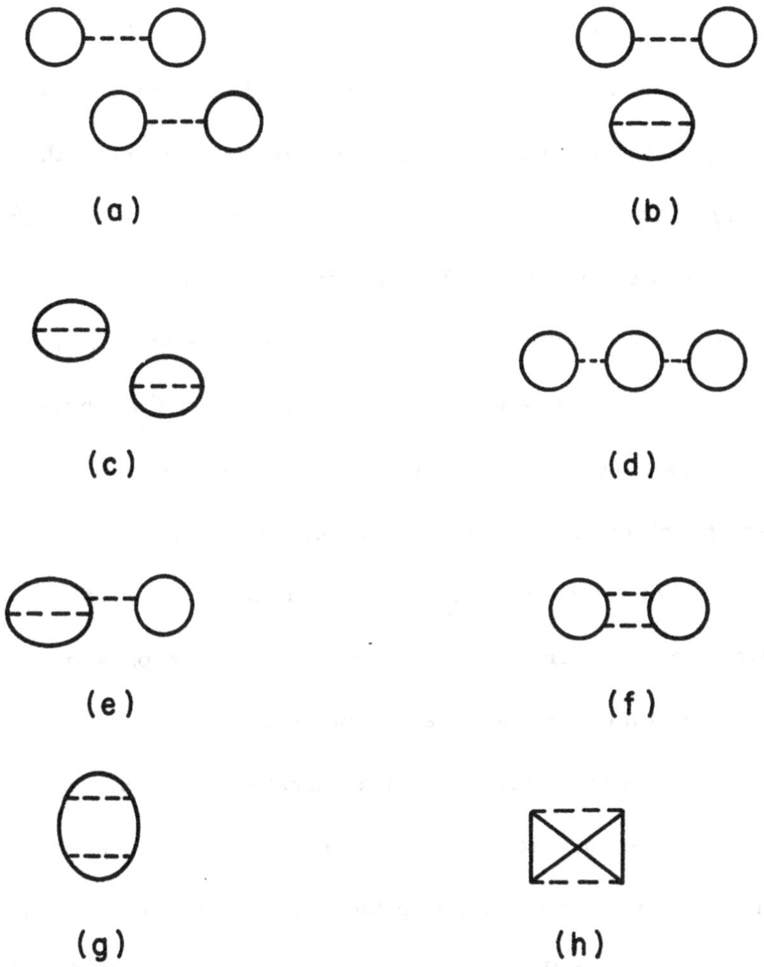

(a) (b) (c) (d) (e) (f) (g) (h)

Figure 9.2. Second-order terms of $\left\langle e^{-\mathcal{H}_1} \right\rangle_o$.

(e) For every closed loop which can be flipped without changing the graph, divide the result of (d) by 2.

The last rule, (d), is to take care of additional symmetry of closed loops, and is best illustrated by examples. In Fig. 9.1b, each loop can be flipped upside down without changing the graph and thus each giving a factor 1/2. This rule takes care of the 1/2 factor in front of each σ^2 in (9.16). If no such flip symmetry exists in a graph, then the 1/2 is always canceled in view of the fact that there are two ways to join to a σ_i in σ_i^2. Most authors prefer to keep track of these extra factors of 2's and define $\mathcal{H}_1$ without the 1/2 for each σ^2. This is a matter of convention. Our rules here are completely in terms of graph symmetry and no more factors are needed.

Thus the factor 1/8 in the first term of (9.15) comes from the permutation symmetry of x and x', and the flipping symmetry of the two loops of Fig. 9.1b. The factor 1/4 comes from the permutation of x and x' and the flipping symmetry of the loop of Fig. 9.1c.

Clearly, there are terms which are represented by disconnected graphs such as Fig. 9.2a and there are terms which are represented by connected graphs. The "linked cluster theorem" tells us that, when we take the logarithm in (9.11), only connected graphs will remain. This fact is easily established as follows.

A general graph has several disconnected pieces. Some of them may be identical. Let C_λ denote the contribution of a certain connected diagram. Thus a general disconnected graph gives

$$(m_1! \ldots m_k!)^{-1} \; C_1^{m_1} \; C_2^{m_2} \ldots C_k^{m_k} \quad . \tag{9.17}$$

The factor in front enables us to divide out the number of permutations which leave the graph unchanged. (The symmetry of each connected piece is taken care of in C_λ.) If we sum over all graphs, i.e., over all $m_1, m_2, \ldots$ in (9.17), taking into account all C_λ's, we get

$$\left\langle e^{-\mathcal{H}_1} \right\rangle_o = \exp(C_1 + C_2 + \cdots) \ , \tag{9.18}$$

where the zeroth-order term, i.e., 1, is included. Taking the logarithm gives simply the sum of all connected graphs.

Equation (9.11) becomes

$$\mathfrak{F} = \mathfrak{F}_o - L^{-d}\left(\sum_\lambda C_\lambda\right) . \tag{9.19}$$

This is the linked cluster theorem. Every C_λ is a "cluster."

The above rules are easily generalized to include the calculation of average values via (9.10). For example, if $Q = \sigma_i(y)\,\sigma_i(z)$, we have the correlation function

$$G(y-z) = \sum_{m=0}^{\infty} \frac{(-)^m}{m!} \langle \mathcal{H}_1^m \, \sigma_i(y)\,\sigma_i(z)\rangle_o \bigg/ \sum_{m=0}^{\infty} \frac{(-)^m}{m!} \langle \mathcal{H}_1^m \rangle_o \,. \tag{9.20}$$

The only new things here are the two σ's. The coordinates y and z are fixed. In addition to the lines drawn according to Rule (a), we also draw lines to join y and z to the dashed lines. Figure 9.3 shows some examples. Note that $\langle \sigma_i \rangle = 0$ by symmetry. The points y and z are connected by a continuous sequence of solid lines. The lines joined directly to the points y and z are called <u>external lines</u> or <u>legs</u>. All other lines are called <u>internal</u>

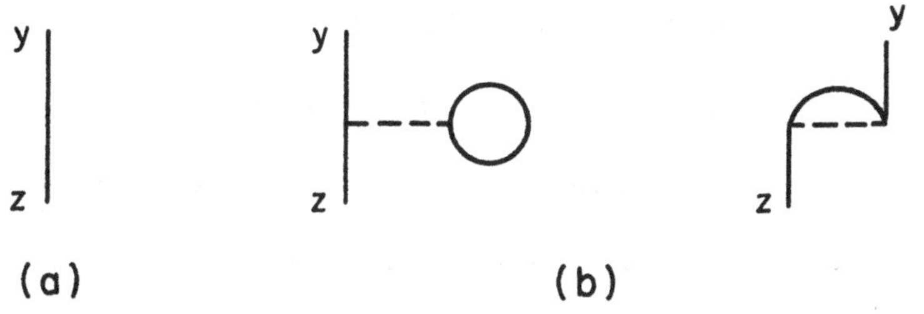

Figure 9.3. (a) Zeroth- and (b) first-order terms of G(y - z).

lines. The numerator of (9.20) is the sum of 2-leg graphs while the denominator is the sum of 0-leg graphs. Note that the sum of 2-leg graphs is the product of the sum of connected 2-leg graphs and the sum of all 0-leg graphs. Thus the 0-leg graphs in the denominator cancel those in the numerator. We are left with the conclusion that G(y - z) is the sum of all connected 2-leg graphs (apart from the trivial term $G_o(y - z)$). Obviously (9.19) says that the free energy (apart from the $\mathfrak{F}_o$ term) is the sum of connected 0-leg graphs.

More generally, we define

$$G_{i_1 i_2 \ldots i_m}(x_1, x_2, \ldots, x_m) \equiv \left\langle \sigma_{i_1}(x_1)\, \sigma_{i_2}(x_2) \ldots \sigma_{i_m}(x_m) \right\rangle_c \tag{9.21}$$

is the sum of m-leg connected graphs. Note that we put subscript c on the average sign to denote "cumulant." It is possible to still have disconnected graphs even after the 0-leg graphs are factored out. Figure 9.4 shows an example. The cumulant average of a product excludes all products of cumulant averages of all subsets. For example,

$$\langle A_1 A_2 \rangle_c = \langle A_1 A_2 \rangle - \langle A_1 \rangle \langle A_2 \rangle ,$$

$$\begin{aligned} \langle A_1 A_2 A_3 \rangle_c = \langle A_1 A_2 A_3 \rangle &- \langle A_1 A_2 \rangle_c \langle A_3 \rangle \\ &- \langle A_2 A_3 \rangle_c \langle A_1 \rangle \\ &- \langle A_1 A_3 \rangle_c \langle A_2 \rangle \\ &- \langle A_1 \rangle \langle A_2 \rangle \langle A_3 \rangle . \end{aligned} \quad (9.22)$$

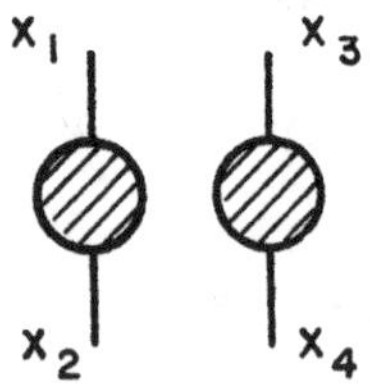

Figure 9.4. A disconnected graph for $\langle \sigma(x_1)\sigma(x_2)\sigma(x_3)\sigma(x_4)\rangle$.

The disconnected pieces in Fig. 9.4 are just cumulants of two subsets of $\sigma(x_1), \ldots, \sigma(x_4)$. Excluding such terms means keeping only graphs which are connected. The subscript c thus also implies connectedness.

Often it is more convenient to use wave vectors as integration variables by Fourier transforming all coordinates. Then every line carries a wave vector (instead of every vertex carrying a coordinate). The sum of all wave vectors that point toward a vertex must be zero (the wave-vector conservation rule). A dashed line with a wave vector q gives a factor

$$u_q = \int d^d x \; e^{-iq \cdot x} \; u(x) \tag{9.24}$$

and a solid line with a wave vector q gives $G_o(q) = (r_o + q^2)^{-1}$. This is the "wave vector representation" instead of the "coordinate representation" used so far. As an example, Fig. 9.5 gives

$$-G_o(k)(2\pi)^{-d} \int d^d q \; G_o(q) \left(u_o \frac{n}{2} + u_{k+q} \right) G_o(k) \tag{9.25}$$

which is the first-order contribution to G(k) [and is just

Figure 9.5. The terms in (9.25).

Fig. 9.3(b) in the wave-vector representation]. All wave vectors not fixed by the conservation law at vertices must be integrated over. Each integral carries a factor $(2\pi)^{-d}$. The wave-vector dependence of u_q will be ignored subsequently, assuming $u(x)$ has a very short range.

That part of a graph for G containing no single isolated solid line is frequently called the "self-energy part" Σ. The graphs in Fig. 9.5 excluding the legs are graphs for the self-energy. Equation (9.25) is a term of $-G_o \Sigma G_o$. The minus sign is a matter of convention. Graphs for G are then obtained by hooking up self-energy graphs as shown in Fig. 9.6. Consequently,

$$G(k) = G_o(k) - G_o(k)\,\Sigma(k)\,G_o(k) + G_o(k)\,\Sigma(k)\,G_o(k)\,\Sigma(k)\,G_o(k) - \cdots$$

$$= (G_o^{-1}(k) + \Sigma(k))^{-1}$$

$$= (r_o + k^2 + \Sigma(k))^{-1} \quad . \qquad (9.26a)$$

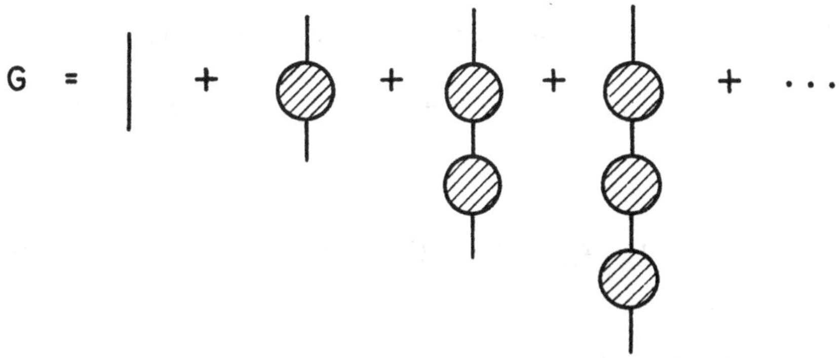

Figure 9.6. The correlation function in terms of the self energy part, Eq. (9.25).

It will turn out to be more convenient in many calculations to modify the definition of G_o slightly. We write

$$G^{-1}(k) = r_o + \Sigma(0) + k^2 + \Sigma(k) - \Sigma(0)$$

$$\equiv r + k^2 + \Sigma(k) - \Sigma(0) \quad . \tag{9.26b}$$

Here

$$r = G^{-1}(0) = r_o + \Sigma(0) \tag{9.27}$$

is just the inverse susceptibility. Now we define

$$G_o(k) = (r + k^2)^{-1} \quad , \tag{9.28a}$$

and calculate everything as a function of r instead of r_o. We must subtract $\Sigma(0)$ from $\Sigma(k)$ when we calculate $G(k)$. At the critical point, $r = 0$ and $r_o = r_{oc}$. Equation (9.27) gives

$$r_o - r_{oc} = r - (\Sigma(0, r) - \Sigma(0, 0)) \quad . \tag{9.28b}$$

A word of caution: if one is interested in calculating the free-energy graphs, then self-energy insertions cannot be correctly summed by the geometric series (9.26). The above rule of using (9.28a) as G_o and subtracting $\Sigma(0)$ from $\Sigma(k)$ will not work for free-energy graphs owing to extra symmetry factors. The reader can easily check this by studying the lowest-order graphs. Since we shall rarely be interested in computing free-energy graphs, no further elaboration on this point is needed here.

3. DIVERGENCE OF THE PERTURBATION EXPANSION AT THE CRITICAL POINT

In this section we show that the expansion parameter of the perturbation expansion outlined above is actually $ur^{(d-4)/2}$ for $d < 4$. The inverse susceptibility r

vanishes as the critical point is approached, and the expansion parameter becomes infinite. Consequently the perturbation expansion diverges. To be specific, consider the calculation of the self energy $\Sigma(k)$ for $k \to 0$. The conlusion is more general.

Every term in the perturbation expansion is a multiple integral over the wave vectors labeling the internal lines. A graph of the mth-order term has m dashed lines representing m powers of u. A graph of (m+1)st order has one more dashed lines. Since each dashed line joins four ends of solid lines, and each solid line has two ends, one additional dashed line implies two additional solid lines, which implies two extra G_o's and two extra wave vectors. The wave-vector conservation law for the wave vectors joined to this additional dashed line leaves one of the two extra wave vectors undetermined and to be integrated over. In short, when the order goes up by one there are two more G_o's and one more integration. Since there are two powers of wave vectors (or one power of r) in each G_o^{-1} and since each wave-vector integral involves d powers of wave vectors, a graph of the (m+1)st order has an extra factor of

$u(r^{1/2})^{d-4}$ compared to a graph of the mth order, apart from a numerical factor after all wave-vector integrals are done. Therefore the larger the order, the larger the contribution if $u \neq 0$, $d < 4$, and r is sufficiently small.

For $d > 4$, the wave-vector integrations will not diverge even if $r = 0$. An (m+1)st-order graph will have an extra factor $u\Lambda^{d-4}$ compared to an mth-order graph. Here Λ is the upper cutoff of the wave-vector integrals. (For $d < 4$, there is also a contribution proportional to $u\Lambda^{d-4}$ besides terms proportional to $ur^{1/2(d-4)}$ discussed above, but such a contribution plays no part in our conclusions.)

For $d = 4$ one expects logarithms of Λ and r instead of $r^{1/2(d-4)}$ and Λ^{d-4}.

The above analysis can be replaced by a simple dimensional argument. It is left to the reader as an exercise to show that the only dimensionless parameters proportional to u are $ur^{1/2(d-4)}$ and $u\Lambda^{d-4}$.

While it is easy to show the divergence of the perturbation expansion when the expansion parameter diverges, there is no proof that the expansion converges if the expansion parameter does not diverge. There has been no established radius of convergence for the expansion.

The above discussion can be generalized when the quantity of interest depends also on a wave vector k, for example, $\Sigma(k)$ with $k \neq 0$. There is then another expansion parameter uk^{d-4}, which diverges for $k \to 0$.

So far we have assumed that $\langle \sigma \rangle = 0$. The perturbation expansion must be modified when $\langle \sigma \rangle \neq 0$; this case will be discussed in Sec. 7.

4. THE 1/n EXPANSION OF CRITICAL EXPONENTS

Although the perturbation expansion diverges for $k \to 0$ and at the critical point, it still can be used as a mathematical tool for extrapolation schemes known as the ϵ expansion and the $1/n$ expansion. The former is conceptually more complicated and will be discussed in the next section.

The logical basis of the $1/n$ expansion of exponents by perturbation method is simply illustrated by the following example. The RG analysis of Chapter VI tells us that the correlation function at T_c for small k behaves like

$$G(k, \mu(T_c)) \propto k^{-2+\eta}(1 + O(k^{-y_2})) \; . \qquad (9.29)$$

[See Eq. (6.20).] The analysis of Chapter VIII tells us that $\eta = 0$ and $y_2 = d-4$ for $n \to \infty$. Suppose that $\eta = O(1/n)$ and $y_2 = d - 4 + O(1/n)$. Then we obtain

$$\begin{aligned} k^2 G(k, \mu(T_c)) \propto 1 &+ \eta \ln k \\ &+ \frac{1}{2}(\eta \ln k)^2 + \frac{1}{3!}(\eta \ln k)^3 + \cdots \\ &+ O(k^{4-d}, n^{-1} k^{4-d} \ln k, \ldots) \end{aligned} \qquad (9.30)$$

if we expand (9.29) formally in powers of $1/n$. If we can calculate Gk^2 by any means to $O(1/n)$ or higher orders in $1/n$, η can be extracted from the ratio of the coefficient of the $\ln k$ term and the constant term. Since η is universal, we can pick the simplest possible model for computing η. Let us pick the Ginzburg-Landau model with $u = O(1/n)$. Then the perturbation expansion becomes a series of powers of $1/n$ (apart from a slight complication to be discussed shortly). Note that $u = O(1/n)$ is our special choice for the sake of getting exponents; this choice is mathematically motivated and has no physical meaning.

Let us illustrate the basic technique of the $1/n$ expansion. We shall always assume that $2 < d < 4$ unless otherwise stated.

At $T = T_c$, we have $r = 0$ and (9.26b) becomes, to $O(1/n)$,

$$G^{-1}(k) = k^2 + \Sigma(k) - \Sigma(0)$$

$$= k^2 (1 - \eta \ln k + \cdots) . \tag{9.31}$$

All we need to do to find η to $O(1/n)$ is to obtain the $k^2 \ln k$ term in $\Sigma(k) - \Sigma(0)$. Since we choose $u = O(1/n)$, can we drop all but the first-order in the perturbation expansion? The answer is no. We argued in the previous section that the expansion parameter is actually uk^{d-4} and therefore we shall get a term proportional to k^{d-4} instead of $\ln k$ from the first-order term. We must note that for every close loop of solid lines, there is a factor n owing to the summation over spin components [see Rule (c) of Sec. 2]. The more loops, the more factors of n. Figure 9.7(a), (b), (c) show graphs for Σ up to $O(1/n)$. A wavy line represents the sum of graphs in Fig. 9.7(d). Whenever we include a u, we must include all "bubble

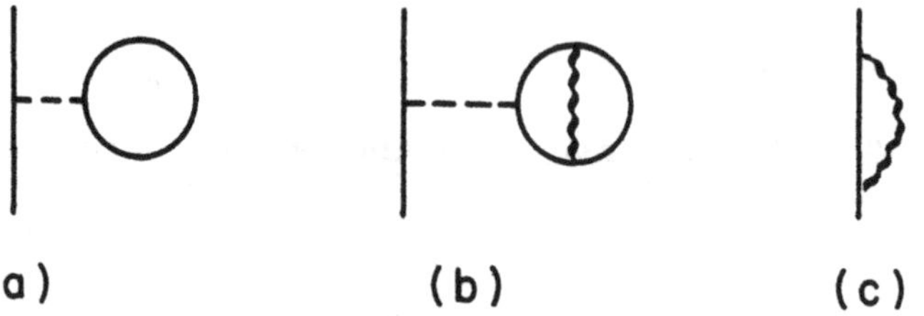

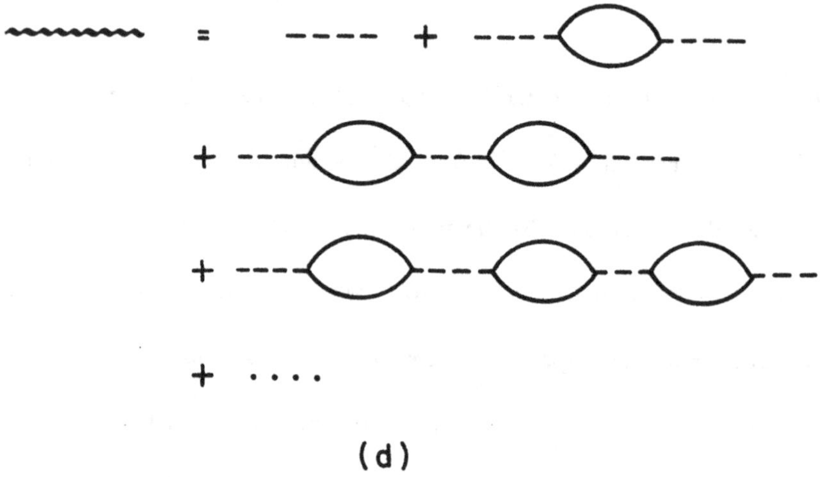

Figure 9.7. (a) $O(1)$ term of Σ.
(b) and (c) $O(1/n)$ terms of Σ.
(d) Definition of the wavy line. Every term in the sum is of $O(1/n)$.

corrections" in Fig. 9.7(d) because each bubble is proportional to n and compensates for the 1/n in the additional u needed to join the bubble to another bubble. The sum of graphs in Fig. 9.7(d) gives a "dressed" u. Thus a wavy line represents

$$-u + u^2 \frac{n}{2} \Pi(k) - u^3 \left[\frac{n}{2} \Pi(k)\right]^2 + \cdots$$

$$= -u \left(1 + u \frac{n}{2} \Pi(k)\right)^{-1} , \qquad (9.32)$$

where $\Pi(k)$ is the bubble given by

$$\Pi(k) = (2\pi)^{-d} \int d^d p \; G_o(p) \; G_o(p+k) \equiv \Pi(r, k^2) \qquad (9.33)$$

which is a function of k^2 and r. Here we set $r = 0$ for the case $T = T_c$.

The integral (9.33) is easily done. Just by dimensional analysis we can write

$$\Pi(k) = \Pi(0,1)\, k^{d-4} \left(1 + O((k/\Lambda)^{4-d})\right) \qquad (9.34)$$

for $k \ll \Lambda$. The constant $\Pi(0, 1)$, obtained by a more detailed calculation, is given in Table 9.1. This table also lists properties of Π for $r \neq 0$.

We note that the contribution of Fig. 9.7(a) and (b) to $\Sigma(k)$ is independent of k. Therefore only Fig. 9.7(c) contributes to $\Sigma(k) - \Sigma(0)$, which is

$$\Sigma_c(k) - \Sigma_c(0)$$

$$= (2\pi)^{-d} \int d^d p \, \frac{u}{\left(1 + u \frac{n}{2} \Pi(p)\right)} \left[\frac{1}{(p+k)^2} - \frac{1}{p^2}\right]. \tag{9.35}$$

The task now is to extract the $k^2 \ln k$ term in this integral. Clearly the $k^2 \ln k$ terms must come from the small-p region of the integral, i.e., the region near $p = 0$. If p is not small, the contribution to the integral divided by k^2 is well defined for $k \to 0$ and cannot give rise to an $\ln k$. If p is small, $\Pi(p) \sim p^{d-4}$ is large and we can drop the 1 compared to the $u \frac{n}{2} \Pi(p)$ in the denominator of (9.35), i.e.,

$$\frac{u}{1 + u \frac{n}{2} \Pi(p)} \approx \frac{2}{n} \Pi(p)^{-1} = \frac{2}{n} \Pi(0,1)\, p^{4-d} \quad . \tag{9.36}$$

Note that u has dropped out. It should drop out in view of the expectation that η must be independent of u. Substituting (9.36) in (9.35) and performing the integral, one indeed finds a $k^2 \ln k$ term

$$\Sigma_c(k) - \Sigma_c(0) = -\eta k^2 (\ln k + \text{const} + O(k^{4-d})) \quad ,$$

$$\eta = 4(4/d - 1)\, S_d/n \quad , \tag{9.37}$$

where S_d is defined in Table 9.1.

The above calculation illustrates fully how the 1/n expansion can be done using the perturbation expansion. The introduction of an artificial small parameter has allowed us to take advantage of the perturbation expansion, which we have shown to be meaningless in describing critical phenomena earlier.

Other exponents have also been calculated using the 1/n expansion. The technical aspects of this procedure are well presented in the literature [see, for example, Ma (1973a)]. A summary of results will be given at the end of this chapter.

Table 9.1

Some useful data for the 1/n expansion calculations

$$\Pi(r, k^2) = \Pi(r/k^2, 1)\, k^{d-4}$$

$$\Pi(r, 1) = \Pi(0, 1)(1 + 4r)^{(d-3)/2}$$

$$+ 2J\left(1 - \frac{1}{2}d\right)^{-1} r^{d/2-1}\, F\left(1, \frac{1}{2}, \frac{1}{2}d; -4r\right)$$

$$= \Pi(0, 1)[1 - 2(3-d)r + O(r^2)]$$

$$+ 2J\left(1 - \frac{1}{2}d\right)^{-1} r^{d/2-1}\, [1 - (4/d)r + O(r^2)]$$

$$\Pi(0, 1) = JB\left(\frac{1}{2}d - 1, \frac{1}{2}d - 1\right)$$

$$\Pi(r, 0) = J r^{d/2-2}$$

$$J = \frac{1}{2} K_d\, \pi\left(\frac{1}{2}d - 1\right) \csc \pi\left(\frac{1}{2}d - 1\right)$$

$$S_d = \frac{1}{2} K_d\, \Pi(0, 1)^{-1} = \frac{\sin \pi\left(\frac{1}{2}d - 1\right) \Gamma(d-1)}{2\pi \Gamma(d/2)^2}$$

$$= \frac{\sin \pi\left(\frac{1}{2}d - 1\right)}{\pi\left(\frac{1}{2}d - 1\right) B\left(\frac{1}{2}d - 1, \frac{1}{2}d - 1\right)}$$

$$K_d = 2^{-d+1} \pi^{-d/2} / \Gamma(d/2)$$

5. THE ϵ EXPANSION OF CRITICAL EXPONENTS

As in the $1/n$ expansion illustrated above, the ϵ expansion of critical exponents by using perturbation methods follows the same steps. Again, let us discuss (9.29), (9.30) and (9.31), and the calculation of η. Instead of choosing $u = O(1/n)$, we now choose $u = O(\epsilon)$. The quantity n is arbitrary here. We then calculate $G(k)$ by perturbation expansion and try to get η from the $k^2 \ln k$ term of $\Sigma(k) - \Sigma(0)$. We know that $\eta = O(\epsilon^2)$ from Chapter VII.

Now we run into a complication which did not occur for the $1/n$ expansion. The exponent y_2 is $-\epsilon + O(\epsilon^2)$ [see Eq. (7.73)]. Consequently the $O(k^{-y_2})$ term in (9.29) will also appear as a series in $\ln k$, i.e., $k^{-y_2} = 1 + \epsilon \ln k + \cdots$, when expanded in powers of ϵ. Note that in the case of the $1/n$ expansion, k^{-y_2} appears as $k^{4-d}\left(1 + O\left(\frac{1}{n} \ln k\right)\right)$. The k^{4-d} dependence is easily identified by inspection. If we calculate $G(k)k^2$ in powers of ϵ, we shall get not only powers of $\ln k$ owing to the k^η, which are what we want, but also powers of $\ln k$

from k^{-y_2}, which we are not interested in. The ln k's from the two origins cannot be separated by inspection in a practical calculation of G(k). A mathematical prescription must be developed to separate them in order to find η.

Wilson (1972) put forth such a mathematical prescription. Operationally, it simply involves matching: Choose a special value of u, which he called $u_o(\epsilon)$, in the Ginzburg-Landau model such that the calculated series of ln k for $G(k)k^2$ exponentiates to match the expected power law behavior k^{η}. In other words, adjust u to a special value so that the coefficient of $(\ln k)^2$ is 1/2 of the square of the coefficient of ln k, that of $(\ln k)^3$ is 1/3! of the cube of that of ln k and so on. Once $u_o(\epsilon)$ is found [$u_o(\epsilon)$ turns out to be of $O(\epsilon)$], one can use the perturbation expansion to calculate other quantities. The associated exponents can be extracted from logarithms. For this special value of $u = u_o(\epsilon)$, the logarithms owing to $O(k^{-y_2})$ do not appear.

The meaning of $u_o(\epsilon)$ and the reason behind its special property was explained by Wilson (1972). His explanation was brief and was often misunderstood, even though everyone learned how to calculate following his prescription. A good deal of confusion in the literature appears to have grown out of such misunderstanding. We shall comment on this point later in Sec. 8.

Let us explain the meaning of $u_o(\epsilon)$ in detail. We need to take for granted the qualitative conclusions of Chapter VI on the structure of the parameter space of the RG.

Recall that for μ near a fixed point μ^* we can expand $\delta\mu = \mu - \mu^*$ in eigenvectors e_j of R_s^L:

$$\mu = \mu^* + \sum_j t_j e_j$$

$$= \mu^* + t_1 e_1 + h e_h + \sum_{j=2}^{\infty} t_j e_j \quad . \tag{9.38}$$

The transformation under R_s is given by

$$R_s \mu = \mu^* \pm (s/\xi)^{1/\nu} e_1 + h s^{y_h} e_h$$

$$+ \sum_{j=2}^{\infty} s^{y_j} t_j e_j \quad , \tag{9.39}$$

where the last term is denoted by $O(s^{y_2})$ for large s in (6.29) because y_2 is the largest of all y_j's in the last sum. We showed in Chapter VI that the power law behavior for the correlation function and other physical quantities follows when the $O(s^{y_2})$ terms are neglected.

If the t_2 in the sum of (9.39) is zero, then the largest term in the last sum of (9.39) will be $s^{y_3} t_3 e_3$ for large s, and the $O(k^{-y_2})$ term in (9.29) will be replaced by $O(k^{-y_3})$. If y_3 is of $O(1)$ instead of $O(\epsilon)$, then k^{-y_3} will not give rise to powers of $\ln k$ which get mixed up with those from k^{η}. In fact $y_3 = -2 + O(\epsilon)$, as we discussed in Chapter VII, and $k^{-y_3} = k^2 (1 + O(\epsilon) \ln k)$, which has an extra k^2 in front. Thus to remove the unwanted logarithms from the ϵ expansion, we need simply to choose the parameters in the Hamiltonian such that $t_2 = 0$. We now define $u_o(\epsilon)$ by the following statement:

<u>The Ginzburg-Landau model with $h = 0$, $r_o = r_{oc}(r = 0)$ and $u = u_o(\epsilon)$ is just that Ginzburg-Landau model represented by the special point μ with $h = t_1 = t_2 = 0$.</u>

Note that $(u_o(\epsilon), r_{oc})$ are not (u_4^*, r_o^*), the values of u_4 and r_o at the fixed point, even though $(u_o(\epsilon), r_{oc})$ happen to have the same values as (u_4^*, r_o^*) to the first order in ϵ. The fixed point $\mathcal{K}^*$ is not of a Ginzburg-Landau form beyond $O(\epsilon)$.

The removal of t_2 is sometimes referred to as "the removal of the slow transient" or the "enlargement of the critical region." Since s^{y_2} with a very small although negative y_2 implies a very slow approach or transient of $R_s\mu$ to μ^* as s increases, the removal of $t_2 s^{y_2}$ is thus the removal of the slow transient. According to (6.47) - (6.50), the critical regions depend on y_2 crucially. If $|y_2|$ is small, so is the critical region. If $t_2 = 0$, all the y_2's in (6.47) - (6.50) will be replaced by a much larger $|y_3|$ and the critical region will be enlarged from $O(10^{-1/\epsilon})$ to $O(1)$. In the case of the $1/n$ expansion, 4-d is not a small quantity and the critical region is already of $O(1)$. Therefore there is no need to set u to any special value.

In principle r_{oc} and $u_o(\epsilon)$ can be obtained by solving the equations $t_1(r_{oc}, u_o(\epsilon)) = 0$ and $t_2(r_{oc}, u_o(\epsilon)) = 0$.

In practice there is no need to solve these equations. One can simply use the matching prescription discussed above. There is no need to make any reference to the RG as far as the practical calculation of exponents in powers of ϵ is concerned. However, without understanding the RG structure first we would not understand the criteria for matching. Wrong criteria lead to meaningless results.

6. SIMPLE ILLUSTRATIVE CALCULATIONS, η AND α

The following examples illustrate the use of graphs and roughly how the ϵ expansion is carried out in practice.

First, let us calculate η to $O(\epsilon^2)$. Assume $u = u_o(\epsilon) = O(\epsilon)$. The second-order contribution to $\Sigma(k)$ is given in Fig. 9.8. Note that the first-order graphs, shown in Fig. 9.5, are independent of k and therefore do not contribute to $\Sigma(k) - \Sigma(0)$. We obtain, following the rules developed earlier,

$$\Sigma(k) = -u^2\left(\frac{n}{2}+1\right)(2\pi)^{-2d}\int d^dp\, d^dq\, G_o(p+k)\, G_o(q)\, G_o(p+q)\,. \tag{9.40}$$

Figure 9.8. Second-order terms of $\Sigma(k)$ which depend on k.

The integral is easily done by going back to the coordinate representation:

$$G_o(x) = (2\pi)^{-d} \int d^d k \; e^{ik \cdot x} \, G_o(k) \; , \tag{9.41}$$

$$\Sigma(k) = -u^2\left(\frac{n}{2} + 1\right) \int d^d x \; e^{-ik \cdot x} \, G_o(x)^3 \; . \tag{9.42}$$

Since we set $r = 0$, we have $G_o(k) = k^{-2}$. Setting $d = 4$, we obtain

$$G_o(x) = (2\pi)^{-2} \; |x|^{-2} \; (1 - J_o(\Lambda |x|)) \; . \tag{9.43}$$

We are only interested in the $k^2 \ln k$ term in $\Sigma(k)$. Let us replace $e^{-ik \cdot x}$ in (9.42) by the expansion

$$e^{-ik \cdot x} = 1 - ik \cdot x - \frac{1}{2}(k \cdot x)^2 + \cdots \tag{9.44}$$

The first term is independent of k and can be ignored. The second term is odd in x and does not contribute to the integral. The third term is equivalent to $-\frac{1}{2}\cdot\frac{k^2|x|^2}{4}$ by spherical symmetry. Thus

$$\Sigma(k)-\Sigma(0)=u^2\left(\frac{n}{2}+1\right)(2\pi)^{-6}\int d^4x\,|x|^{-6}\,\frac{k^2|x|^2}{8}+\cdots\,. \tag{9.45}$$

The integral of (9.45) is logarithmically divergent. The expansion (9.44) is in fact illegitimate. However, for extracting $k^2 \ln k$ it does serve the purpose. The upper cutoff of the integral is $\sim 1/k$ given by $e^{-ik\cdot x}$ and the lower cutoff is $\sim\Lambda^{-1}$ given by the J_0 in (9.43). We obtain

$$\begin{aligned}
k^{-2}G^{-1}(k^2) &= 1+(\Sigma(k)-\Sigma(0))/k^2\\
&= 1-\left(\frac{u}{16\pi^2}\right)^2\left(\frac{n}{2}+1\right)(\ln k+\text{const})+O(k^2)\\
&= k^{-\eta}\left(1+O\left(k^{-y_2}\right)+O\left(k^{-y_3}\right)\right).
\end{aligned} \tag{9.46}$$

Note that the ln k term in (9.46) is the sum of $-\eta \ln k$

from $k^{-\eta}$ and the contribution from $O(k^{-y_2})$ as we explained in the previous section. The next task is to calculate $u_o(\epsilon)$. When we set $u = u_o(\epsilon)$ in (9.46) the $O(k^{-y_2})$ becomes zero and we obtain

$$\eta = (u_o(\epsilon)/16\pi^2)^2 \left(\frac{n}{2} + 1\right) . \qquad (9.47)$$

There are several ways of obtaining $u_o(\epsilon)$ by matching. One way is to calculate the $(\ln k)^2$ term of $k^{-2} G^{-1}$ to $O(\epsilon^3)$. Since there is no $O(\epsilon^3)$ term in the matching formula

$$k^{-\eta} = 1 - \eta \ln k + \frac{1}{2} (\eta \ln k)^2 + \cdots \qquad (9.48)$$

because $\eta = O(\epsilon^2)$, we can set the coefficient of $(\ln k)^2$ to $O(\epsilon^3)$ to zero and solve for u. We then obtain $u_o(\epsilon)$ to $O(\epsilon)$. To compute $G^{-1}k^{-2}$ to $O(\epsilon^3)$ we not only have to calculate to $O(u^3)$ in the perturbation expansion, but also have to obtain the $O(\epsilon u^2)$ terms in (9.40) by expanding the ϵ occurring in $(2\pi)^{-(4-\epsilon)} \int d^{(4-\epsilon)} p d^{(4-\epsilon)} q$. This program is not difficult to carry out, but there are simpler ways to obtain $u_o(\epsilon)$ to $O(\epsilon)$ by using matching formulas other than (9.48). The most convenient one is that involving exponents that are already known to $O(\epsilon)$. It was first used by Wilson (1972) and is obtained as follows.

First we define the correlation function of four spins:

$$G(k_1 k_2 k_3 k_4, \mu) = \int d^d x_1 \, d^d x_2 \, d^d x_3 \, e^{-ik_i x_1 - ik_2 \cdot x_2 - ik_3 \cdot x_3}$$

$$\times \langle \sigma_i(x_1) \, \sigma_i(x_2) \, \sigma_j(x_3) \, \sigma_j(0) \rangle_c$$

$$= L^d \left\langle \sigma_{ik_1} \, \sigma_{ik_2} \, \sigma_{jk_3} \, \sigma_{jk_4} \right\rangle_c \quad , \qquad (9.49a)$$

where $k_4 = -k_1 - k_2 - k_3$. This correlation function is represented by graphs of four legs. Now we divide it by $G(k_1)\, G(k_2)\, G(k_3)\, G(k_4)$ and take the limit of small k's to obtain

$$\Gamma_4(\mu) \equiv - \lim_{k_i \to 0} \; G(k_1 \ldots k_4, \mu) \Big/ \prod_{i=1}^{4} G(k_i, \mu). \qquad (9.49b)$$

This division removes the contribution of the four legs (or "amputates the legs"). The transformation formula (5.16) for $G(k, \mu)$ is easily generalized to correlation functions involving more spins. We obtain

$$G(k_1 \ldots k_4, \mu) = s^{d+4-2\eta}\, G(sk_1 \ldots sk_4, \mu') \ . \qquad (9.50)$$

Note that $\lambda_s^2 s^d = s^{2-\eta}$ in (5.16) and $L^d = L' s^d$ in (9.49a). We thus obtain from (9.49b)

$$\Gamma_4(\mu) = s^{d-4+2\eta}\, \Gamma_4(\mu') \ . \qquad (9.51)$$

The generalization of (6.19) using the same arguments gives

$$\Gamma_4 \propto \xi^{d-4+2\eta}\left(1 + O(\xi^{y_2})\right)$$

$$\propto r^{(\epsilon-2\eta)/(2-\eta)}\left(1 + O\left(r^{-y_2/(2-\eta)}\right)\right) \qquad (9.52)$$

since $1/r \propto \xi^{2-\eta}$. To the first order in ϵ, $\eta = 0$, $y_2 = -\epsilon$; we therefore obtain

$$\Gamma_4 \propto \left(1 + \frac{1}{2}\epsilon \ln r + \text{const } t_2 y_2 \ln r + O(\epsilon^2)\right) , \qquad (9.53)$$

which involves no unknown exponent.

The lowest-order graph for Γ_4 is just Fig. 9.9(a),

which gives simply u. The next-order graphs are shown by Fig. 9.9(b). We obtain, using $G_o(p) = 1/(r + p^2)$,

$$\Gamma_4 = u\left(1 - \left(\frac{n}{2} + 4\right) u(2\pi)^{-d} \int d^d p \, G_o(p)^2\right)$$

$$= u\left(1 + \left(\frac{n}{2} + 4\right) \frac{u}{16\pi^2} (\ln r + \text{const}) + O(\epsilon^2)\right). \quad (9.54)$$

Comparing (9.54) and (9.53), we see that in order to set $t_2 = 0$ and remove the $\ln r$ coming from $O(\xi^{y_2})$, we need to choose $u = u_o(\epsilon)$ with

$$u_o(\epsilon) = \epsilon(8\pi^2)\Big/\left(\frac{n}{2} + 4\right) + O(\epsilon^2) . \quad (9.55)$$

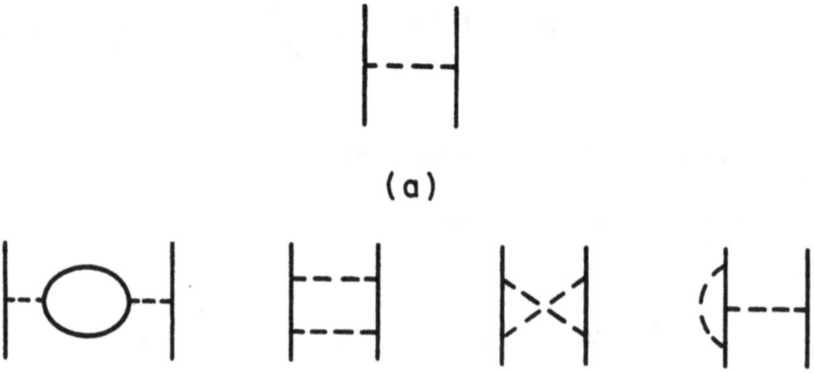

(a)

(b)

Figure 9.9. Graphs for Γ_4: (a) $O(u)$, and (b) $O(u^2)$.

Substituting (9.55) in (9.47), we obtain

$$\eta = \frac{(n+2)\epsilon^2}{2(n+8)^2} + O(\epsilon^3) \; . \tag{9.56}$$

The exponent γ can be obtained to $O(\epsilon)$ by evaluating Fig. 9.5 with $k = 0$ and using the formula

$$r_o - r_{oc} = r - \Sigma(0, r) + \Sigma(0, 0)$$

$$\propto r^{1/\gamma} \; . \tag{9.57}$$

The algebra is left as an exercise for the reader. The result is

$$\gamma = 1 + \frac{n+2}{2(n+8)}\epsilon + O(\epsilon^2) \; . \tag{9.58}$$

We now proceed to calculate the specific heat exponent α.

The specific heat can be obtained from

$$C \propto -\partial^2 F/\partial r_o^2$$

$$\propto \frac{1}{4}\int d^d x \, \langle \sigma^2(x)\, \sigma^2(0) \rangle_c \; . \tag{9.59}$$

The zeroth-order term is shown in Fig. 9.10a which gives

$$\frac{n}{2}\,\Pi = -\frac{n}{32\pi^2}\ \ln r\left[1 - \frac{\epsilon}{4}\ln r + \epsilon A\right] + B \tag{9.60}$$

where A and B are cutoff-dependent constants of O(1). To O(1) one would conclude that $C \propto \ln r$, which is consistent with $\alpha = 0 + O(\epsilon)$. The O(u) terms are shown in Fig. 9.10b, which gives $-u\frac{n}{2}\left(\frac{n}{2}+1\right)\Pi^2$. Combining it with (9.60), we obtain

$$C \propto B - \frac{n}{32\pi^2}\ln r\left[1 + \epsilon A' + \left(\frac{u}{16\pi^2}\left(\frac{n}{2}+1\right) - \frac{\epsilon}{4}\right)\ln r\right] \tag{9.61}$$

where A′ is a constant of O(1). The expected behavior of C (see Sec. VI. 3), is

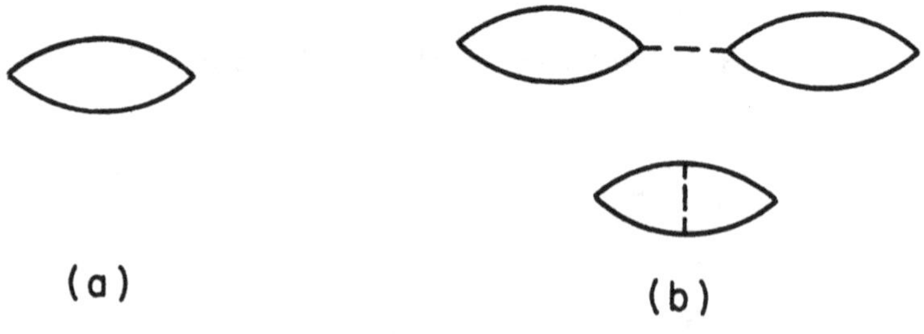

Figure 9.10. Graphs for $\langle\sigma^2(x)\,\sigma^2(0)\rangle_c$: (a) O(1), and (b) O(u).

$$C \propto \text{const} + |r_o - r_{oc}|^{-\alpha}$$

$$\propto \text{const} + r^{-\alpha/\gamma}$$

$$\propto \text{const} - \frac{\alpha}{\gamma} \ln r \left[1 - \frac{\alpha}{2\gamma} \ln r + \cdots\right] \quad (9.62)$$

Comparing (9.61) and (9.62), we obtain

$$\frac{u_o(\varepsilon)}{8\pi^2} \left(\frac{n}{2} + 1\right) - \frac{\varepsilon}{2} = -\frac{\alpha}{\gamma} \quad . \quad (9.63)$$

Note that (9.62) and (9.63) do not have any specified overall constant factor. We can only compare what is in the brackets. Using (9.55) we obtain

$$\alpha = \frac{1}{2}\left(\frac{4-n}{n+8}\right)\varepsilon + O(\varepsilon^2) \quad . \quad (9.64)$$

The simplicity of the $O(\varepsilon)$ calculation shown above does not remain for higher-order calculations. The number of graphs increases rapidly with the order. Counting graphs is a trivial task compared to performing the integrals. We shall not elaborate on higher-order calculations here.

7. THE PERTURBATION EXPANSION IN THE PRESENCE OF A NONZERO $\langle \sigma \rangle$

Recall that in the Gaussian approximation discussed in Chapter III, the spin fluctuation [i.e., the deviation of the spin configuration $\sigma(x)$ from the most probable configuration $\tilde{\sigma}(x)$] was treated by the quadratic, or harmonic, approximation. The perturbation expansion is just an extension of the Gaussian approximation to include the anharmonic terms in $\mathcal{H}$. The most probable configuration $\tilde{\sigma}(x)$ is the configuration that minimizes $\mathcal{H}[\sigma]$, as was explained in Chapter III. For $T > T_c$ and $h = 0$, we have $\tilde{\sigma}(x) = 0$ so that $\sigma - \tilde{\sigma} = \sigma$. The perturbation expansion discussed in this chapter has assumed that $\tilde{\sigma} = 0$.

For $T < T_c$ or $h \neq 0$, $\tilde{\sigma}(x) \neq 0$. The perturbation expansion must be modified. The Gaussian approximation with $\tilde{\sigma} \neq 0$ was discussed in Chapter III. To set up the perturbation expansion we extend the Gaussian approximation by keeping in $\mathcal{H}$ anharmonic terms of $\sigma - \langle \sigma \rangle$ as well. In the Gaussian approximation $\tilde{\sigma}(x)$ is just the average spin $\langle \sigma \rangle$. Beyond the Gaussian approximation the most probable configuration $\tilde{\sigma}$ is no longer the same as the average spin $\langle \sigma \rangle$, but it is more convenient to formulate the perturbation expansion in terms of $\sigma - \langle \sigma \rangle$ than in terms of $\sigma - \tilde{\sigma}$.

Let the field h be along the 1 direction so that the magnetization is long the 1 direction, i.e., $\langle \sigma_i(x) \rangle = m\delta_{i1}$. Write $\sigma_1' = \sigma_1 - m$ and $\mathcal{H}[\sigma]$ is then expressed in terms of σ_1' and $\sigma_\perp \equiv (0, \sigma_2, \sigma_3 \ldots \sigma_n)$ as

$$\begin{aligned}
\mathcal{H} = \frac{1}{2} L^d \left(r_o m^2 + \frac{1}{4} u m^4 \right) - L^d hm \\
+ \frac{1}{2} \int d^d x \Big[(\nabla \sigma_1')^2 + (\nabla \sigma_\perp)^2 + \left(r_o + \frac{1}{2} u m^2 \right) \sigma_\perp^2 \\
+ \left(r_o + \frac{3}{2} u m^2 \right) \sigma_1'^2 \\
+ u m \sigma_1' (\sigma_1'^2 + \sigma_\perp^2 + m^2) \\
+ \frac{1}{4} u (\sigma_\perp^2 + \sigma_1'^2)^2 + (r_o m - h) \sigma_1' \Big] .
\end{aligned} \tag{9.65}$$

This Hamiltonian is manifestly asymmetric under rotations in the spin space. The field h has selected a special direction. The fact that h is the only source of the rotation asymmetry leads to some important mathematical consequences which we shall mention before going into the details of the perturbation expansion.

Define

$$G_1(k) = \langle |\sigma'_{1k}|^2 \rangle ,$$

$$G_\perp(k)\, \delta_{ij} = \langle \sigma_{ik}\, \sigma_{j-k} \rangle , \quad i, j \neq 1 . \qquad (9.66)$$

If we change the field h to $h + \delta h$, the average spin m will change into $m + \delta m$. If $\delta h \parallel h$, the ratio $\delta m/\delta h$ is called the longitudinal susceptibility, which is related to G_1 by

$$(\delta m/\delta h)_1 = G_1(0) . \qquad (9.67)$$

If $\delta h \perp h$, the ratio is called the transverse susceptibility related to $G_\perp$ by

$$(\delta m/\delta h)_\perp = G_\perp(0) . \qquad (9.68)$$

Now we note that applying an infinitesimal field δh perpendicular to h is the same as rotating h by an infinitesimal angle $\delta h/h$. As a result the original m must be rotated by the same angle, i.e.,

$$\delta h/h = \delta m/m \quad , \tag{9.69}$$

which implies that $(\delta m/\delta h)_\perp = m/h$, i.e.,

$$G_\perp(0) = m/h \quad . \tag{9.70}$$

This result fixes the small k limit of $G_\perp(k)$. Of special interest is the case $h \to 0$ and $m \neq 0$, which occurs for $T < T_c$. It follows from (9.70) that

$$G_\perp^{-1}(0) = 0 \quad , \tag{9.71}$$

for $T < T_c$, $h \to 0$. Equations (9.70) and (9.71) are very important mathematical relations independent of the perturbation expansion. Of course for $T > T_c$, m vanishes when h vanishes and $G_\perp^{-1}(0) = G_1^{-1}(0) \neq 0$ is just the inverse susceptibility. Physically (9.71) simply says that it requires no work to rotate m below T_c if $h = 0$. Of course for $n = 1$ there is no $\perp$ component.

Now we return to (9.65). Note that (9.65) is simply

$\mathcal{H}[\sigma]$ written in terms of m, σ_1', and $\sigma_\perp$ without specifying what m is. We need to specify $m = \langle \sigma \rangle$, or, equivalently,

$$\langle \sigma_1' \rangle = 0 \quad . \tag{9.72}$$

Equation (9.72) is an equation obtained by calculating $\langle \sigma_1' \rangle$ using (9.65). It gives an equation from which we can solve for m. It is equivalent to the equation $\partial F(m)/\partial m = 0$, where F(m) is the free energy of $\mathcal{H}$ given by (9.65) at a fixed m.

The perturbation expansion now must include interactions involving m and must differentiate between the $\perp$ and the 1 components, in addition to including those features mentioned earlier in this chapter.

We summarize the new features as follows.

(a) Define $G_{o\perp}$ and G_{o1} as

$$G_{o1}(k) = (r_1 + k^2)^{-1} \quad ,$$

$$G_{o\perp}(k) = (r_\perp + k^2)^{-1} \quad , \tag{9.73}$$

where $r_\perp \equiv G_\perp^{-1}(0) = h/m$ and $r_1 \equiv G_1^{-1}(0)$. If a solid line is labeled by 1, it represents G_{o1}, otherwise it represents both G_{o1} and $G_{o\perp}$.

(b) Figure 9.11 shows new interactions. Each dotted line represents a factor m. The first graph is a constant $\frac{1}{8}um^4$ for the free energy. Each m is counted as of $O(u^{-1/2})$. [We already encountered this fact in Chapter III. See Eq. (3.18). Note that $a_4 = 8u$.] Thus in Fig. 9.11 the terms (1), (2), (3), (4), and (5) are, respectively, of the order u^{-1}, $u^{-1/2}$, 1, 1, and $u^{1/2}$. The term (6) = $O(u)$ is of course the interaction included in the perturbation expansion discussed earlier.

(c) The sum of graphs with one external line (the "tadpole" graphs) shown in Fig. 9.12 simply gives $\langle\sigma_1'\rangle$, which is zero. Thus (9.72) is just the equation (sum of tadpole graphs) = 0, from which we can solve for m.

Note that if we remove a dotted line from the graphs for $F(m)$ in all possible ways, then we simply obtain graphs for $\partial F(m)/\partial m$. These graphs are just the tadpole graphs with the leg removed. Thus the equation $\langle\sigma_1'\rangle = 0$ is equivalent to $\partial F(m)/\partial m = 0$.

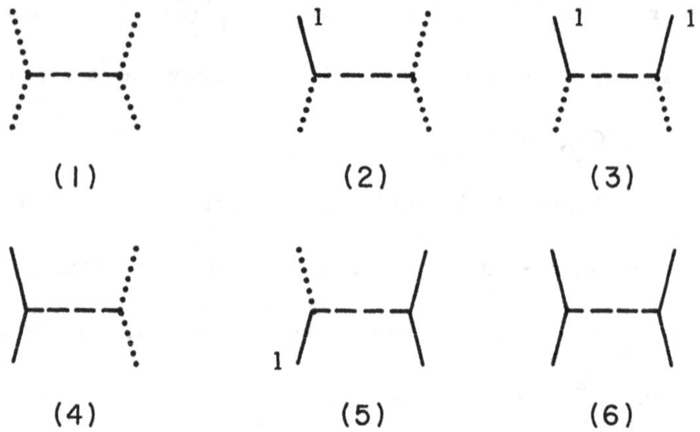

Figure 9.11. Interaction terms involving powers of m, (1) - (5). The last one, (6), is just $u(\sigma^2)^2$.

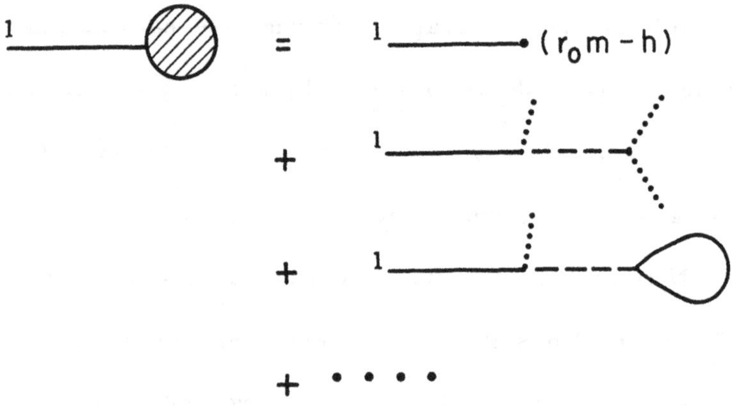

Figure 9.12. Graphs for $\langle \sigma_1' \rangle$

(d) r_1 can be obtained as a function of r_o via

$$r_1 = r_o + \Sigma_1 (k = 0, \; r_1 , \; r_\perp) \; ,$$

$$r_\perp = h/m \; , \tag{9.74}$$

where Σ_1 is the sum of self energy graphs for G_1. The equation $\langle\sigma_1'\rangle = 0$ written in terms of h, m, and r_o is then the equation of state relating h, m, and r_o. An equivalent equation of state is, for $n \geq 2$, given by $r_\perp = h/m$:

$$h/m = r_o + \Sigma_\perp (k = 0, \; r_1 , \; h/m) \; , \tag{9.75}$$

where r_1 must be expressed in terms of r_o, h, and m.

As an illustrative example consider the case of large n. We choose $u = O(1/n)$ for simplicity. Since $m = O(u^{-1/2})$, we have $m = O(n^{1/2})$. The lowest-order tadpole graphs are shown in Fig. 9.12. The equation $\langle\sigma_1'\rangle = 0$ to this order is

$$-r_o m + h - \frac{1}{2} um^3 - \frac{1}{2} umn(2\pi)^{-d} \int d^d p \, (r_\perp + p^2)^{-1} = 0 \; . \tag{9.76}$$

At the critical point we have $r_o = r_{oc}$, $h = 0$, $m = 0$, and $h/m = 0$. Divide (9.76) by m and subtract its value at the critical point. We obtain

$$r_{oc} - r_o + \frac{h}{m} - \frac{1}{2} um^2$$
$$- \frac{1}{2} un(2\pi)^{-d} \int d^d p \left[\frac{1}{r_\perp + p^2} - \frac{1}{p^2} \right] = 0 \ . \tag{9.77}$$

This is the equation of state to O(1) upon substituting $r_\perp = h/m$. For small $r_\perp$ the integral in (9.77) is easily evaluated. We obtain

$$r_{oc} - r_o + \frac{h}{m} - \frac{1}{2} um^2 + (h/m)^{d/2-1} \frac{1}{2} unJ \Big/ \left(\frac{d}{2} - 1\right) = 0 \ . \tag{9.78}$$

The constant J is given in Table 9.1.

When $h = 0$ and $r_o < r_{oc}$, (9.78) gives

$$m \propto (r_{oc} - r_o)^{1/2} \ , \tag{9.79}$$

which implies that $\beta = 1/2$. For $r_o = r_{oc}$ we have

$$m^2 \propto (h/m)^{d/2 - 1} ,$$

$$m \propto h^{(d-2)/(d+2)} , \tag{9.80}$$

which implies that $\delta = (d-2)/(d+2)$. These values of β and δ are consistent with the values $1/\nu = d-2$, $\eta = 0$, and the scaling laws to the leading order in $1/n$.

The equation of state (9.76) can also be obtained from (9.75). The algebra is left as an exercise for the reader. The transverse correlation function $G_\perp(k)$ is

$$G_\perp(k) = (k^2 + h/m)^{-1} \tag{9.81}$$

to $O(1)$. Note that $G_\perp$ is independent of r_o when $h = 0$. This means that $\langle \sigma_\perp^2 \rangle = (2\pi)^{-d} \int d^d p \; G_\perp(p)$ is independent of r_o and hence the transverse spin fluctuations do not contribute to the specific heat to $O(1)$:

$$C \propto -\frac{\partial}{\partial r_o} \langle \sigma^2 \rangle$$

$$= -\frac{\partial}{\partial r_o} \left(m^2 + \langle \sigma_\perp^2 \rangle + \langle \sigma_1'^2 \rangle \right)$$

$$= -\frac{\partial}{\partial r_o} m^2 (1 + O(1/n))$$

$$= \frac{1}{u}(1 + O(1/n)) \; . \qquad (9.82)$$

On the other hand, above the critical point and $h = 0$, the $r_\perp$ in (9.77) must be replaced by an r which is nonzero. One obtains $r_o - r_{oc} \propto r^{d/2 - 1}$ and

$$C \propto -\frac{\partial}{\partial r_o} \langle \sigma^2 \rangle$$

$$\propto n(r_o - r_{oc})^{\frac{4-d}{d-2}} + \text{const} \; , \qquad (9.83)$$

which gives $\alpha = (d-4)/(d-2)$. Thus (9.82) seems to give a different exponent, i.e., $\alpha' = 0$ for $r_o < r_{oc}$. This interpretation is not quite right, however. The $\frac{1}{u} O(1/n)$ term in (9.82) can in fact be shown to be proportional to $(r_{oc} - r_o)^{-\alpha}$ with $\alpha = (d-4)/(d-2)$. This situation illustrates the extreme care needed in interpreting results in $1/n$ and ϵ expansions. A calculation to a certain order may not give the right exponent to that order.

The behavior of the longitudinal correlation function

$G_1(k)$ is very different. Figure 9.13 shows the graph for Σ_1 to O(1).

We note that the first two graphs give just $\Sigma_\perp$. Therefore

$$G_1^{-1}(k) = r_\perp + um^2 \Big/ \left(1 + \frac{n}{2}\, u\Pi\,(r_\perp, k)\right)$$

$$\approx r_\perp + \frac{2m^2}{n}\,\Pi^{-1}(r_\perp, k) \ , \qquad (9.84)$$

where $\Pi(r, k)$ is given in Table 9.1. In particular for $k = 0$ we have

$$G_1^{-1}(0) = r_1 = \frac{2m^2}{nJ}\,(h/m)^{2-d/2} + O(h/m) \ , \qquad (9.85)$$

which vanishes for $h \to 0$. This means that $(\partial h/\partial m)_T$ below T_c vanishes for $T < T_c$ on the coexistence curve

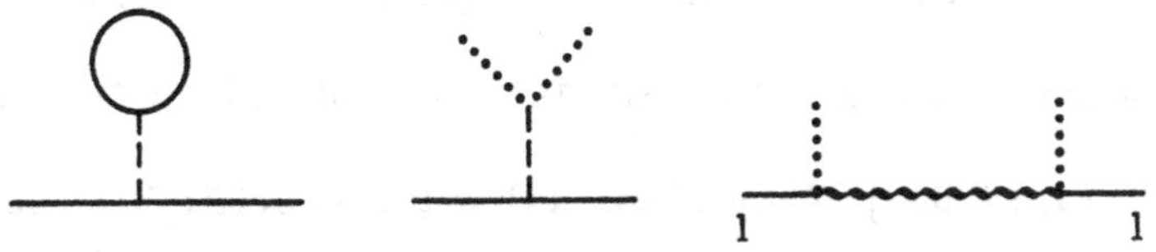

Figure 9.13. Self energy graphs of O(1) for large n.

for large n. [Brezin, Wallace, and Wilson (1972) and Brezin and Wallace (1973) showed that this conclusion should hold for any $n \geq 2$.]

8. REMARKS

It must be kept in mind that the calculation of exponents by the ϵ or the $1/n$ expansion is an extrapolation procedure based on the assumption that the results of the RG analysis are correct. Mathematically the series in $\ln k$ can converge only when $|\ln k|$ is not too large. Exponents such as η are defined by $G(k)k^2 \propto k^{\eta}$ for $k \to 0$, i.e., when $|\ln k|$ is very large. We rely heavily on what the RG tells us, i.e., that the $\ln k$ term should show what η is.

Sometimes there may be extra logarithmic terms which are unrelated to the exponent or the "slow transient" mentioned earlier. A classic example is the specific heat calculation of Abe and Hikami (1973) to $O(1/n)$. [See also Ma (19 3a, 1975).] They showed that the specific heat behaves like

$$C/n \propto [(1+B_\ell)(r_o - r_{oc})^{-\alpha} - B_\ell (r_o - r_{oc})^{\ell-1}$$

$$+ \text{(other integral powers of } (r_o - r_{oc}))] , \quad (9.85')$$

where $\ell \geq 2$ is any integer and B_ℓ is a constant. They found that B_ℓ is nonzero only when $d = 2 + 2/\ell$, where α_o, the O(1) term of α, is $\alpha_o = (d-4)/(d-2) = 1 - \ell$. which is an integer. Consequently if one calculates C/n to O(1) the B_ℓ terms cancel. To O(1/n) and at $d = 2 + 2/\ell$, the B_ℓ terms appear as a logarithmic term ($\alpha = \alpha_o + \alpha_1$):

$$B_\ell (r_o - r_{oc})^{\ell-1} \left[(r_o - r_{oc})^{-\alpha_1} - 1 \right]$$

$$= -\alpha_1 B_\ell (r_o - r_{oc})^{\ell-1} \ln (r_o - r_{oc}) ,$$

which is in addition to the $-\alpha_1 (r_o - r_{oc})^{\ell-1} \ln (r_o - r_{oc})$ term of interest. Of course, without knowing this peculiar appearance of B_ℓ, one would conclude from the extra log term that α_1 changes by an anomalous factor $1 + B_\ell$ at $d = 2 + 2/\ell$. This was in fact the first interpretation. Later Abe and Hikami (1973) and Wilson (private communication) established the specific heat equation (9.85′). Clearly one must always think twice before asserting that a certain logarithmic term reflects a part of an exponent. There are various cases where extra logarithmic terms cannot be interpreted as exponents directly.

We conclude this section by mentioning two widespread and often misleading statements which grew out of a lack of understanding or carelessness.

(a) That the mean field theory and the Gaussian approximation are good away from T_c. The fact is that most many-body systems are strongly interacting and the mean field theory and the Gaussian approximation are <u>not good at any temperature</u>. The mean field and the Gaussian approximation give good descriptions of critical behaviors only for a <u>weakly interacting</u> system over a temperature range in which the Ginzburg parameter ζ_T is very small. [See Sec. III.6.]

(b) That the interaction is weak for d near 4. Of course the interaction strength is a special property of the particular system of interest. It may be weak or it may be strong. Just because we <u>arbitrarily choose</u> $u = u_o(\varepsilon) = O(\varepsilon)$

for the sake of computing exponents, the thought that somehow the interaction strength in general should be of $O(\epsilon)$ has become widespread. A consequence of thinking that $u = O(\epsilon)$ is thinking that $u = 0$ at $d = 4$. This of course leads to the false conclusion that the critical behavior for $d = 4$ is trivial and given by the Gaussian approximation. In fact the critical behavior for $d = 4$ is far from trivial and will not be pursued here. As was mentioned in Sec. VII. 5, the nontrivial fixed point approaches the trivial one as $d \to 4$. At $d = 4$ the degeneracy of the two fixed points makes the critical behavior very complicated.

9. THE RG IN THE PERTURBATION EXPANSION

The study of the RG by means of the perturbation expansion was already set up in Chapter VII [see Eqs. (7. 13) - (7. 16)], and applied to the case of small ϵ [see Secs. VII. 5 and 6], without the aid of graphs. Now we have developed the language of graphs. The mathematical procedure in Chapter VII can be put in a more convenient and general form (with no change of the substance, of course), which is the subject of this section.

The transformation of $\mathcal{H}[\sigma]$ into $\mathcal{H}'[\sigma]$ under the RG is defined by

$$\mathcal{H}'[\sigma] + AL^d = -\ln\left[\int \delta\phi \; e^{-\mathcal{H}[\phi + \sigma']}\right]_{\sigma'_k \to s^{1-\eta/2}\sigma_{sk}} , \tag{9.86}$$

where σ' and ϕ are, respectively, the long- and short-wavelength parts of σ as defined by (7.16) or some other procedure. Now we write

$$\mathcal{H}[\phi + \sigma'] = \mathcal{H}[\sigma'] + \mathcal{H}_o[\phi] + \mathcal{H}_1[\phi, \sigma'] . \tag{9.87}$$

The first term involves no ϕ; $\mathcal{H}_o[\phi]$ is quadratic in ϕ involving no σ', and $\mathcal{H}_1[\phi, \sigma']$ is the rest. Using the definition (7.16) we have

$$\mathcal{H}_o[\phi] = \frac{1}{2} \sum_{\Lambda/s < q < \Lambda} \sum_{i=1}^{n} |\sigma_{iq}|^2 (r_o + q^2) . \tag{9.88}$$

$\mathcal{H}_1[\phi, \sigma']$ is more complicated. If $\mathcal{H}[\sigma]$ includes the quartic $u(\sigma^2)^2$ and the six-power $u_6(\sigma^2)^3$ terms as shown in Fig. 9.14a, $\mathcal{H}_1$ will have the terms shown in

Fig. 9.14b. Note that we use a thin solid line to denote σ' and a thick solid line to denote ϕ. Thus $\mathcal{H}_1[\phi, \sigma']$ has odd as well as even powers of ϕ.

Substituting (9.87) in (9.86) we obtain

$$\mathcal{H}'[\sigma] = \left[\mathcal{H}[\sigma'] - \ln\left\langle e^{-\mathcal{H}_1[\phi,\sigma']}\right\rangle_o\right]_{\sigma'_k \to \sigma_{sk}\, s^{1-\eta/2}}$$

$$+ \ln\left\langle e^{-\mathcal{H}_1[\phi,0]}\right\rangle_o \quad , \tag{9.89}$$

$$AL^d = -\ln\left(\int \delta\phi\, e^{-\mathcal{H}_o[\phi]}\right) - \ln\left\langle e^{-\mathcal{H}_1[\phi,0]}\right\rangle_o \quad , \tag{9.90}$$

where $\langle \cdots \rangle_o$ means averaging over $P_o[\phi] \propto \exp(-\mathcal{H}_o[\phi])$, keeping σ' fixed. The major task of finding $\mathcal{H}'[\sigma]$ is thus the evaluation of $-\ln\left\langle e^{-\mathcal{H}_1}\right\rangle_o$, which is just the free energy for ϕ with σ' fixed. In graph language we have

$$\ln\left\langle e^{-\mathcal{H}_1[\phi,\sigma']}\right\rangle_o - \ln\left\langle e^{-\mathcal{H}_1[\phi,0]}\right\rangle_o$$

= (the sum of all connected graphs with thin external lines and thick internal lines) . (9.91)

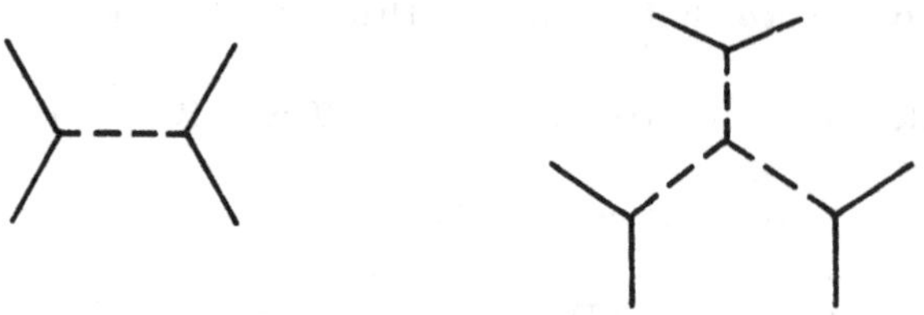

(a)

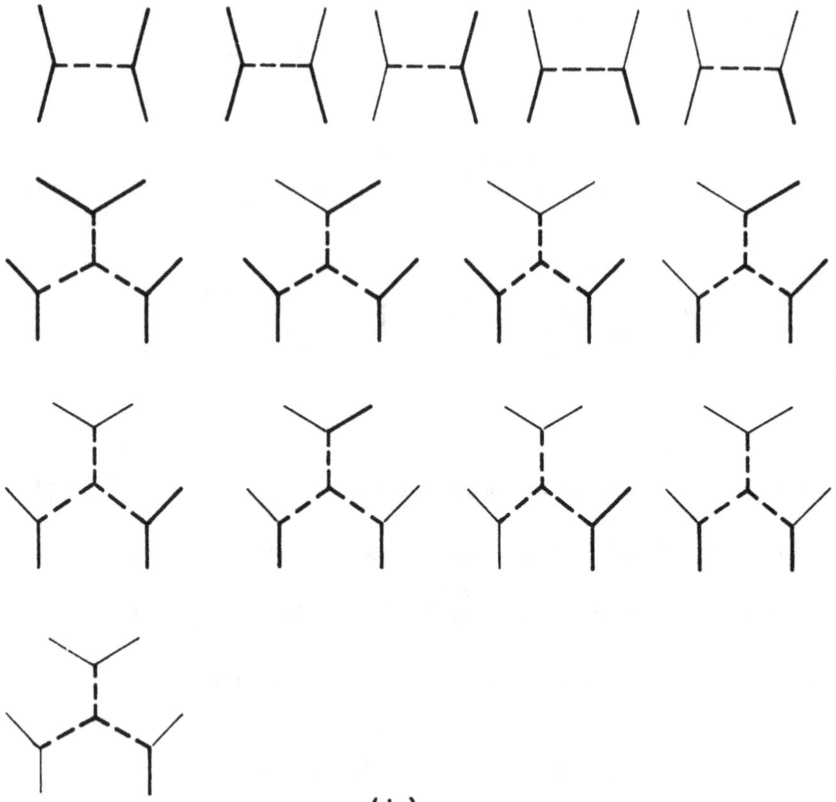

(b)

Figure 9.14. (a) $u(\sigma^2)^2$ and $u_6(\sigma^2)^3$ in $\mathcal{H}[\sigma]$.
(b) Terms in $\mathcal{H}_1[\phi, \sigma']$.

The thin external lines represent the fixed σ'. Each thick internal line represents a factor $\mathcal{G}_o(x - x') = \langle \phi(x)\, \sigma(x') \rangle_o$. [The subtraction in (9.91) simply removes those graphs without any external line, i.e., those terms which are σ'-independent.]

As an illustration we give in Fig. 9.15 the $O(u^2)$ graphs which represent the terms (b) through (g) of (7.61). The change of scale and further algebra are already discussed in Chapter VII.

Let us summarize the general definition of $R_s \mu = \mu'$ in graph language in the wave-vector representation. Given $\mu = (u_1, u_2, u_3 \ldots)$ representing

$$\mathcal{H} = \sum_{m=1}^{\infty} L^{d-md/2} \sum_{k_1 \ldots k_{m-1}} \sum_{i_1 \ldots i_m}$$

$$\times\, u_{m i_1 \ldots i_m}(k_1 \ldots k_{m-1})\, \sigma_{i_1 k_1} \ldots \sigma_{i_m k_m} \,, \qquad (9.92)$$

where $k_m = -k_1 - k_2 - \ldots - k_{m-1}$, we define σ' and ϕ as earlier, and $\delta_{ij} \mathcal{G}_o^{-1}(k) = u_{2ij}(k)$ for $\Lambda/s < k < \Lambda$ and zero otherwise. There are two steps in defining R_s:

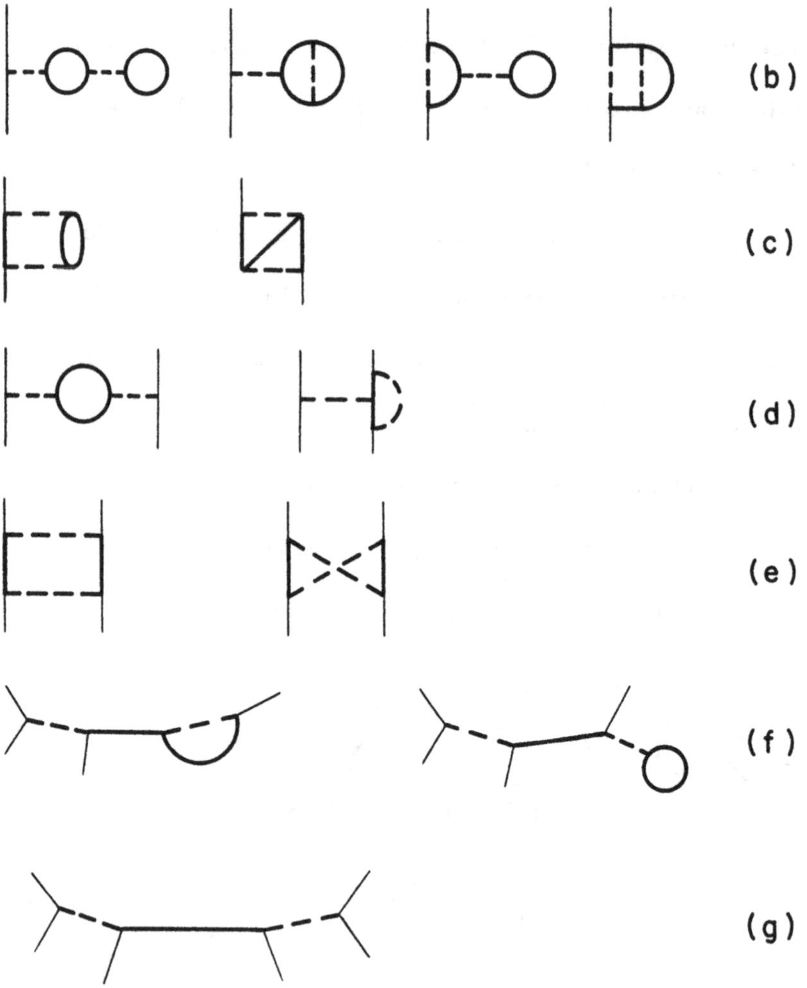

Figure 9.15. Graphs representing the terms (b) - (g) of (7.61).

Step (i): Calculate $\bar{u}_m$ defined by

$$\bar{u}_{mi_1\ldots i_m}(k_1\ldots k_{m-1}) = u_{mi_1\ldots i_m}(k_1\ldots k_{m-1})$$

$$\text{- (the sum of connected graphs of } m \text{ external lines)} \quad , \tag{9.93}$$

where all internal lines of the graphs must have wave vectors in the shell $\Lambda/s < k < \Lambda$, and all external lines must have wave vectors $k < \Lambda/s$.

Step (ii): Obtain

$$u'_{mi_1\ldots i_m}(k_1\ldots k_{m-1})$$

$$= \bar{u}_{mi_1\ldots i_m}(k_1/s\ldots k_{m-1}/s)\; s^{d+m(2-\eta-d)/2} \quad , \tag{9.94}$$

which gives $\mu' = (u'_1, u'_2, \ldots) = R_s\mu$. The exponent η is chosen to assure the existence of a fixed point μ^*.

The above definition is formal and makes no reference to the ϵ or $1/n$ expansion. In practice it is a convenient definition to remember for various small-ϵ calculations of the RG such as that in Chapter VII. More examples and generalizations will appear in subsequent chapters.

Note that in (9.92) we have included odd as well as even powers of σ and also arbitrary indices. This has been done only to remind the reader that there are infinite varieties of Hamiltonians other than the ones we have been studying. In all our earlier discussions we have assumed complete rotation symmetry except for the magnetic field h. The effect of anisotropy will be our next subject.

10. ANISOTROPIC PARAMETERS AND COMMENTS ON THE LIQUID-GAS CRITICAL POINT

Anisotropic parameters are those parameters in $\mathcal{H}$ which are not invariant under rotations, as, for example, when the field h is a vector in the spin space. Since we are already familiar with the effect of h, we shall assume that $h = 0$ in the subsequent discussion unless otherwise stated.

In a real crystal lattice the spin interactions are not completely rotation-invariant because the lattice itself is not. One has to include in $\mathcal{H}$, in addition to the invariant terms studied so far, the terms

$$\mathcal{H}_A = \sum_\lambda a_\lambda \int d^d x \, D_\lambda \tag{9.95}$$

where a_λ are the anisotropic parameters and D_λ are in general powers and products of σ and derivatives of σ. Since all our previous conclusions about critical behaviors assumed that $\mathcal{H}_A = 0$, it is crucial to find out whether the presence of $\mathcal{H}_A$ will change those conclusions. Let us phrase this problem in the language of the RG.

Let the set of parameters μ include the anisotropic parameters. Let μ_n^* be the completely n-component-spin-rotation-invariant fixed point, which has been studied so far, and s^{y_j} be the eigenvalues of the linearized RG transformation R_s^L in the neighborhood of μ_n^*. [See Eqs. (6.3) - (6.11).] Assume that a_λ are small and the terms in $\mathcal{H}_A$ are so arranged that a_λ are among the t_j's in (6.9), i.e., $a'_\lambda = a_\lambda s^{y_\lambda}$, where y_λ are the exponents associated with a_λ. Then the crucial question is whether y_λ's are positive or negative, or zero. If one or more a_λ with $y_\lambda \geq 0$ is nonzero, then μ_n^* is unstable and the previously described critical behaviors would be strongly modified. (See the discussion in Sec. VII. 3.) One has to calculate the y_λ associated with the anisotropic parameters of physical interest. The study of these exponents has been done mostly for small ϵ or for small $1/n$. We shall briefly summarize three examples. Details of calculations will be omitted.

(a) Spin tensor interaction

Consider the parameters $a_{\tau ij}$, $i, j = 1, \ldots, n$, transforming in the same manner as a traceless spin tensor under rotations in the n-dimensional spin space. The exponent y_τ associated with a_τ can be obtained by calculating a correlation function involving the tensor

$$\tau_{ij} = \sigma_i \sigma_j - (1/n)\sigma^2 \delta_{ij} , \tag{9.96}$$

for example

$$\int d^d x \, d^d x' \langle \tau_{ij}(0)\, \sigma_i(x)\, \sigma_j(x')\rangle \propto r^{-1-y_\tau/(2-\eta)} . \tag{9.97}$$

(Recall that $r \propto \xi^{2-\eta}$.) The result for small $1/n$ is

$$y_\tau = 2(1 - 8\,d^{-1}\,S_d/n) + O(n^{-2}) \ , \qquad (9.98)$$

where the quantity S_d is defined in Table 9.1. [See Ma
(1974b), Sec. 5E. Note that the summary equation (1.13)
has a misprint in y_τ. Equation (5.16) in the text is correct.] For small ϵ one has

$$y_\tau = 2\left[1 - \frac{1}{n+8}\,\epsilon + \frac{n^2 - 18n - 88}{4(n+8)^3}\,\epsilon^2 + O(\epsilon^3)\right] \qquad (9.99)$$

[taken from λ_2 of Aharony (1975), Eq. (3A.10)].

   Clearly $y_\tau > 0$ and a_τ is relevant. The simplest
example of a nonzero a_τ is the uniaxial perturbation term
($a_{11} = a$, $a_{ij} = 0$ otherwise)

$$a(\sigma_1^2 - \sigma^2/n) \qquad (9.100)$$

in $\mathcal{K}$ which makes the 1 direction special. Fisher and
Pfeuty (1972) studied this perturbation to $O(\epsilon)$. If $a < 0$,
$R_s\mu$ will approach the fixed point μ_1^* for $n = 1$ because
there is an easy axis, the 1 axis, of magnetization. If
$a > 0$, $R_s\mu$ will approach the fixed point μ_{n-1}^* because
the 1 axis is a "hard axis," and the system will have the

critical behavior of an (n-1)-component spin system.

(b) Cubic interaction

A term in $\mathcal{H}$ which is invariant under the permutation of spin components but not invariant under spin rotations is called a cubic interaction term by convention. The simplest example is

$$a_c \sum_{i=1}^{n} \int \sigma_i^4 \, d^d x \quad . \tag{9.101}$$

(Note that $\Sigma_i \sigma_i^2 = \sigma^2$ is rotation invariant.)

The exponent y_c associated with a_c and related questions was first studied by Wallace (1973). One can find y_c to $O(1/n)$ by calculating the correlation function

$$\int d^d x_1 \ldots d^d x_4 \langle \sigma_i^4(0)\, \sigma_i(x_1) \ldots \sigma_i(x_4) \rangle_c$$

$$\propto r^{-4-(y_c-4+d-2\eta)/(2-\eta)} \quad , \tag{9.102}$$

where the subscript c under $\langle \ldots \rangle$ denotes cumulant average. The result is

$$y_c = 4 - d - (16/n)(1+2/d)\, S_d + O(n^{-2}) \tag{9.103}$$

[see Ma (1975), Sec. V.6], which is positive for large enough n. The study for small ϵ gives

$$y_c = \frac{n-4}{n+8}\epsilon + \frac{5n^2 + 14n + 152}{(n+8)^3}\epsilon^2 + O(\epsilon^3) \ . \qquad (9.104)$$

[See Ketly and Wallace (1973) and Aharony (1975), Sec. V.] This result indicates that for sufficiently small n, y_c is negative, the isotropic fixed point is stable, and a_c is therefore irrelevant. For large enough n, $y_c > 0$ and a_c becomes relevant and the symmetric fixed point becomes unstable. The borderline case is at $n_c = 4 - 2\epsilon + O(\epsilon^2)$ where $y_c = 0$. The reader is referred to Wallace (1973) and the papers quoted above for further details.

(c) <u>The vector term $\sum_i a_i \, \delta\mathcal{H}/\delta\sigma_i$</u>

Here $\mathcal{H}$ denotes the isotropic Hamiltonian, the Ginzburg-Landau Hamiltonian, for example. This is an example of a special kind of term called "redundant" (Wegner, 1974b), and not of physical interest. It can be removed by a change of variable

$$\sigma_i \to \sigma_i + a_i \qquad (9.105)$$

(assuming a_i is very small). Thus a_i will not appear in the free energy. One can calculate the correlation function involving $\delta \mathcal{H}_1 / \delta \sigma_i$ and the associated exponent $y_{\sigma 1}$ by the expansion methods discussed above and find that

$$y_{\sigma 1} = \frac{1}{2} (d - 2 + \eta) \quad . \tag{9.106}$$

This exponent actually implies that the correlation functions only have a short-range behavior at the critical point. For further discussions see Wilson and Kogut (1974) Appendix, Wegner (1974b), and Ma (1974b).

The last example brings us to the subject of the liquid-gas critical point for which the order parameter does not have the inversion symmetry (symmetry of $\mathcal{H}$ under $\sigma \to -\sigma$) characterizing the magnetic systems. There are other cases such as the binary-fluid critical points which may also lack such symmetry. An important question is whether the exponents of the liquid-gas critical points are the same as those of the $n = 1$ magnetic critical points. Experimental data so far seem to indicate some small differences, but there has been no clear theoretical understanding.

The asymmetry can be described by including odd powers of the order parameter σ in $\mathcal{H}$. Then one asks how the previous RG analysis is modified by such odd powers. Hubbard and Schofield (1973) looked into this question. They showed that a Ginzburg-Landau-like Hamiltonian with all powers of σ can be derived for the fluid system. For d near 4, they argued that powers beyond the fourth could be safely dropped. Furthermore, they observed that the σ^3 term can be removed by a change of variable $\sigma \to \sigma + a$ as in Eq. (9.106) with a proper choice of the constant a. Eliminating the σ^3 term also eliminates the difference between the RG analysis of the liquid-gas system and that of the magnetic system. Therefore Hubbard and Schofield concluded that the liquid-gas critical exponents should be the same as those of the $n = 1$ magnetic critical exponents at least in the ϵ expansion. At this moment we cannot regard this conclusion as final. While the σ^3 term can be removed by a change of variable $\sigma \to \sigma + a$, other powers $\sigma^5, \sigma^7 \ldots$ etc., which are important for $d = 3$ cannot be removed. The asymmetry is not something superficial. So far there has been no serious attempt to study the RG with asymmetry. We must remind ourselves that our knowledge of the RG so far is quite limited to the neighborhood of the symmetric fixed point. Global properties have not been well analyzed. The liquid-gas critical point still remains an open problem.

11. TABLES OF EXPONENTS IN ϵ AND $1/n$ EXPANSIONS

In Table 9.2 we list the results of ϵ-expansion calculations of some exponents. An advanced and detailed discussion on the ϵ expansion is given by Wilson and Kogut (1974), which also includes an exhaustive list of references. A study of higher-order terms in the ϵ expansion and convergence problems has been done by Nickel (1974). Brezin, Wallace, and Wilson (1972) and Brezin and Wallace (1973) have studied the equation of state in the ϵ expansion. There is a large literature on the ϵ expansion using the Callan-Symansik equation, starting with the work of Brezin, LeGuillo, and Zinn-Justin (1973). Some of the higher-order terms listed in Table 9.2 were calculated using the Callan-Symansik-equation approach. We

Table 9.2

Exponents for the isotropic n-components spin fixed point in the ϵ expansion

$$\alpha = -\frac{(n-4)}{2(n+8)}\epsilon - \frac{(n+2)^2}{4(n+8)^3}(n+28)\epsilon^2 - \frac{(n+2)}{8(n+8)^5}\{n^4 + 50n^3 + 920n^2 + 3472n + 4800 - 192(5n+22)(n+8)T\}\epsilon^3 + O(\epsilon^4)$$

$$\beta = \frac{1}{2} - \frac{3}{2(n+8)}\epsilon + \frac{(n+2)(2n+1)}{2(n+8)^3}\epsilon^2 + \frac{(n+2)}{8(n+8)^5}\{3n^3 + 128n^2 + 488n + 848 - 48(5n+22)(n+8)T\}\epsilon^3 + O(\epsilon^4)$$

$$\gamma = 1 + \frac{(n+2)}{2(n+8)}\epsilon + \frac{(n+2)}{4(n+8)^3}(n^2 + 22n + 52)\epsilon^2 + \frac{(n+2)}{8(n+8)^5}\{n^4 + 44n^3 + 664n^2 + 2496n + 3104 - 96(5n+22)(n+8)T\}\epsilon^3 + O(\epsilon^4)$$

$$\delta = 3 + \epsilon + \frac{1}{2(n+8)^2}\{n^2 + 14n + 60\}\epsilon^2 + \frac{1}{4(n+8)^4}\{n^4 + 30n^3 + 276n^2 + 1376n + 3168\}\epsilon^3 + \frac{1}{16(n+8)^6}\{2n^6 + 96n^5 + 1778n^4 + 12760n^3 + 50280n^2 + 147136n + 263040 + 768(n+2)(n+8)(5n+22)T\}\epsilon^4 + O(\epsilon^5)$$

$$\eta = \frac{(n+2)}{2(n+8)^2}\epsilon^2 + \frac{(n+2)}{8(n+8)^4}(-n^2 + 56n + 272)\epsilon^3 + \frac{(n+2)}{32(n+8)^6}\{-5n^4 - 230n^3 + 1124n^2 + 17920n + 46144 - 768(5n+22)(n+8)T\}\epsilon^4 + O(\epsilon^5)$$

$$\nu = \frac{1}{2} + \frac{(n+2)}{4(n+8)}\epsilon + \frac{(n+2)}{8(n+8)^3}(n^2 + 23n + 60)\epsilon^2 + \frac{(n+2)}{32(n+8)^5}\{2n^4 + 89n^3 + 1412n^2 + 5904n + 8640 - 192(5n+22)(n+8)T\}\epsilon^3 + O(\epsilon^4)$$

$$y_2 = -\epsilon + \frac{3(3n+14)}{(n+8)^2}\epsilon^2 + O(\epsilon^3)$$

$$y_T = 2\left[1 - \frac{1}{n+8}\epsilon + \frac{n^2 - 18n - 88}{4(n+8)^3}\epsilon^2\right] + O(\epsilon^3)$$

$$y_c = \frac{n-4}{n+8}\epsilon + \frac{5n^2 + 14n + 152}{(n+8)^3}\epsilon^2 + O(\epsilon^3)$$

The first five exponents are from the collection of Wilson and Kogut (1974) Table 8.1. The last three are available in Aharony (1975). The constant T is 0.60103.

shall not elaborate on this approach except to remark, for those readers who have a field-theory background, that the Ginzburg-Landau model is mathematically equivalent to an unrenormalized $\lambda\phi^4$ (Euclidean) model in quantum field theory. It is important to note that the $\lambda\phi^4$ theory is, in the conventional renormalization scheme, super-renormalizable and $\lambda = 0$ is a stable fixed point for the Callan-Symansik equation, and hence does not describe the relevant physics of critical phenomena. In this sense the statement that "quantum field theory = classical statistical mechanics" has little meaning. To extract what is relevant, Brezin and coworkers and other authors had to use an unconventional renormalization scheme in which the bare coupling constant goes to infinity with the cutoff Λ as $\lambda \propto \Lambda^{4-d}$. The $\lambda\phi^4$ theory becomes renormalizable in this new scheme.

In Table 9.3 we list exponents in the $1/n$ expansion. Abe and coworkers have carried out the $1/n$ expansion farther than anyone else. Their technique is slightly different from the perturbation method discussed in this chapter. The reader is referred to their papers for further information.

In addition to their use in calculating exponents and the equations of state, the ϵ and the $1/n$ expansions have been used for studying the correlation function $G(k)$ as a function of k and ξ. [See Fisher and Aharony (1973) and Aharony (1973).]

Table 9.3

Exponents for the isotropic n-component-spin fixed point in the 1/n expansion

$$\eta = 4(4/d - 1)\, S_d/n + O(n^{-2}) \;,$$

$$\eta = 8(3\pi^2 n)^{-1} - (8/3)^3 (\pi^4 n^2)^{-1} + O(n^{-3}), \quad d = 3, \text{ [Abe (1973)]}$$

$$\gamma = \left(\frac{1}{2} d - 1\right)^{-1} (1 - 6\, S_d/n) + O(n^{-2})$$

$$\alpha = (d-4)(d-2)^{-1} (1 + 8(1-d)(4-d)^{-1} S_d/n) + O(n^{-2})$$

$$y_2 = d - 4 + \frac{8}{d}(4-d)(d-1)^2\, S_d/n + O(n^{-2})$$

$$y_\tau = 2 - 16 d^{-1} S_d/n + O(n^{-2})$$

$$y_c = 4 - d - 16(1 + 2/d)\, S_d/n + O(n^{-2})$$

$$S_d = \frac{\sin\pi\left(\frac{1}{2} d - 1\right)}{\pi\left(\frac{1}{2} d - 1\right) B\left(\frac{1}{2} d - 1, \frac{1}{2} d - 1\right)} = \frac{\sin\pi\left(\frac{1}{2} d - 1\right) \Gamma(d-1)}{2\pi\Gamma(d/2)^2}$$

Data taken from Ma (1973a, 1974b)

$$S_{4-\epsilon} = \epsilon/2 - \epsilon^2/4 + O(\epsilon^3) \;,$$

$$S_3 = 2/\pi^2 \;,$$

$$S_{2+\delta} = \delta/4 + O(\delta^3) \;.$$

X. THE EFFECT OF RANDOM IMPURITIES AND MISCELLANEOUS TOPICS

SUMMARY

We discuss the effect of random impurities in some detail, with emphasis upon basic concepts and ideas. The RG approach needs to be formulated slightly differently when quenched impurities are present. We also include a brief section on the use of graphs in studying quenched impurities. The self-avoiding random walk problem is then discussed, and its connection to the critical behavior of the $n = 0$ Ginzburg-Landau model is demonstrated.

1. RANDOM IMPURITIES

Some impurities always exist in any real material. The study of critical phenomena must pay attention to their effects. There are many forms of impurities, whose effects are very different. For example, in a ferromagnetic crystal, there may be a fraction of sites occupied by atoms which cannot produce a magnetic moment. If this fraction is larger than a certain value, ferromagnetism is completely suppressed. If the fraction is small, then a decrease in T_c is expected. Critical exponents may be modified. Another example is that there may be random distortions of the lattice causing preferred directions for spin orientations which are randomly distributed. The superfluid transition in liquid He^4 in a porous medium is yet another example.

Theoretical studies of random impurity effects on various phenomena began many years ago. The motion of electrons in disordered solids, the problem of percolation, the Ising model with spins on random sites, etc., constitute a vast literature, which we shall not be able to review here. We shall only touch upon the RG approach to the study of effects of small amounts of impurity on critical behavior.

The effects of impurities on critical behavior are

expected to be very important for the following reason. Suppose we mix a few impurities into a system near its critical point, in effect turning on a small perturbation. The response of the system to the perturbation is described by various susceptibilities and correlation functions. Near the critical point, some of these quantities are very large and are singular functions of the temperature. This means that a small amount of impurity can produce a large effect near the critical point, thereby altering the critical behavior of the pure system substantially. The critical exponents may be modified. Singular structures of certain quantities may be smeared out. Even the critical point itself may disappear. These effects are profound and still not well understood.

Before we begin any specific analysis, let us set straight some terminology.

By convention, impurities are classified according to the way they are distributed in the host system, either as annealed impurities or as quenched impurities.

Annealed means that the impurities are in thermal equilibrium with the host system. Let φ_i denote the variables specifying the impurity configuration. At thermal equilibrium, the joint probability distribution for φ and the spin configuration σ of the host system is given by

$$P[\varphi, \sigma] = Z^{-1} e^{-\mathcal{H}[\varphi, \sigma]} , \tag{10.1}$$

$$Z = \int \delta\varphi \ \delta\sigma \ e^{-\mathcal{H}[\varphi, \sigma]} ,$$

where the integral is taken over all impurity and spin configurations, and $\mathcal{H}[\varphi, \sigma]$ is the effective Hamiltonian for the whole system — spins and impurities. To obtain the effective Hamiltonian for σ alone, we can integrate out φ as we did previously with all uninteresting variables:

$$e^{-\mathcal{H}[\sigma]} = \int \delta\varphi \ e^{-\mathcal{H}[\varphi, \sigma]} \tag{10.2}$$

The free energy is simply

$$\mathfrak{F} = -\ln Z . \tag{10.3}$$

Of course, to reach thermal equilibrium one has to wait for a time long compared to the <u>relaxation time</u>, which is determined by dynamical processes which redistribute the impurities. In fluid systems, the relaxation time is short. In solids, often the relaxation time is very long compared to the time of observation of phenomena of interest. Then the impurities must be considered fixed, with a distribution prescribed by the mechanism which introduced the impurities. For example, suppose that impurities are stirred into the system above the melting point of the crystal. Then the system is cooled down and crystallizes with impurities mixed in. If the impurities can hardly move, then their distribution is fixed at the time of solidification, i.e., at the melting temperature. At lower temperatures we have to speak of the behavior of spins for a given impurity distribution. Of course, if we wait for many, many years, thermal equilibrium will eventually be reached between the impurities and the host. But before that, the impurities are fixed and are called quenched.

To discuss quenched impurities, we need to be more general. Let $P[\varphi]$ denote the probability distribution for the quenched impurity configuration, and $\mathcal{H}[\varphi|\sigma]$ denote the Hamiltonian for the spins at the given impurity configuration φ. Then the conditional probability distribution $P[\varphi|\sigma]$, i.e., the probability distribution of σ

given φ, is

$$P[\varphi|\sigma] = Z^{-1}[\varphi]\, e^{-\mathcal{H}[\varphi|\sigma]}$$

$$Z[\varphi] = \int \delta\sigma\ e^{-\mathcal{H}[\varphi|\sigma]} \quad . \tag{10.4}$$

Here $Z[\varphi]$ is the partition function of σ at the given impurity configuration φ. The joint probability distribution for φ and σ is then

$$P[\varphi, \sigma] = P[\varphi]\, P[\varphi|\sigma]$$

$$= P[\varphi]\, Z^{-1}[\varphi]\, e^{-\mathcal{H}[\varphi|\sigma]} \quad . \tag{10.5}$$

The probability distribution for σ alone can again be found by integrating out φ:

$$P[\sigma] = \int \delta\varphi\ P[\varphi, \sigma]$$

$$= \int \delta\varphi\ P[\varphi]\, Z^{-1}[\varphi]\, e^{-\mathcal{H}[\varphi|\sigma]} \quad . \tag{10.6}$$

Here it is absolutely crucial to note that $Z^{-1}[\varphi]$ is not a constant but a function of φ.

At a given φ, the free energy for the spins is

$$\mathfrak{F}[\varphi] = -\ln Z[\varphi] \quad . \tag{10.7}$$

Then we average it over $P[\varphi]$ to find the free energy of interest

$$\mathfrak{F} = \langle \mathfrak{F}[\varphi] \rangle = -\int \delta\varphi \; P[\varphi] \ln Z[\varphi] \tag{10.8}$$

One might feel uneasy speaking of the average of a free energy since free energy is not a mechanical variable (i. e., it is intrinsically statistical and cannot be represented by an operator in quantum mechanics). Still, Eq. (10.8) produces the right averages as a good free energy should. For example, the average of σ is

$$m = \langle \sigma \rangle = \int \delta\varphi \; P[\varphi] \langle \sigma \rangle_\varphi \tag{10.9}$$

where $\langle \sigma \rangle_\varphi = \partial \mathfrak{F}[\varphi]/\partial h$ is the average of σ at a given φ. Clearly, (10.9) is given by

$$m = \partial \mathfrak{F}/\partial h \tag{10.10}$$

with $\mathfrak{F}$ given by (10.8).

Note that the formulas (10.4) - (10.10) are all identities of probability theory. They are true for any probability distribution of φ and σ. In particular they are true if the impurities are annealed. The annealed impurities simply have the particular probability distribution

$$P[\varphi] = \int \delta\sigma \; P[\varphi, \sigma]$$

$$= \int \delta\sigma \; Z^{-1} \, e^{-\mathcal{H}[\varphi, \sigma]} \tag{10.11}$$

via (10.1). In some literature one finds the statement that for annealed impurities the free energy is $-\ln \langle Z[\varphi] \rangle$, i.e., average the partition function first, then take the logarithm. This is misleading unless we arbitrarily modify the definition of the average. If we use the average over the correct distribution (10.11), then $\langle Z[\varphi] \rangle$ has no meaning. We have to write $\mathcal{H}[\varphi, \sigma]$ as $\mathcal{H}_o[\varphi] + \mathcal{H}_1[\varphi, \sigma]$ with $\mathcal{H}_o$ depending on φ only. Then (10.1) becomes

$$Z = \int \delta\varphi \; e^{-\mathcal{H}_o} \int \delta\sigma \; e^{-\mathcal{H}_1[\varphi, \sigma]} \, . \tag{10.12}$$

If we formally regard the first integral as an averaging process over the probability

$$P_o[\varphi] \propto e^{-\mathcal{H}_o} \, , \tag{10.13}$$

which is not the probability distribution function of φ, and the second integral as a partition function $Z_1[\varphi]$, then Z is $\langle Z_1[\varphi]\rangle_o$, formally the average of Z_1 over P_o. It is not $\langle Z[\varphi]\rangle$. It is unfortunate that many authors interchange the averaging over impurity distribution and the taking of the logarithm as a way to account for the difference between the annealed impurities and quenched impurities.

The only difference between the annealed impurities and the quenched impurities is the difference in the probability distribution $P[\varphi]$. The annealed impurities follow the spins of the host system and vice versa, and $P[\varphi]$ depends strongly on the spins. The quenched impurities stand firm with $P[\varphi]$ not affected by the spins. The spins distribute themselves to fit the condition set up by the quenched impurities.

We can calculate the effective Hamiltonian $\mathcal{H}[\sigma]$ by integrating out φ, i.e.,

$$\int \delta\varphi \; P[\varphi, \sigma] \propto e^{-\mathcal{H}[\sigma]} , \tag{10.14}$$

regardless of whether the impurities are annealed or

quenched. For the annealed case we can use Eq. (10.2). The total Hamiltonian $\mathcal{H}[\varphi, \sigma]$ is (assuming that there is no long-range force among φ or σ) a smooth function of temperature, σ and φ. The integration over φ in (10.2) is expected to produce a smooth $\mathcal{H}[\sigma]$. Of course, it is possible that the interaction is such that φ becomes the order parameter of some critical point or behaves in some collective manner. Here we imagine that the amount of impurity is sufficiently small so that such things do not happen. The $\mathcal{H}[\sigma]$ so produced is expected to behave qualitatively the same as those block Hamiltonians discussed in previous chapters.

For the quenched impurities, we need to use (10.6) for calculating $\mathcal{H}[\sigma]$:

$$e^{-\mathcal{H}[\sigma]} \propto \int \delta\varphi \; P[\varphi] \; Z^{-1}[\varphi] \; e^{-\mathcal{H}[\varphi|\sigma]} \; . \qquad (10.15)$$

Again, we imagine that the amount of impurity is small. The quantity $Z[\varphi]$, the partition function of the spins at a given small amount of fixed impurity, is not a smooth function of temperature or φ near the critical point. Consequently, we may not assume that $\mathcal{H}[\sigma]$ is a smooth function, and the RG approach would not be useful in studying $\mathcal{H}[\sigma]$. It is easier not to calculate $\mathcal{H}[\sigma]$, but to study $P[\varphi]$ and $\mathcal{H}[\varphi|\sigma]$ directly.

On the other hand, the $P[\varphi]$ for the annealed case, given by (10.11), is strongly dependent on the behavior of σ and is expected to be a singular function of the temperature. It is therefore not convenient to study $P[\varphi]$ and $\mathcal{H}[\varphi|\sigma]$. It is far easier to study $\mathcal{H}[\sigma]$.

Finally, we remark that quenched impurities are never the only impurities. There are always other impurities which are not fixed and can adjust themselves to the spins and the quenched impurities. Modifications of the impurities by short-wavelength spin fluctuations which we want to integrate out must be considered. On a large scale, impurities are never completely quenched. They are dressed by all kinds of things which vary with the temperature and other parameters. This point will be taken up in the next section.

2. THE RG APPROACH TO NONMAGNETIC IMPURITIES

We now apply the RG machinery to the study of quenched impurities.

Recall that the Kadanoff transformation, i.e., the coarse graining procedure, is a central concept. This procedure eliminates certain short-wavelength modes. The eliminated modes affect the remaining modes through their effect on the parameters specifying the Hamiltonian for the remaining modes. The parameters reflect the environment, or the reservoir, seen by the remaining modes. In thermal equilibrium, all we need is one Hamiltonian with definite parameters. In the presence of quenched impurities, it is no longer desirable to use the effective Hamiltonian for the spins alone as we mentioned earlier. It is easier to keep track of the probability distribution of possible environments, i.e., probability distribution $P[\varphi]$ of the parameters φ in $\mathcal{H}[\varphi | \sigma]$. Now we must note that the parameters φ, which will be called random fields, depend not only on the impurity configuration (which we denoted by φ before), but also on all the eliminated modes. They can be regarded as the "dressed impurities," i.e., the combined effect of impurities and the eliminated modes. Qualitatively, the RG approach will concentrate on how the impurities are "dressed" upon repeated coarse graining. If the effect of impurities is diminished upon dressing, the critical behavior will not be affected qualitatively. If the effect is amplified, the critical behavior will be affected qualitatively. There are borderline cases, too.

Let us consider some explicit calculations in the case of nonmagnetic impurities, which do not affect the rotation symmetry of the spin system. We begin with a Ginzburg-Landau form

$$\mathcal{H}[\varphi\,|\,\sigma] = \frac{1}{2}\int d^dx \left[\varphi_r(x)\,\sigma^2(x) + \frac{1}{4}\varphi_u(x)\,\sigma^4(x) + \varphi_c(x)\,(\nabla\sigma(x))^2\right], \tag{10.16}$$

which is just the old Ginzburg-Landau form with random fields φ_r, φ_u, and φ_c replacing the old r_o, u, and c. The probability distribution of φ is $P[\varphi]$. We can imagine that $P[\varphi]$ is a probability distribution on the space of φ, which is the extension of our old parameter space to include nonuniform parameters.

We shall restrict our attention to cases where there is no long-range correlation in $\varphi(x)$ and all averages are translationally invariant. Let us define

$$\langle \varphi_r(x) \rangle = r_o ,$$

$$\langle \varphi_u(x) \rangle = u ,$$

$$\langle \varphi_c(x) \rangle = c . \tag{10.17}$$

No long-range correlation means

$$\langle \varphi_i(x)\, \varphi_j(x') \rangle_c = 0$$

$$\langle \varphi_i(x)\, \varphi_j(x')\, \varphi_\ell(x'') \rangle_c = 0 \tag{10.18}$$

and all higher cumulants vanish unless all $x, x', x'', \ldots$ are within a distance b, the spin block size, of one another.

In previous discussions we spoke of the transformation by R_s of a point μ to another point μ'. Likewise we can use the same definition of R_s to obtain $\varphi' = R_s\varphi$ in the extended space of φ. The only difference is that now parameters are position dependent. Now a probability distribution $P[\varphi]$ can be thought of as describing a cloud of points in this space. Each point moves under R_s. As a result, the cloud flows. The center of mass and the shape of the cloud change. In a manner analogous to the study of gas flow, one derives formulas for the change of the center of mass and the shape of the cloud. The quantities r_o, u, c defined by (10.17) play the role of the center of mass of the cloud $P[\varphi]$ in the abstract φ space. The "shape" is characterized by various moments of correlation functions of φ:

$$\int d^dx \, \langle \varphi_i(x) \, \varphi_j(x') \rangle_c = \Delta_{ij} \; ,$$

$$\int d^dx \; d^dx' \langle \varphi_i(x) \, \varphi_j(x') \, \varphi_\ell(x'') \rangle_c = \Delta_{ij\ell} \; ,$$

$$\int d^dx (x - x')^2 \, \langle \varphi_i(x) \, \varphi_j(x') \rangle_c = \Delta_{ij}^{(2)} \; , \qquad (10.19)$$

etc.

We can use these quantities to characterize $P[\varphi]$ and define

$$\tilde{\mu} = (r_o, u, c; \Delta_{ij}, \Delta_{ij\ell}, \ldots) . \qquad (10.20)$$

This defines the parameter space representing various $P[\varphi]$.

The transformation $R_s \tilde{\mu} = \tilde{\mu}'$ is defined by the following procedure.

First, we perform steps (i) and (ii) as we did before on $\mathcal{H}[\varphi|\sigma]$, treating $\varphi_r, \varphi_u, \varphi_c$ as we treated r_o, u, and c, keeping in mind that φ_r, φ_u, and φ_c depend on x. We then get a new Hamiltonian $\mathcal{H}'$. The coefficients of $1/2\,\sigma^2$, $1/8\,\sigma^4$, and $1/2(\nabla\sigma)^2$ in $\mathcal{H}'$ are φ_r', φ_u', and φ_c', respectively. They depend on φ_r, φ_u, and φ_c. We average these coefficients over $P[\varphi]$ and call the results r_o', u', and c'. Then we compute the cumulants of φ_i' via Eqs. (10.19), for example

$$\int d^d x \, \langle \varphi_i'(x)\, \varphi_j'(x') \rangle_c = \Delta_{ij}' , \qquad (10.21)$$

to obtain Δ_{ij}', $\Delta_{ij\ell}'$, etc. Then we have

$$\tilde{\mu}' = (r_o', u', c'; \Delta_{ij}', \Delta_{ij\ell}', \ldots) = R_s \tilde{\mu} . \qquad (10.22)$$

The transformation $\mu' = R_s \mu$ is thereby defined. The above procedure must be generalized when more parameters are needed.

Let us summarize the results of calculation to $O(\varepsilon)$. Some details will be discussed in Section 4.

To $O(\varepsilon)$, it turns out to be sufficient to keep track of φ_r, u, and c only. Furthermore, only $\Delta_{rr} \equiv \Delta$ plays a role and all other moments need not be considered if we are interested in $O(\varepsilon)$ only. Thus we simply write

$$\tilde{\mu} = (r_o, u, c; \Delta) . \qquad (10.23)$$

The result of carrying out the above procedures for $R_s \tilde{\mu} = \tilde{\mu}'$ is given by

$$r_o' = s^{2-\eta} \left\{ r_o + \left[\left(\frac{n}{2} + 1 \right) u - \Delta \right] \times K_4 \left(\frac{\Lambda^2}{2c} (1 - s^{-2}) - \frac{r_o}{c} \ln s \right) \right\} + O(\varepsilon u, \varepsilon \Delta, u^2, \Delta^2) , \qquad (10.24a)$$

$$u' = s^{\epsilon - 2\eta} \left\{ u - \frac{u}{c^2} \left(\left(\frac{n}{2} + 4 \right) u - 6\Delta \right) K_4 \ln s \right\} , \qquad (10.24b)$$

$$\Delta' = s^{\epsilon - 2\eta} \left\{ \Delta - \frac{\Delta}{c^2} \left((n+2)u - 4\Delta \right) K_4 \ln s \right\} , \qquad (10.24c)$$

$$c' = s^{-\eta} c . \qquad (10.24d)$$

These are among the results of Lubensky (1975). If we have $\Delta = 0$, we will get $\Delta' = 0$, and the formulas discussed in Chapter VII. We clearly have all the fixed points obtained there. Choosing $\eta = 0$, we get the Gaussian fixed point and the stable fixed point given by (7.70) and (7.71). We also find a fixed point with $\Delta^* > 0$ for $n < 4$. These fixed points and some of their properties are summarized in Table 10.1. The projection of the fixed points and flow lines in the (u, Δ) plane is shown in Fig. 10.1. Note that u' and Δ' are determined by u and Δ only, and not affected by r_o to $O(\epsilon^2)$, as (b) and (c) of Eq. (10.24) show.

There is apparently another fixed point at $u = 0$, $\Delta = -\epsilon/4K_4$. This fixed point has no physical meaning [and is called the "unphysical fixed point" by Lubensky

Table 10.1

Fixed points and exponents to $O(\epsilon)$ allowing quenched nonmagnetic impurities

Fixed Point	Gaussian	Pure	Random
u^*	0	$\frac{2\epsilon}{(n+8)K_4}$	$\frac{\epsilon}{2(n-1)K_4}$
Δ^*	0	0	$\frac{\epsilon(4-n)}{8(n-1)K_4}$
$y_1 = 1/\nu$	2	$2-\left(\frac{n+2}{n+8}\right)\epsilon$	$2-\frac{3n\epsilon}{8(n-1)}$
y_2	ϵ	$-\epsilon$	$-\epsilon$
y_3	ϵ	$\left(\frac{4-n}{n+8}\right)\epsilon$	$\frac{(n-4)\epsilon}{4(n-1)}$
$2y_1-d = \alpha/\nu$	ϵ	$\left(\frac{4-n}{n+8}\right)\epsilon$	$\frac{(n-4)\epsilon}{4(n-1)}$
Stability	Unstable	Stable $n > 4$ only	Stable for $n < 4$ unphysical for $n > 4$

Information concerning $O(\epsilon^2)$ can be found in Lubensky (1975). For $n = 1$,

$$K_4\Delta^* = \left(\frac{6}{53}\epsilon\right)^{1/2}, \quad u^* = \frac{4}{3}\Delta^*,$$

$$\eta = -\epsilon/106,$$

$$y_1 = 1/\nu = 2 - \left(\frac{6}{53}\epsilon\right)^{1/2}.$$

Note that there is quantitative discrepancy between these results obtained through Lubensky (1975) and those obtained by Khmelnitsky (1975).

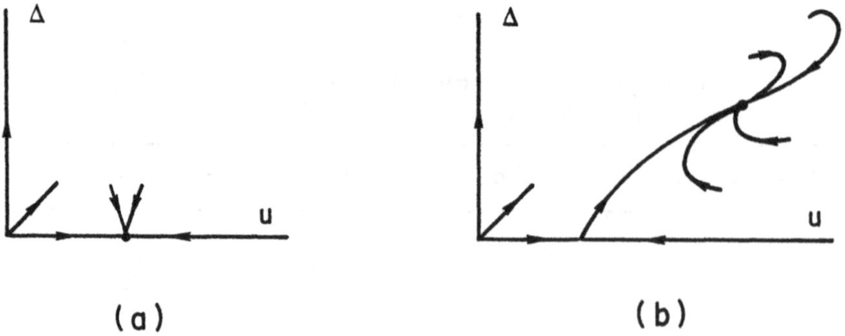

Figure 10.1. Fixed points on the (u, Δ) plane.
(a) $\alpha < 0$. The pure fixed point is stable.
(b) $\alpha > 0$. The pure fixed point is unstable and a new fixed point appears with $\Delta^* \neq 0$.

(1975)] since Δ, Δ' must be positive by construction.

Recall that

$$\Delta = \int d^d x \, \langle \varphi_r(x) \, \varphi_r(x') \rangle_c$$

$$= L^{-d} \left\langle \left(\int d^d x \, \varphi_r(x) \right)^2 \right\rangle_c > 0 \, , \qquad (10.25)$$

$$\Delta' = L^{-d} \left\langle \left(\int d^d x \, \varphi'_r(x) \right)^2 \right\rangle_c > 0 \, ,$$

by (10.19) and (10.21).

Table 10.1 tells us that for $n > 4$, the pure fixed point is stable. The critical behavior therefore is not affected by the kind of quenched impurity discussed here. When $1 < n < 4$, the pure fixed point is unstable ($y_3 > 0$), but the random fixed point is stable. The critical exponents are different from those given by the pure fixed point. For $n = 1$, the Ising model, the pure fixed point is unstable but (10.24) gives no other finite fixed point for $n = 1$. Khmelnitsky (1975) showed that there is a fixed point for $n = 1$ with u^* and Δ^* of order $\sqrt{\epsilon}$. This fixed point can be obtained when the next-order terms in Δ and u are included in (10.24). Such terms are available in Lubensky (1975), giving the values shown in Table 10.1. Further implications of this fixed point remain to be understood.

Clearly, y_3 (which appears in the linearized RG formula $\delta\Delta' = s^{y_3}\delta\Delta + \cdots$), is just $2y_1 - d = \alpha/\nu$, where $\alpha = 2 - d\nu$ is the specific heat exponent. A fixed point is unstable when $\alpha > 0$. This result is quite general and not restricted to the first order in ϵ, as is easily shown by the following arguments.

First, consider a fixed point with $\Delta^* = 0$, i.e., without impurity. Now switch on a small $\delta\varphi_r(x)$, which means a small change in r_o. As a result $\delta\varphi_r$ transforms under R_s like δr_o. We know that

$$\delta r_o' = s^{y_1} \delta r_o + \cdots \qquad (10.26)$$

where $\cdots$ means terms proportional to s^{y_i} with $y_i < 0$. (Recall that y_1 is assumed to be the only positive one of all y_i's for rotationally symmetric scaling fields.) Therefore we must have

$$\delta\varphi_r'(x/s) = s^{y_1} \delta\varphi_r(x) + \cdots \qquad (10.27)$$

It follows from the definition of Δ and Δ' that

$$\Delta' = \int d^d x' \langle \delta\varphi_r'(x') \, \delta\varphi_r'(y'') \rangle$$

$$= s^{2y_1 - d} \int d^d x \langle \delta\varphi_r(x) \, \delta\varphi_r(y) \rangle + \cdots$$

$$= s^{2y_1 - d} \Delta + \cdots . \qquad (10.28)$$

The above arguments are easily generalized to fixed points with $\Delta^* \neq 0$. We simply add a small new random field $\delta\phi_r$ to the φ_r which has the property $\int d^d x \langle \varphi_r(x) \varphi_r(y) \rangle_c = \Delta^*$. Assume that

$$\langle \varphi_r(x)\, \delta\phi_r(y) \rangle = 0 \tag{10.29}$$

Therefore Δ^* is changed to $\Delta^* + \delta\Delta$ with

$$\delta\Delta = \int d^d x \,\langle \delta\phi_r(x)\, \delta\phi_r(y) \rangle \tag{10.30}$$

as a result of $\delta\phi_r$. Again $\delta\phi_r$ transforms like δr_o. Replacing $\delta\varphi_r$ by $\delta\phi_r$, Δ by $\delta\Delta$ in (10.28), we get

$$\delta\Delta' = s^{2y_1 - d}\, \delta\Delta + \cdots \tag{10.31}$$

We therefore conclude that, in the presence of nonmagnetic impurities, a stable fixed point must have the property

$$2y_1 - d < 0 \quad , \tag{10.32}$$

i.e., $\alpha < 0$.

This result is generalized to other kinds of impurities in the next section. Qualitative, Eq. (10.32) says that impurities remove the divergence of the specific heat. The random field $\varphi_r(x)$ couples to $\sigma^2(x)$, which is roughly the energy density or entropy density. This random field destroys the long-range correlation of $\sigma^2(x)$ and thereby suppresses the divergence of $\int d^d x \, \langle \sigma^2(x) \sigma^2(y) \rangle$, which is proportional to the specific heat.

From a physical viewpoint, the important role of the specific heat exponent α is obvious. The random field φ_r can be regarded as a random perturbation of local temperature since r_o is linear in temperature near the critical point. The linear response to a perturbation of temperature is the specific heat. If $\alpha > 0$, the response blows up as $T \to T_c$, implying instability. If $\alpha < 0$, the response is finite.

3. FIXED POINT STABILITY CRITERIA AND OTHER IMPURITIES

The criterion (10.32) is a special case of the more general criteria, which are established by the same arguments as those leading to (10.32),

$$y_i + y_j - d < 0 \tag{10.33a}$$

$$y_i + y_j + y_\ell - 2d < 0 \tag{10.33b}$$

etc.

for the stability of a fixed point in the presence of random impurities. These criteria simply state that the parameters Δ_{ij}, $\Delta_{ij\ell}$, etc., defined by (10.19), must be irrelevant. Each φ_i of (10.19) gives y_i and $d^d x$ gives -d in (10.33). Clearly that Eq. (10.33a) which has the largest y_i, y_j is the most important equation. The others are more easily satisfied. For nonmagnetic impurities, y_1 is the largest and thus (10.32) is the most important criterion. For small ϵ the random fields φ_u and φ_c play no part in critical behavior because Δ_{rc}, Δ_{ru}, Δ_{cc}, Δ_{uu}, and Δ_{cu} are all irrelevant. Note that in calculating the fixed point to $O(\epsilon^2)$, these parameters must be taken into account along with other irrelevant parameters. (Note again that irrelevant parameters need not be zero at a fixed point. R_s will drive them toward their fixed point values, which are in general nonzero.)

The criteria (10.33) can be used to test the stability of fixed points when quenched impurities of various kinds are added to the system. The examples below illustrate the significance of these criteria, and some special features of the impurity effects.

First, we discuss the effect of a random tensor field.

For certain amorphous solids, the perfect crystal lattice does not extend very far, but its lattice orientations vary from place to place randomly. In other words, the solid is a collection of very small pieces of crystal, each of which has the same structure but a random orientation. The spins in the solid interact like those discussed in previous chapters except that now they may have the preference of lining up along an axis determined by the orientation of the lattice. Such a preference can be described by a term

$$\int d^d x \, (\varphi(x) \cdot \sigma(x))^2 = \sum_{i,j=1}^{n} \int d^d x \, \varphi_i \varphi_j \sigma_i \sigma_j \qquad (10.34)$$

in $\mathcal{H}$, where the vector $\varphi(x)$ lies along the preferred axis at x. The effect of (10.34) with φ^2 = constant has been studied in some detail by Aharony (1975b).

Note that

$$\sum_{i,j} \varphi_i \varphi_j \sigma_i \sigma_j = \sum_{i,j} \left(\varphi_i \varphi_j - \frac{1}{n} \delta_{ij}\right) \sigma_i \sigma_j + \varphi^2 \sigma^2 . \qquad (10.35)$$

Clearly, φ^2 plays the role of φ_r discussed earlier, if it is not a constant but a random field. The first term is a special case of the traceless tensor field φ_τ :

$$\sum_{i,j} \varphi_{\tau ij} \, \sigma_i \, \sigma_j \tag{10.36}$$

We can use the criterion (10.33) to determine whether the pure fixed point is stable against a nonzero random traceless tensor field. The exponent y_τ is known for small ϵ or for small $1/n$. Tables 9.2 and 9.3 give

$$y_\tau = 2\left(1 - \frac{\epsilon}{n+8} + O(\epsilon)^2\right)$$

$$= 2(1 - 8\, S_d/nd) + O(n^{-2}) \, . \tag{10.37}$$

Substituting them in (10.33), we get

$$2y_\tau - d = \frac{n+4}{n+8}\, \epsilon + O(\epsilon^2)$$

$$= 4 - d - (16/d)\, S_d/n + O(n^{-2}), \tag{10.38}$$

which suggests that for $d < 4$, the criterion (10.33a) cannot be satisfied and the pure fixed point is therefore unstable.

We next consider a random magnetic field $\varphi_h(x)$, i.e., a term

$$- \int \varphi_h(x) \cdot \sigma(x)\, d^d x \qquad (10.39)$$

in $\mathcal{H}$. We assume that

$$\langle \varphi_h(x) \rangle \equiv h = 0 \ . \qquad (10.40)$$

Since $y_h = \frac{1}{2}(d + 2 - \eta)$, we have

$$2y_h - d = 2 - \eta \qquad (10.41)$$

Clearly (10.33a) cannot be satisfied unless $\eta > 2$. For a fixed point of small η, the instability is very serious.

In fact, the effect of a random magnetic field is so strong for the cases $n \geq 2$ that there can no longer be a finite magnetization for $d \leq 4$ no matter how weak the field is, as long as it is nonzero. [See Imry and Ma (1975).] We shall not prove this statement but give some strong evidence.

It is evident that, if φ_h is sufficiently strong, i.e., stronger than the field produced by neighboring spins, the spin configuration will simply follow the field. Since $\langle \varphi_h \rangle = 0$, we must have $m = \langle \sigma \rangle = 0$ at any temperature for any n. If the field is very weak, the situation is less obvious.

Suppose that we start with $\varphi_h = 0$ and $m \neq 0$. Then we turn on a very weak φ_h. For $n \geq 2$, the component of φ_h perpendicular to m will produce a magnetization with Fourier components which can be expressed in terms of the transverse susceptibility $G_\perp(k)$:

$$\langle \sigma_k^\perp \rangle_\varphi \equiv m_k^\perp = G_\perp(k) \, \varphi_{hk}^\perp \tag{10.42}$$

for the given φ_h to the first order in φ_h. Now we calculate the correlation function averaged over the random field configurations:

$$\langle m^\perp(x) \, m^\perp(x') \rangle = (2\pi)^{-d} \int d^d k \; G_\perp(k)^2 \langle |\varphi_{hk}^\perp|^2 \rangle \, e^{ik\cdot(x-x')} \; . \tag{10.43}$$

Since $G_\perp(k) \propto k^{-2}$ for $m \neq 0$, (10.43) diverges as long as $\lim_{k \to 0} \langle |\varphi_{hk}|^2 \rangle \neq 0$, no matter how small, for $d \leq 4$. Since $m^\perp(x)$ is never infinite, we have a contradiction. The conclusion is that the assumption $m \neq 0$ is wrong. Therefore

$m = 0$. The above argument is not quite a proof because only the lowest order in φ_h is accounted for. It has some resemblance to the argument given in Sec. III.7, where we discussed the Hohenberg-Mermin-Wagner theorem.

For $n = 1$, the above argument does not apply. It seems clear that for sufficiently low temperatures and sufficiently weak φ_h, m must not vanish since $G(k)$ for the pure system of $n = 1$ does not diverge for $k \to 0$. The fact that $m \neq 0$ for sufficiently low temperatures and $m = 0$ for high temperatures of course does not imply that there must be a fixed point. The pure fixed point with small η is never stable for any n and d. It turns out that for $d > 4$ there is a stable fixed point analogous to the random fixed point discussed in Sec. 2. For $d \leq 4$, $n \geq 2$, we have no ferromagnetic transition at all as argued above. For $n = 1$, $d \leq 4$, the situation is not yet understood.

Random impurities produce spatial nonuniformity and sometimes anisotropy in coordinate or spin spaces (even if statistically there is still uniformity and isotropy). As a result, at a given configuration φ, certain averages do not vanish, in contrast to what we are used to dealing with in previous chapters. For example, if φ is not uniform, then the correlation function

$$C(k) \equiv \langle |\sigma_k|^2 \rangle \tag{10.44}$$

is no longer the same as the cumulant

$$G(k) \equiv \langle |\sigma_k - \langle \sigma_k \rangle_\varphi |^2 \rangle$$

$$= \langle |\sigma_k|^2 \rangle - \langle |\langle \sigma_k \rangle_\varphi |^2 \rangle \tag{10.45}$$

because $\langle \sigma_k \rangle_\varphi$ may not vanish. The limit $k \to 0$ of $G(k)$ gives the susceptibility $\partial m/\partial h$, but the limit of $C(k)$ is entirely something else.

A more subtle example is that

$$K(k) = \int d^d x \, e^{-ik \cdot x} \left(\langle \sigma^2(x) \sigma^2(0) \rangle - \langle \sigma^2 \rangle^2 \right) \tag{10.46}$$

is very different from

$$F(k) = \int d^d x \, e^{-ik \cdot x} \left\langle (\sigma^2(x) - \langle \sigma^2(x) \rangle_\varphi)(\sigma^2(0) - \langle \sigma^2(0) \rangle_\varphi) \right\rangle \tag{10.47}$$

since

$$\left\langle \langle \sigma^2(x) \rangle_\varphi \langle \sigma^2(0) \rangle_\varphi \right\rangle \neq \langle \sigma^2(x) \rangle \langle \sigma^2(0) \rangle \ . \quad (10.47')$$

The specific heat is $F(0)$, not $K(0)$.

4. COMMENTS ON GRAPHS

The graph technique developed on Chapter IX is easily generalized to account for the presence of random fields. As an illustration, let us consider the model

$$\mathcal{K}[\varphi | \sigma] = \frac{1}{2} \int d^d x \left[r_o \sigma^2 + \frac{1}{4} u \sigma^4 + c(\nabla \sigma)^2 + \varphi_r \sigma^2 \right] \ . \quad (10.48)$$

Here we assume that $\varphi_r(x)$ has a Gaussian distribution with

$$\langle \varphi_r(x) \rangle = 0 \ ,$$

$$\langle \varphi_r(x) \varphi_r(x') \rangle = \Delta \delta(x - x') \ . \quad (10.49)$$

For a given φ_r, the perturbation expansion in powers of u and φ_r can be made. The only new element added is the fixed external field φ_r. We shall represent it by a dotted line joining two solid lines as shown in Fig. 10.2a.

The free energy and various cumulant averages can be computed at a fixed φ_r. They are still represented by various connected graphs now dressed with φ_r lines. Examples are given in Fig. 10.2b for a term of free energy, 10.2c for a term in spin susceptibility, and 10.2d for a term in the cumulant

$$\langle \sigma^2(x)\,\sigma^2(x')\rangle_\varphi - \langle \sigma^2(x)\rangle_\varphi \langle \sigma^2(x')\rangle_\varphi \,. \qquad (10.50)$$

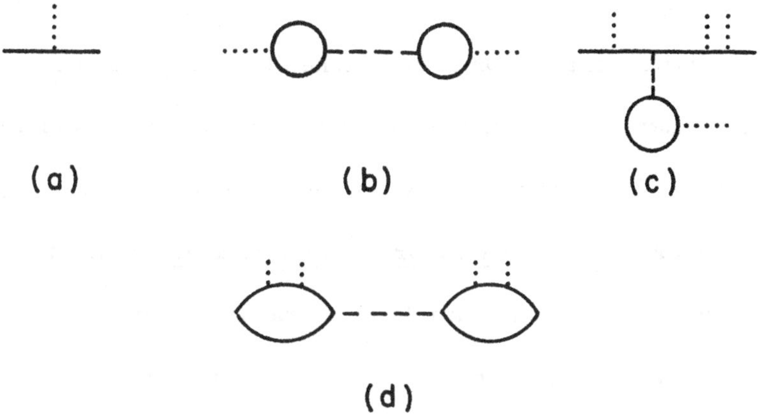

Figure 10.2. Graphs with fixed external field φ_r.
(a) Basic graph element showing the coupling $1/2\ \sigma^2 \varphi_r$.
(b) A graph for the free energy.
(c) A graph for the susceptibility.
(d) A graph for $\langle \sigma^2(x)\,\sigma^2(x')\rangle_c$.
Dashed lines represent the interaction u and dotted lines φ_r.

Remember that cumulant averages are always represented by connected graphs, with or without an external field.

Now we need to average over the probability distribution of φ_r. The Gaussian property (10.49) allows us to write the average of a product of φ_r's as a product of pairwise averages. Graphically, we keep track of such pairwise averages by pairing the dotted lines with small circles. Figure 10.3 shows the averages of the terms in Fig. 10.2b, c, d. Every circled line gives a factor Δ according to (10.49).

Thus, in addition to the dashed lines representing u, we have dotted circled lines giving the effect of the random field φ_r. By construction, when the circled lines are cut, a graph for the free energy or a cumulant average must remain connected. For example, the graphs shown in Fig. 10.4 are not allowed. Note that the definition of cumulant is another source of confusion. Here we mean the cumulant defined by averages over $e^{-\mathcal{K}[\varphi|\sigma]}$ at a fixed φ. The average over φ is then taken. An example is (10.47). If the cumulant is defined with respect to the full average, e.g., (10.46), then the situation is quite different.

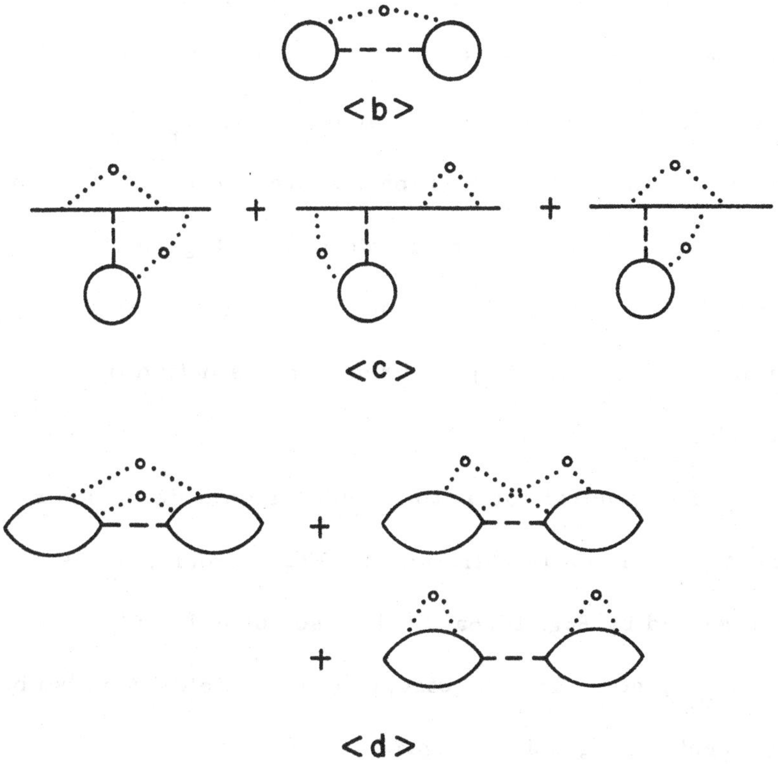

Figure 10.3. Averages of the graphs Fig. 10.3b, c, d. Each circled dotted line represents a factor Δ.

Figure 10.4. These are not graphs for cumulant averages or the free energy, because they will not remain connected upon cutting all circled dotted lines.

If the random field φ_r does not follow a Gaussian distribution as assumed in (10.49), we would need to keep track of all cumulants, for example, $\langle \varphi_r(x)\, \varphi_r(x')\, \varphi_r(x'') \rangle_c$. The generalization of the graphic representation to account for higher cumulants is straightforward. Figure 10.5 shows some examples. In addition to joining pairs by a circle, we need also to join three or more dotted lines by a circle.

The generalization to include random fields other than φ_r is also straightforward. For example, φ_u is represented by Fig. 10.6a. It is also possible to have φ_{u_6}, φ_{u_8}, etc. The anisotropic random fields can also be represented by graphs. Figure 10.6b, c shows some examples. Finally, the cumulants of random fields may

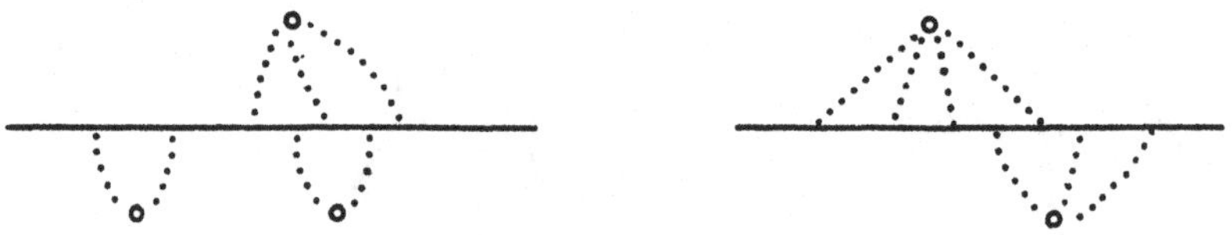

Figure 10.5. Graphs for G(k) when φ_r has nonzero higher cumulants. The third cumulant is represented by three dotted lines joined to a circle; the fourth cumulant is represented by four dotted lines joined by a circle.

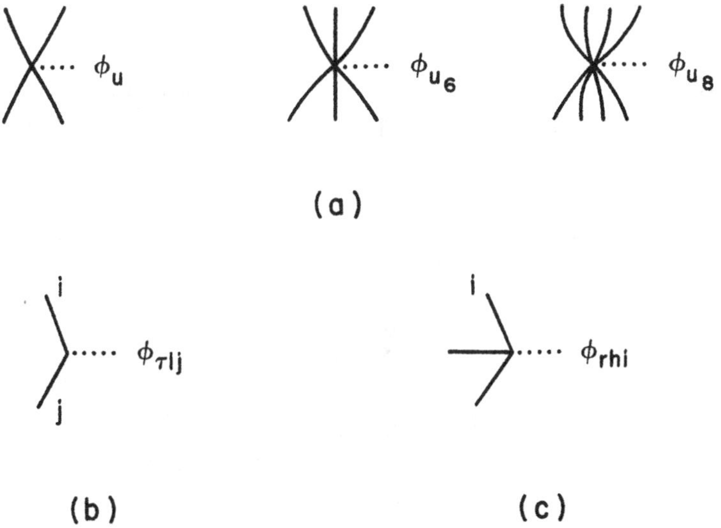

Figure 10.6. (a) Random u, u_6, and u_8.
(b) Random tensor field, [see (10.36)].
(c) This graph represents a term $\varphi_{rhi}\,\sigma_i\,\sigma^2$ in $\mathcal{H}$.

not be simply proportional to a δ-function like (10.49). Such complications must also be kept track of.

We now discuss briefly the graphic representation of the RG. First, we follow the rules in Chapter IX to obtain $\varphi' = R_s \varphi$. For example, Fig. 10.7a shows the graphs contributing to φ'_r up to $O(u\varphi_r, \varphi_r^3)$. General graphs for φ'_r are those with two external solid lines and any number of dotted lines as shown in Fig. 10.7b.

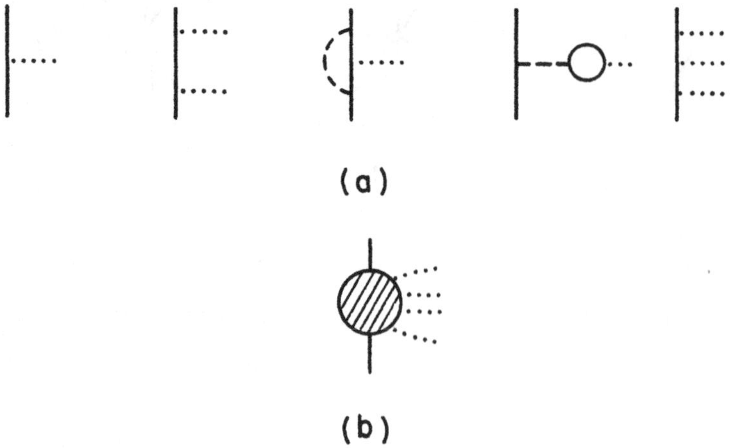

Figure 10.7. Graphs for φ_r'.
(a) Graphs up to $O(u\varphi_r, \varphi_r^3)$.
(b) Form of a general graph for φ_r'. Note that all solid lines except the two external legs carry wave vectors in the shell $\Lambda/s < q < \Lambda$.

Then we compute the cumulants of φ_r' and calculate the new Δ's following (10.19). This completes the RG transformation $\tilde{\mu}' = R_s \tilde{\mu}$ defined by (10.22) and the procedure described above (10.22). Figure 10.8a shows the transformed Δ to $O(\Delta, u\Delta, \Delta^2)$. It is just Fig. 10.7a multiplied by itself and then averaged over φ_r. A general graph for the transformed Δ is shown in Fig. 10.8b. The graph for r_o' to $O(\Delta)$ is given in Fig. 10.9. It is just

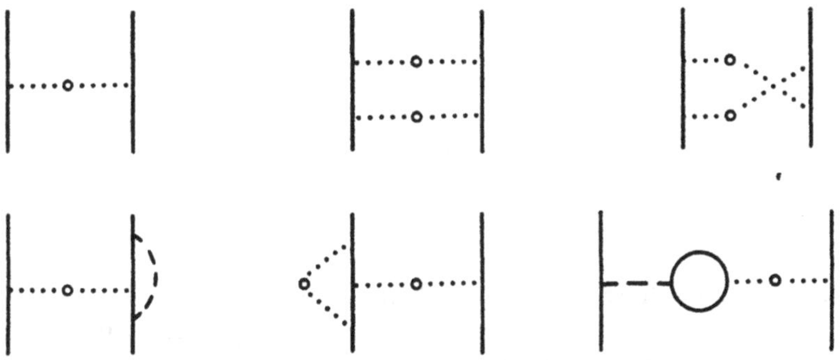

(a)

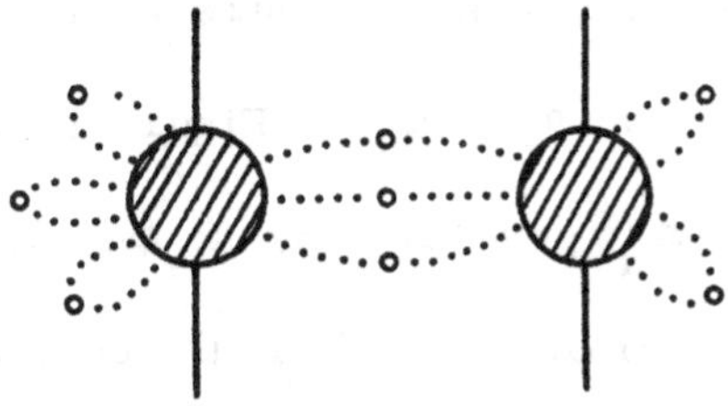

(b)

Figure 10.8. (a) Graphs for Δ' up to $O(u\Delta, \Delta^2)$.
(b) The general form of a graph for Δ'. Δ' is obtained by squaring the sum of all graphs of the form of Fig. 10.7b, and then averaging over the probability distribution of φ_r.

those in Fig. 10.7a averaged over φ_r. Figures 10.8 and 10.9 give the formulas (10.24c) and (10.24a), respectively. The contribution to r_o' from the nonrandom parameters is of course also included in (10.24a).

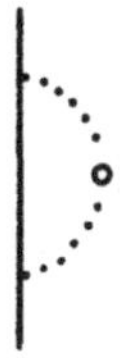

Figure 10.9. The $O(\Delta)$ contribution to r_o'. Note that all except the second graph in Fig. 10.7a average to zero.

$R_s\varphi$ also generates random fields of forms other than those included in (10.48). Figure 10.10a shows the graphs for φ_u' to $O(u\varphi_r, u\varphi_r^2)$. Figure 10.10b shows a general graph for φ_u' generated by $R_s\varphi$. The average of Fig. 10.10a contributes to u' and accounts for the $O(u\Delta)$ term in (10.24b). If one assumes that Δ, u, and r_o are all of $O(\epsilon)$, then (10.24) gives the fixed points and exponents which are consistent with such assumptions. New parameters such as the third cumulant of φ_r' shown in Fig. 10.11, can be consistently ignored to this order.

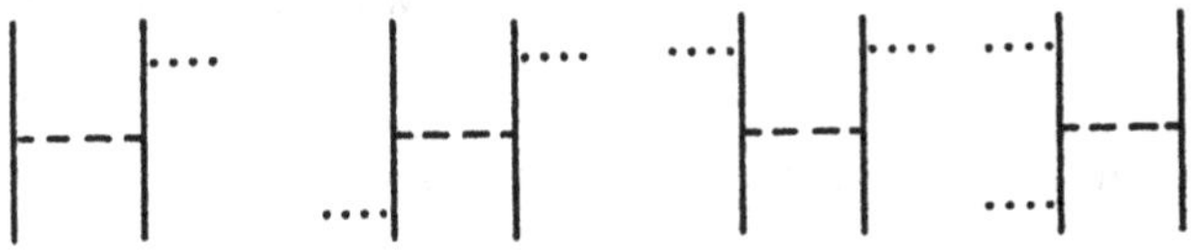

(a)

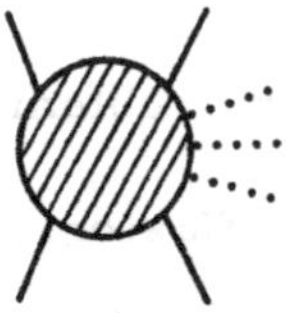

(b)

Figure 10.10. (a) Graphs for φ'_u, $O(\varphi_r u)$, and $O(\varphi_r^2 u)$.
(b) General form of a graph for φ'_u.

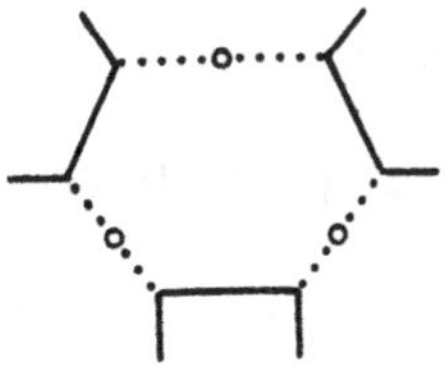

Figure 10.11. The leading contribution to the third cumulant of φ'_r. It is of $O(\Delta^3)$.

5. THE SELF-AVOIDING RANDOM WALK PROBLEM

Consider a particle which moves a step at a time. Each step is assumed to be very small and is taken randomly. If the probability for each step is entirely independent of the previous history and the location of the particle, then we have the well known problem of random walk, or Brownian motion, or diffusion. If the additional feature is added that the particle tends to avoid the places it has been before, we have the self-avoiding random walk problem. We shall study it here because it is formally equivalent to a Ginzburg-Landau model of $n = 0$. This equivalence was pointed out by DeGennes (1972). This problem is also equivalent to the long polymer problem. [See Fisher (1965), for example, for detailed discussion and references.]

The quantity of final interest is $P(x, t)$, the probability distribution of the particle at time t given that the particle is at $x = 0$ at $t = 0$, i.e., $P(x, 0) = \delta(x)$.

There are different possible paths for the particle to go from 0 to x in a time interval t. Let us denote a path by $z(\tau)$. The probability that the particle will take the path $z(\tau)$ is a product of the probabilities for all the steps making up the path. Let this probability be formally denoted by $dP(z \mid x, t)$. Then we have

$$P(x, t) = \int dP(z \mid x, t) \tag{10.51}$$

where the integral is a formal notation for summing over paths $z(\tau)$ which begin at $(0, 0)$ and end at (x, t).

In the absence of any self-avoiding tendency, we have simply the diffusion process, denoted by the subscript o:

$$P_o(x, t) = \int dP_o(z \mid x, t) \ .$$

P_o satisfies the diffusion equation

$$\begin{aligned} \frac{\partial}{\partial t} P_o &= c \nabla^2 P_o \ , \\ P_o(x, 0) &= \delta(x) \ , \end{aligned} \tag{10.52}$$

where c is the diffusion coefficient. The solution of (10.52) is well known:

$$P_o(x, t) = (4\pi c t)^{-d/2} \ e^{-x^2/4ct} \ . \tag{10.53}$$

The root-mean-square distance traveled over a period t is thus proportional to $t^{1/2}$.

The problem is to find out how the self-avoiding tendency alters (10.53).

The self-avoiding tendency reduces the probability for paths which visit a place more than once. We shall assume that, for a given path z, the probability of a particle's visiting $z(\tau)$ and $z(\tau')$ when $z(\tau)$ is very close to $z(\tau')$ is reduced by a factor proportional to

$$e^{-v(z(\tau) - z(\tau'))\,\Delta\tau\,\Delta\tau'} \tag{10.54}$$

where $v(z - z') > 0$ if $|z - z'|$ is shorter than some very small distance and zero otherwise; $\Delta\tau$ and $\Delta\tau'$ are the time intervals during which the particle stays at $z(\tau)$ and $z(\tau')$, respectively. This factor (10.54) is a phenomenological description of the self-avoiding tendency. The total reduction factor is the product of factors of the form (10.54) for all τ, τ' between 0 and t, which is

$$e^{-\int_0^t d\tau \int_0^\tau d\tau' \, v(z(\tau) - z(\tau'))} = e^{-\frac{1}{2}\int_0^t d\tau \int_0^t d\tau' \, v(z(\tau) - z(\tau'))} . \tag{10.55}$$

Summing over all paths, we obtain

$$P(x, t) = N \int dP_o(z \mid x, t)\, e^{-\frac{1}{2}\int_o^t d\tau\, d\tau'\, v(z(\tau) - z(\tau'))} \tag{10.56}$$

where N is a normalization factor. The task is then to evaluate this integral. Rather than attempt this here, we shall show how this task can be related to solving the $n = 0$ case of the Ginzburg-Landau model.

Expanding (10.56) in powers of v, we obtain

$$P(x, t) = N \int dP_o(z \mid x, t) \sum_{n=0}^{\infty} \frac{(-)^n}{n!} \int_o^t d\tau_1\, d\tau_1' \ldots d\tau_n\, d\tau_n'$$

$$\times\, 2^{-n} v(z(\tau_1) - z(\tau_1'))\, v(z(\tau_2) - z(\tau_2')) \ldots v(z(\tau_n) - z(\tau_n')) \,. \tag{10.57}$$

Now we order every possible set of values of $\tau_1, \tau_1' \ldots \tau_n, \tau_n'$, in the integrand and rename them $t_1, \ldots, t_{2n}$:

$$0 < t_1 < t_2 < t_3 \ldots < t_{2n} < t \,, \tag{10.58}$$

and define

$$x_i = z(t_i) , \quad i = 1, \ldots, 2n \quad . \tag{10.59}$$

Between any of the time intervals (t_i, t_{i+1}), the particle random walks according to the diffusion equation. The integration over paths in (10.57) can be carried out first for a fixed set of x_i; we simply obtain

$$P_o(x - x_{2n}, t - t_{2n}) \ldots P_o(x_2 - x_1, t_2 - t_1) \, P_o(x_1, t_1) \, . \tag{10.60}$$

We then integrate over $x_1 \ldots x_{2n}$. Now every term can be represented by a graph. We start from the origin. We draw a line from the origin to x_1, then a line from x_1 to $x_2, \ldots,$ and finally a line from x_{2n} to x. Every line represents a factor P_o in (10.60). Now we pair up the points $x_1 \ldots x_{2n}$ and join them pairwise by dashed lines representing the -v's in (10.57). Figure 10.12 shows an example. It should be obvious that the $P(x, t)$ is the sum of all graphs with a continuous solid line going from 0 to x and decorated with dashed lines with ends attaching to the solid line. Such graphs are just those for the

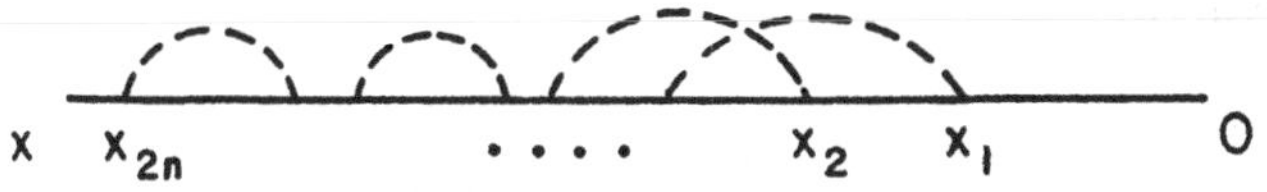

Figure 10.12. Graph representation of a sequence of random walks from 0 to x_1, x_1 to $x_2, \ldots, x_{2n}$ to x. Every dashed line represents a factor $-v(x_i - x_j)$.

correlation function G(k) of a Ginzburg-Landau model discussed in Chapter IX, provided that all closed loops of solid lines in graphs for G(k) are excluded. Excluding closed loops is equivalent to setting $n = 0$. We have thus shown that graphs for P(x, t) are graphs for an $n = 0$ Ginzburg-Landau theory. To relate P(x, t) to G(k) quantitatively, we take the Fourier and Laplace transform of P(x, t):

$$N^{-1} \int d^d x \; e^{-ik \cdot x} \int_0^\infty dt \; e^{-r_o t} \; P(x, t) \equiv \Gamma(k, r_o) \; . \quad (10.61)$$

Then (10.52) gives (note that $N = 1$ for $v = 0$),

$$\Gamma_o(k, r_o) = (r_o + ck^2)^{-1} \; , \qquad (10.62)$$

which is just what we called $G_o(k)$. Since v(x) has a

short range,

$$\int d^d x \; e^{-ik\cdot x} \; v(x) \;=\; u \tag{10.63}$$

is roughly a constant independent of k. Then graph by graph we can show that

$$\Gamma(k, r_o) = [G(k)]_{n=0} \; , \tag{10.64}$$

$$P(x,t) = N(2\pi)^{-d} \int d^d k \; e^{ik\cdot x} \int_{-i\infty}^{i\infty} \frac{dr_o}{2\pi i} \, e^{r_o t} \, \Gamma(k, r_o) \; . \tag{10.65}$$

Suppose that we take for granted that our knowledge about the critical behavior of the Ginzburg-Landau model of $n \geq 1$ at least applies qualitatively, even when $n = 0$. Then there should be a critical value r_{oc} of r_o, and singular behaviors of $\Gamma(k, r_o)$ for small k and $r_o - r_{oc}$ are expected. We should have

$$\begin{aligned} \Gamma(0, r_o) &\propto (r_o - r_{oc})^{-\gamma} \; , \\ \Gamma(k, r_{oc}) &\propto k^{-2+\eta} \; , \\ \Gamma(k, r_o) &= k^{-2+\eta} \, f(k\xi) \; , \end{aligned} \tag{10.66}$$

for sufficiently small k and $r_o - r_{oc}$, where

$$\xi = (r_o - r_o)^{-\nu} \tag{10.67}$$

and where f is some function independent of $r_o - r_{oc}$ and k except through the combination $k\xi$. From (10.66) and (10.67) we can deduce some properties of $P(x, t)$ for large x and t. When we let $r_o - r_{oc} = \lambda$ and use λ as an integration variable, we obtain from (10.65)

$$P(x, t) = N' t^{\gamma - 1 - \nu d} \tilde{f}(xt^{-\nu}) , \tag{10.68}$$

$$y^{2 - \eta - d - 1/\nu} \tilde{f}(y) = (2\pi)^{-d} \int d^d k' \times \int_{-i\infty}^{i\infty} \frac{d\lambda'}{2\pi i} e^{ik' \cdot \hat{x} + \lambda' k'^{1/\nu} y^{-1/\nu}} f(\lambda'^{-\nu}) \tag{10.69}$$

where N' is the normalization constant to assure that $\int d^d x \; P(x, t) = 1$. Note that a factor $e^{-r_{oc} t}$ appears in (10.65) as we change the integration variable from r_o to λ via $\lambda = r_o - r_{oc}$. This factor, being independent of x, drops out when we normalize the probability distribution.

In view of (10.68), the characteristic distance is t^ν at the time t. Recall that the root-mean-square distance in the absence of any self-avoiding tendency is $t^{1/2}$, as given by (10.53).

Another quantity of interest is $P(0, t)$ for large t, which tells how the probability decays at the place where the particle started. We note that in the Ginzburg-Landau model

$$(2\pi)^{-d} \int d^d k \;\; G(k) = \frac{1}{n} \langle \sigma^2 \rangle$$

$$\propto \; (r_o - r_{oc})^{1-\alpha} + \text{const} \qquad (10.70)$$

which implies that

$$(2\pi)^{-d} \int d^d k \; \Gamma(k, 0) = N^{-1} \int_0^\infty dt \; e^{-r_o t} \; P(0, t)$$

$$\propto \; (r_o - r_{oc})^{1-\alpha} + \text{const} \qquad (10.71)$$

which in turn implies that

$$P(0, t) \; \propto \; t^{-2+\alpha} + \text{const} \;\; \delta(t) \qquad (10.72)$$

for large t. We write a $\delta(t)$ term only to remind the reader that there are short time variations in $P(0,t)$ not derivable from the kind of qualitative arguments here.

What we need now are the values of these exponents for $n=0$. So far the only place where we can get an estimate of them is the small ϵ results. We set $n=0$ in Table 9.1 and list the results in Table 10.2. If we set $\gamma = 1$ and $\nu = 1/2$ in (10.68), the result is consistent with (10.53). That is, Eq. (10.53) is just the Gaussian approximation for (10.68).

We see that ν is greater than 1/2. Thus t^{ν} gives a larger distance than $t^{1/2}$ predicted by the diffusion

Table 10.2

Exponents for $n=0$ to $O(\epsilon^2)$
$\eta = 0.016\,\epsilon^2$
$\gamma = 1 + 0.125\,\epsilon + 0.051\,\epsilon^2$
$\alpha = 0.250\,\epsilon - 0.055\,\epsilon^2$
$\nu = 0.5 + 0.063\,\epsilon + 0.030\,\epsilon^2$

equation. Intuitively, this is expected. A tendency to avoid places visited before should make the probability distribution spread out more. Again, if we put $\alpha = 2 - d/2$, the value under the Gaussian approximation, we get $P(0,t) \propto t^{-d/2}$, which is consistent with (10.53). Note that $-2 + \alpha = -d\nu$. The ν given by Table 10.2 is larger than 1/2. Thus the self-avoiding tendency makes the probability at $x = 0$ decay faster.

The self-avoiding random walk problem is mathematically identical to the problem of a long flexible polymer chain. The chain is analogous to the path of the random walk and the length of the chain plays the role of the time t. The relative position between the two ends of the chain plays the role of x. The atoms making up the chain are not penetrable. This makes the chain behave like a self-avoiding path. In view of the above results, the distance between the two ends of the polymer chain is expected to be proportional to the length of the chain raised to the power ν.

A related problem to the self-avoiding walk is the self-attracting walk. Suppose that, instead of avoiding places visited earlier, the particle tends to walk back to these places. This tendency can be described also by (10.54), but with a negative v. We can use the same argument to conclude that this problem is equivalent to the Ginzburg-Landau model of $n = 0$, but now with $u < 0$. (Of course, we need to require that $u > 0$ in order that the Ginzburg-Landau model has any meaning for positive integral values of n. For $n = 0$, such a requirement is not needed.) The important question is whether the critical behavior of such a model is similar to that derived from models with $u > 0$. The answer is, in view of various results, that it is not. Consequently, our knowledge of critical phenomena is not helpful in understanding the self-attracting random walk problem unless sophisticated methods of analytic continuation is invented to utilize the $u > 0$ results for $u < 0$ cases.

An interesting aspect of the self-attracting walk problem is that it is also equivalent to the problem of electronic motion in a random potential. The equivalence can be established graphically. Let $\Gamma(x, t)$ be the solution to the electron Schrödinger's equation

$$i \frac{\partial \Gamma}{\partial t} = \left(- \frac{\hbar^2}{2m} \nabla^2 + \varphi(x) \right) \Gamma \qquad (10.74)$$

where $\varphi(x)$ is the random potential and Γ satisfies the initial condition $\Gamma(x, 0) = \delta(x)$. We define

$$G(k, E) = i \int_0^\infty e^{iEt} \langle \Gamma(x, t) \rangle \, dt \, , \qquad (10.75)$$

where the average $\langle \dots \rangle$ is taken over the probability distribution of φ. In the absence of $\varphi(x)$, $G = G_o$,

$$G_o(k, E) = \frac{1}{-E + k^2} \qquad (10.76)$$

where we have set $\hbar^2/2m = 1$ for simplicity. In the presence of φ, we solve (10.74) in powers of φ, and get the usual Born series,

$$G = G_o - \langle G_o \varphi G_o \rangle + \langle G_o \varphi G_o \varphi G_o \rangle + \cdots \qquad (10.77)$$

as shown in Fig. 10.13a at a given φ. Each solid line is a factor G_o and each dotted line a factor φ. Now we assume that φ has a Gaussian distribution

$$\langle\varphi\rangle = 0 ,$$

$$\langle\varphi(x)\,\varphi(x')\rangle = \Delta\delta(x-x') . \qquad (10.78)$$

Figure 10.13b shows the average of Fig. 10.13a. Dotted lines are paired up to form circled lines, each of which gives a factor Δ, just as in our treatment of random fields earlier in this chapter. Clearly Fig. 10.13b looks just like Fig. 10.12. In fact, if we identify $G_o(k, E)$ with $(r_o+k^2)^{-1}$, i.e., $-E \rightarrow r_o$, Δ with $-u$, then $G(k, E)$ becomes $G(k)$

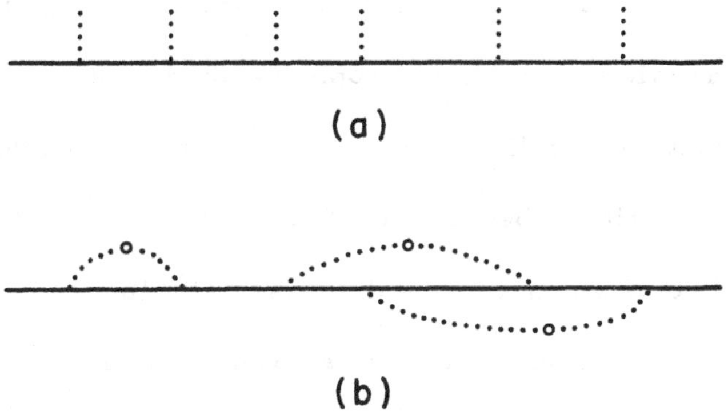

Figure 10.13. (a) A term in the Born series for the electron propagator Γ in the potential φ.
(b) To obtain the average of the Born series over the distribution of φ, we pair up the dotted lines in all possible ways to form circled lines.

of the $n = 0$ Ginzburg-Landau model with $u = -\Delta$, i.e., $\Gamma(k, r_o)$ of the random walk problem with a negative v.

If the random potential has a probability distribution which is more complicated than a Gaussian, then the problem will be equivalent to an $n = 0$ Ginzburg-Landau model with more complicated interactions.

6. OTHER NON-IDEAL FEATURES OF REAL SYSTEMS

Our discussion on the RG approach to critical phenomena thus far can be simply summarized as follows.

(i) Given the system of interest, we approximate the Hamiltonian by an isotropic n-component short-range interacting spin Hamiltonian. The tentative conclusion is that the critical behavior of this system is given by the fixed point which we have discussed all along. The critical exponents associated with this fixed point are determined by numerical approximations or the ϵ and $1/n$ expansion schemes as discussed in Chapters VII, VIII, and IX. Such critical behaviors will be referred to as "ideal."

(ii) Then we turn on the parameters ignored in the above approximation for the Hamiltonian, and refer to them as "non-ideal." We have discussed anisotropy and impurities as examples. If these parameters are irrelevant at the ideal fixed point, the critical behaviors will remain ideal, provided that these parameters are sufficiently small. If they are relevant, then the ideal fixed point is unstable and the critical behaviors will be those described by another fixed point, or there may no longer be any critical point.

In a real system there are often other non-ideal features besides anisotropy and impurities. One has to check all of them and determine the stable fixed point, if any, in order to find the correct critical behaviors. Many of them are unexplored. In the following we shall only mention briefly a few relatively simple and important non-ideal features.

(a) Dipole interaction

The interaction between magnetic dipoles in a ferromagnet is important owing to its long range. In three dimensions we have the familiar form

$$\mathcal{H}_{\text{dipole}} = -g^2 \int d^3x\, d^3x' \sum_{ij} \sigma_i(x)\, \sigma_j(x') \frac{\partial^2}{\partial x_i \partial x_j} \left(\frac{1}{|x - x'|} \right) \tag{10.79}$$

for the dipole interaction energy, where g is a constant. The subscripts denote spin components and coordinate components $(n = d = 3)$. The most important effect of the dipole interaction is that the shape of the sample becomes very important when an external magnetic field is turned on. The net magnetic pole layer on the surface screens out a part of the applied field in the sample. In theory and in the experimental data for ferromagnetic critical behavior, the applied field h is meant to be the true applied field plus the field due to the surface pole layer (or "the field corrected for demagnetization").

Apart from this surface effect, the strong angular dependence of the dipole interaction is such as to make it effectively short-ranged. The parameter g^2 in (10.79) is relevant and there is a new fixed point for dipolar critical behaviors, but the new exponents are found to be close to the ideal ones — so close that experimental detection of the difference would be difficult. The papers by Fisher and Aharony (1973), Aharony (1973), Bruce and Aharony (1974), and references therein should be consulted for details.

(b) Lattice vibrations

The ideal spin systems which we have been discussing are rigid mechanically. If the elastic nature of the lattice is taken into account, many complications arise. Several authors have investigated this problem, e.g., Sak (1974), Wegner (1974), Imry (1974), Rudnick et al. (1974), Aharony and Bruce (1974), etc. The conclusions, however, are not clear. The boundary condition, i.e., whether the pressure or the volume and shape are held fixed, the lattice symmetry, the anharmonic terms in the elastic part of the Hamiltonian, etc., all seem to be of qualitative importance. No one has put everything together in a reasonably clear fashion.

(c) Effect of gravity

In doing experiments near the liquid-gas critical point, we cannot ignore the pressure gradient in the sample due to gravity. Only a small layer of the sample can be at the critical pressure, and the system is no longer uniform. A discussion of the effects of gravity on certain experiments is given by Splittorff and Miller (1974).

(d) Gauge fields

The coupling of the magnetic vector potential, which is an example of a gauge field, to the superconductive order parameter (the Cooper pair amplitude) changes the critical point to a first-order transition. [See Halperin, Lubensky, and Ma (1973).] Very qualitatively speaking, the presence of a nonzero order parameter strongly affects the long-wavelength thermal fluctuations of the magnetic field (essentially the Meissner effect). The Hamiltonian for the order parameter, after the magnetic fluctuations are integrated out, acquires a nonanalytic term as a result and leads to a first-order transition. Similar situations appear in some critical points observed in liquid crystals.

There are many interesting topics which we have left out. Among them are the bicritical, tricritical, and multicritical phenomena, displacive phase transitions in crystals, surface phase transitions, various other formulations of the RG, various approximation schemes, etc. They are materials for advanced study. We shall conclude this chapter by mentioning several general references for advanced topics.

Wilson and Kogut (1974) is an exhaustive review of the RG and the ϵ expansion. It contains a complete list of references.

A new review by Wilson (1975) contains his numerical RG study of the two-dimensional Ising model, as well as a detailed exploration of his RG solution of the Kondo problem.

A book edited by Domb and Green (1976) (which is the 6th volume of the "Phase Transitions and Critical Phenomena" series), contains several advanced reviews.

A book by Toulouse and Pfeuty (1975) provides an elegant exploration of a wide range of topics.

Conference proceedings edited by Gunton and Green (1973) contain short lectures and dialogues reflecting the state of knowledge at that time. Much progress has been made since then, but many fundamental questions raised there still remain unsolved.

XI. INTRODUCTION TO DYNAMICS

SUMMARY

Elementary concepts of Brownian motion and kinetic equations are reviewed in preparation for studying dynamic (time-dependent) critical phenomena. The motion of a harmonic oscillator is analyzed to illustrate basic concepts such as mode-mode coupling, elimination of fast modes, and correlation and response functions. The van Hove theory is discussed.

1. INTRODUCTION

Now we turn our attention to dynamic, i.e., time-dependent, phenomena near critical points, a subject which is still in its developing stage.

In studying static phenomena, we were interested in consequences of the thermal equilibrium probability distribution of the configurations of spins or other quantities. The problem of statics is essentially a problem of statistics. The study of dynamics is a much more complicated problem. We need to know the time evolution of configurations, how physical quantities change under external time-dependent disturbances, and how the equilibrium probability distribution is reached after disturbances are turned off. Dynamic phenomena are much richer in variety than static phenomena, including, for example, diffusion, wave propagation, damping, inelastic scattering of neutrons or light, etc. In the study of critical dynamics, we are mainly interested in the time variations of the large-scale fluctuations of the order parameter and other slowly varying physical quantities near the critical point. Qualitatively it is easy to understand why the order parameter varies slowly in time. Imagine a spin system. Near its critical point, configurations with large spin patches are favorable. In each patch there is a net fraction of spins pointing in the same direction. As we mentioned in Chapter I, thermal agitations will flip spins at random. Owing to the large sizes of the patches, it would take a long time for the thermal agitations to turn a whole patch of spins around. In a more formal language, we say that long wavelength modes of spin fluctuations have very long relaxation times. Often this is referred to as "critical slowing down." There are also other reasons for long relaxation times, mainly conservation laws, for various quantities. Observed dynamic phenomena near critical points are all characterized by long relaxation times. A theory of critical dynamics must explain how these long relaxation times come about in terms of small-scale interactions among spins, how they depend on the temperature, and how they are affected by conservation laws and other features. Of course, if we were able to derive critical dynamics from microscopic models via first principles, then we would have a theory. The criteria for dynamic models are subject to the same kind of discussion as those for static ones in Sec. II. 1 (which should be read again if forgotten), except that many new complications arise for dynamics. In statics, the model provides a basis for statistics, which is essentially the counting of configurations. In going from model (1) to model (5) of Sec. II. 1, the configurations of spins are more and more coarse-grained, but there is no essential difference among the models. For dynamics, we need to do much more than count configurations — we need equations of motion for studying time evolutions. At the level of models (1) and (2), the equations of motion are furnished by time-dependent Schrödinger equations. Because such equations are time-reversal invariant and do not directly describe any dissipation, they are impractical for studying critical dynamics. At the level of models (4) and (5), the coarse-grained spin configurations will be described by time-irreversible equations of motion, which contain dissipative terms. These terms are a result of random thermal agitation over a scale of b, the size of a block. Such equations will be called kinetic equations. The derivation of kinetic equations from Schrodinger's equations for various systems is a very difficult task, which has still not been completed and to which a very large literature has been devoted. We shall not take a microscopic approach. Instead, we shall write down phenomenological kinetic equations, which are constructed under plausible assumptions and general restrictions. These assumptions and restrictions will be motivated by microscopic considerations. Such kinetic equations will serve as the basis for our study of critical dynamics.

In the remainder of this chapter we shall explain the intuitive basis for the construction of kinetic equations and the van Hove theory. Much of this material can be found in textbooks on elementary kinetic theory. We include it here not only for the sake of completeness, but also in order to emphasize the fact that basic ideas are not very complicated.

In the chapters following this we shall apply renormalization group ideas to the study of dynamics of some simple model systems. Dynamic scaling and the extent of universality will be explored within the framework of the RG. The approach here will be along the lines taken by Halperin, Hohenberg, and Ma (1972, 1974) and Ma and Mazenko (1974, 1975). There has been a great deal of recent work in this area by many other authors. Advances made before the use of the RG form a much larger literature. Formulations of dynamic scaling, mode-mode coupling, and other ideas were put forth by Kadanoff, Kawasaki, Halperin, Hohenberg, Ferrell, and many others. A review of experimental results can be found in Stanley's book (1971).

2. BROWNIAN MOTION AND KINETIC EQUATIONS

The kinetic equations which we shall use later for critical dynamics are generalizations of equations for the Brownian motion. Generally kinetic equations describe the time evolution of a set of physical quantities of interest. There are two distinct mechanisms for time evolution, (a) regular or organized motion, and (b) random or disorganized motion. The regular motion follows time-reversal invariant laws of dynamics and is all one has in a microscopic theory. It will be termed the "mode-mode coupling" part of kinetic equations. The random motion is the result of processes not explicitly included in the kinetic equation. It generates statistical distributions for quantities of interest and is responsible for time-irreversible effects.

It will be described phenomenologically by decay rates and noises in kinetic equations. We shall illustrate these basic features by considering the Brownian motion of a harmonic oscillator.

Let us forget about critical phenomena for the moment. Consider a harmonic oscillator with the familiar Hamiltonian

$$H = \frac{p^2}{2m} + \frac{1}{2} Kx^2 \quad . \tag{11.1}$$

Let us introduce the notation

$$q_1 = p, \quad q_2 = x \quad . \tag{11.2}$$

We call q_1 and q_2 modes. We shall refer to the (q_1, q_2) plane as the phase space. The motion of the oscillator is thus represented by the motion of a point in the phase space. In the absence of any other force, the velocity (v_1, v_2) of the point in phase space is

$$\begin{aligned} v_1 &= -Kq_2 \quad , \\ v_2 &= q_1/m \quad , \end{aligned} \tag{11.3}$$

which traces out an ellipse. Now suppose that the oscillator is immersed in a viscous fluid, i.e., in contact with a thermal reservoir. The effect of the reservoir can be approximately accounted for by a damping on the oscillator and a random force, as given by the phenomenological kinetic equations:

$$\frac{\partial q_1}{\partial t} = v_1 - \frac{\Gamma_1}{T}\frac{\partial H}{\partial q_1} + \zeta_1(t) \qquad (11.4a)$$

$$\frac{\partial q_2}{\partial t} = v_2 \qquad (11.4b)$$

where Γ_1/T is a constant and $(-\Gamma_1/T)\,\partial H/\partial q_1 = -\Gamma_1 q_1/mT$ is simply a frictional force. It gives a velocity in phase space pointing in the direction of decreasing energy. $\zeta_1(t)$, a random function of time, is the random force of "noise." What is the reason for defining the damping term with an extra factor $1/T$? The reason is to have the dimensionless combination H/T instead of H appear in the kinetic equation. Experience has told us that H and T most often appear together as H/T. The average value and the correlation function of the noise are assumed to be

$$\langle \zeta_1(t) \rangle = 0$$

$$\langle \zeta_1(t)\, \zeta_1(t') \rangle = 2\, D_1\, \delta(t - t') \qquad (11.5)$$

where the average is taken over the assumed Gaussian probability distribution of ζ_1, and $2D_1$ is a constant. Equation (11.5) simply says that the random forces at different times are not correlated and have zero average at all times.

Equations (11.4) and (11.5) define the motion completely. They are just the Langevin equations for the oscillator. If (q_1, q_2) is at a given point at time $t = 0$, (11.4) tells us that the average value of (q_1, q_2) will go around in an elliptic orbit whose size diminishes as a result of damping. Eventually, the average value of (q_1, q_2) will become zero. The cause of an elliptic orbit is the velocity (v_1, v_2). The energy is transferred from kinetic energy $q_1^2/2m$ to potential energy $\frac{1}{2} K q_2^2$ back and forth as a result. This (v_1, v_2) is a prototype of the mode-mode coupling which Kawasaki introduced in his study of critical dynamics. Here (v_1, v_2) "couples the q_1-mode and q_2-mode."

Of course, only the <u>average</u> value of (q_1, q_2) diminishes. The motion becomes more random but never ceases because the random force $\zeta_1(t)$ in (11.4) keeps on acting. As $t \to \infty$, the oscillator comes into thermal equilibrium with the reservoir and we get an equilibrium probability distribution for (q_1, q_2). It is of interest to know how the probability distribution evolves in time.

Let $P(q_1, q_2, t)\, dq_1 dq_2$ be the probability of finding the oscillator in the area $dq_1 dq_2$ at (q_1, q_2). Equations (11.4) and (11.5) lead to a Fokker-Planck equation for $P(q_1, q_2, t)$, which can be written in an appealing form:

$$\frac{\partial P}{\partial t} + \sum_i \frac{\partial J_i}{\partial q_i} = 0 , \tag{11.6a}$$

with

$$J_i = v_i P - \frac{\Gamma_i}{T} \frac{\partial H}{\partial q_i} P - D_i \frac{\partial P}{\partial q_i} , \tag{11.6b}$$

with v_i given by (11.3), $\Gamma_2 = 0$ for our harmonic oscillator. Clearly (11.6a) is just a continuity equation in the phase space (like the usual $\partial P/\partial t + \nabla \cdot J = 0$). The probability current J_i has three terms. The first one is generated by the nonrandom velocity v_i. The second is generated by the velocity toward lower energy due to damping, and the third is a diffusion current toward lower probability. We have written (11.6) in a generalized form applicable to many cases. A more careful derivation of the Fokker-Planck equation can be found in textbooks. The coefficient of P in J_i and that of $-\partial P/\partial q_i$ can be calculated via

$$\langle \Delta q_i \rangle / \Delta t \quad , \tag{11.7a}$$

$$\frac{1}{2} \langle \Delta q_i \ \Delta q_i \rangle / \Delta t \quad , \tag{11.7b}$$

respectively. Here Δq_i is the change of q_i in a time interval Δt, which is large compared to the time over which noise is correlated, but still small compared to the time over which q_i changes appreciably. If one uses the generalization of (11.4) and (11.5), namely,

$$\frac{\partial q_i}{\partial t} = v_i - \frac{\Gamma_i}{T} \frac{\partial H}{\partial q_i} + \zeta_i(t) \quad , \tag{11.8a}$$

$$\langle \zeta_i(t) \ \zeta_j(t') \rangle = 2 D_i \, \delta_{ij} \, \delta(t - t') \quad , \tag{11.8b}$$

with $\langle \zeta_i(t) \rangle = 0$ understood, one easily shows that (11.7a) and (11.7b) are just Γ_i/T and D_i, respectively.

Equations (11.6) and (11.8) provide a phenomenological description of Brownian motion in a phase space of any number of coordinates. The question of how realistic this description is must be answered by more microscopic studies.

Let us note a few general properties of (11.8). The velocity vector v_i must be orthogonal to the "gradient vector" $\partial H/\partial q_i$ if the total energy H is conserved in the absence of the damping and random force:

$$0 = \frac{dH}{dt} = \sum_i \frac{\partial H}{\partial q_i} \frac{dq_i}{dt} = \sum_i \frac{\partial H}{\partial q_i} v_i \qquad (11.9)$$

which is simply the dot product of the vectors $\partial H/\partial q_i$ and v_i. The Liouville theorem says that

$$\sum_i \frac{\partial v_i}{\partial q_i} = 0 , \qquad (11.10)$$

i.e., the vector v_i is divergence-free in phase space. For a stationary probability distribution, i.e., $\partial P/\partial t = 0$,

(11.6a) tells us that

$$\sum_i \frac{\partial J_i}{\partial q_i} = 0 \ . \qquad (11.11)$$

This equation is satisfied if

$$P \propto e^{-H/T} \qquad (11.12)$$

and J_i is calculated via (11.6b) plus the relation

$$\Gamma_i = D_i \ , \qquad (11.13)$$

which is the so-called Einstein relation. Note that (11.9) and (11.10) are crucial in guaranteeing (11.12) to be a stationary solution of the Fokker-Planck equation. Since we want the equations of motion to generate the equilibrium probability distribution (11.12) for $t \to \infty$, it is necessary to satisfy (11.13).

3. RELAXATION TIMES

Suppose that a quantity such as q_i assumes a certain value at time $t = 0$. Subsequently the value of $q_i(t)$ becomes uncertain (and can be described by a probability distribution). After some time the average value of $q_i(t)$ will approach its thermal equilibrium value. This some time is the relaxation time (often referred to as the decay time), of the mode q_i. Some modes have short relaxation times, some have long ones.

We can solve equations such as (11.8) and obtain relaxation times. In general, the forms of H and v_i as well as the values of Γ_i all play interrelated roles in determining relaxation times. As a simple illustration, let us go back to the kinetic equations (11.4) for the harmonic oscillator.

Clearly the thermal equilibrium average value of q_1 and q_2 is zero. To obtain relaxation times, we solve for the average values of $q_i(t)$ given the initial values $q_i(0)$. Now suppose that we neglect v_i in (11.4). Then the modes q_1 and q_2 are decoupled, and we get

$$\frac{\partial}{\partial t}\langle q_1(t)\rangle = -(\Gamma_1/mT)\langle q_1(t)\rangle ,$$

$$\frac{\partial}{\partial t}\langle q_2(t)\rangle = 0 . \tag{11.14}$$

This means a relaxation time mT/Γ_1 for q_1. Since

$\langle q_2 \rangle$ never changes according to (11.14), the relaxation time for q_2 is infinite, which is contrary to what we expect of a harmonic oscillator. Of course, we know the motion of a harmonic oscillator too well. The mode-mode coupling terms v_i cannot be ignored. In fact, we know how to solve equations with v_i easily by taking linear combinations of (11.4). Let

$$q_{\pm} = \tau_{\mp} q_1 + q_2 \ . \tag{11.15}$$

Then we obtain from (11.4)

$$\frac{\partial}{\partial t} q_{\pm} = -q_{\pm}/\tau_{\pm} + \tau_{\mp} \zeta_1 \tag{11.16}$$

where $\tau_{\pm}$ are the relaxation times given by

$$1/\tau_{\pm} = \Gamma_1/(2\,mT) \pm ((\Gamma_1/2\,mT)^2 - K/m)^{1/2} \ . \tag{11.17}$$

Let us assume that $\Gamma_1/Tm \gg (K/m)^{1/2}$. Then

$$1/\tau_+ \approx (\Gamma_1/T)/m \ ,$$
$$1/\tau_- \approx K(T/\Gamma_1) \ . \tag{11.18}$$

Thus we have a "fast mode" q_+ with a shorter relaxation time and a "slow mode" q_- with a longer one. These simple results illustrate the important role of the mode-mode coupling terms v_i in (11.4).

4. ELIMINATION OF FAST MODES

In our study of static critical phenomena, the elimination of short-wavelength modes played an important part, especially in the renormalization group. In critical dynamics, we are concerned mainly with modes with long relaxation times. The elimination of modes with short relaxation times will again play an important part. In statics, the elimination was effected by integrating out the unwanted modes in the probability distribution. In dynamics, the elimination process is effected very differently. As a simple illustration, let us return to the harmonic oscillator equations.

If we make use of (11.16), then the elimination of the fast mode is trivial, namely, simply dropping the equation for q_+. This simplicity is a special property of the harmonic oscillator and does not appear in more general problems. It is more instructive to examine (11.4). Let us eliminate q_1 by solving (11.4a) for q_1 and substitute it in (11.4b):

$$\frac{\partial}{\partial t} q_2 = q_1/m$$

$$= -\frac{K}{m} \int_{-\infty}^{t} dt' \, e^{-\gamma(t-t')} q_2(t') + \zeta'(t) \quad , \qquad (11.19)$$

$$\zeta'(t) = \int_{-\infty}^{t} dt' \, e^{-\gamma(t-t')} \zeta_1(t')/m \quad , \qquad (11.20)$$

where $\gamma \equiv \Gamma_1/(Tm) \approx 1/\tau_+$. Over a period of time long compared to τ_+ but still short compared to τ_-, $q_2(t)$ can be taken as approximately constant. Equations (11.19) and (11.20) reduce to

$$\frac{\partial}{\partial t} q_2 \approx -\frac{\Gamma'}{T} \frac{\partial H}{\partial q_2} + \zeta'(t) \quad , \qquad (11.21)$$

$$\Gamma'/T \equiv (m\gamma)^{-1} = T/\Gamma_1 \quad , \qquad (11.22)$$

$$H = \frac{1}{2} K q_2^2 \quad . \qquad (11.23)$$

From (11.20) and (11.5), we obtain

$$\langle \zeta'(t)\, \zeta'(t') \rangle = 2D'(\gamma/2)\, e^{-\gamma|t-t'|}$$

$$\approx 2D' \,\delta(t-t') \quad , \qquad (11.24)$$

$$D' \equiv D_1 (T/\Gamma_1)^2 \quad . \qquad (11.25)$$

Equations (11.21) and (11.24) have the form of (11.8). It is easy to verify that $\Gamma' = D'$. What we have just done is eliminate q_1, the momentum of the oscillator, which now acts as a noise for the remaining variable q_2, the coordinate. Note that (11.21) is equivalent to the equation for q_- in (11.16) as far as variations over a time long compared to $\tau_+ = 1/\gamma$ are concerned. Equation (11.21) does not describe the details of q_2 over a time scale τ_+. Apart from such details, q_2 is the same as q_-.

The elimination process also reveals the limitation of the general form (11.8) of kinetic equations. In (11.8), q_i are the modes of interest. The effects of all unwanted modes are included in the noise ζ_i. The property (11.8b) of the noise is a good approximation only if the relaxation times of the unwanted modes are much shorter than the time scales of interest for q_i. [This condition allows us to replace $(\gamma/2)\, e^{-\gamma|t-t'|}$ by $\delta(t-t')$ in (11.24).]

Of course there are many other important features of elimination processes which are not revealed by the above simple illustration. We shall eventually be interested in elimination processes involving virtually an infinite number of modes coupled nonlinearly, namely coarse graining processes like those we discussed in statics. Sometimes the eliminated modes cannot be simply described as noise. There may be combinations of them which appear as important new modes. An example is the heat diffusion mode, which will be discussed later. (See Sec. XIII. 2.)

The relaxation time $(\Gamma' K/T)^{-1} = (TK/\Gamma_1)^{-1}$ given by (11.21) is just τ_- given by (11.18). Note that K is the curvature of the potential well $\frac{1}{2} K q_2^2$ (see p. 445, Fig. 11.1). The flatter the well, the smaller the K and hence the longer the relaxation time. The flatter well has a smaller driving force toward the center. It also gives a bigger spread in the equilibrium probability distribution for q_2, namely,

$$\langle q_2^2 \rangle \propto T/K \ . \tag{11.26}$$

This is consistent with $\langle q_2^2 \rangle \sim D' \tau_-$, which says that it takes a time τ_- for the diffusion process to achieve the equilibrium distribution for q_2 .

5. RESPONSE FUNCTIONS AND CORRELATION FUNCTIONS

Response functions describe how the system behaves under very weak perturbation. Suppose that some perturbation causes H to change into H′

$$H'/T = H/T - \sum_i f_i q_i \quad . \tag{11.27}$$

Then $-\partial H/\partial q_i$ changes into $-\partial H/\partial q_i + f_i T$. We note that $-\partial H/\partial q_i$ is a gradient vector in the phase space pointing in the direction of steepest decrease of energy. It is convenient to regard it as a force. Thus f_i is the perturbing force. As a result of f_i, average values of various quantities are modified. If f_i is very small, then the modification is approximately linear in f_i. We write

$$\langle q_i(t) \rangle = \sum_j \int G_{ij}(t-t')\, f_j(t')\, dt' + O(f^2)\ . \qquad (11.28)$$

Here we assume that $\langle q_i(t) \rangle = 0$ in the absence of the perturbation. In terms of Fourier transforms, we have

$$\langle q_i(\omega) \rangle = \sum_j G_{ij}(\omega)\, f_j(\omega) + O(f^2)\ . \qquad (11.29)$$

G_{ij} are called the response functions. As a simple example, let us add a force f to $-(\partial H/\partial q)/T$ in (11.21) (dropping the subscript 2 for simplicity). We get

$$\frac{\partial q}{\partial t} = -(\Gamma' K/T) q + \Gamma' f\ . \qquad (11.30)$$

The solution is trivial:

$$q(\omega) = G(\omega)\, f(\omega)\ ,$$

where

$$G(\omega) = (K/T - i\omega/\Gamma')^{-1} \qquad (11.31)$$

is the response function.

Now we leave out the perturbation and define the correlation function C_{ij} as

$$C_{ij}(t-t') = \langle q_i(t)\, q_j(t') \rangle \quad , \tag{11.32a}$$

or in Fourier transforms

$$2\pi\,\delta(\omega+\omega')\, C_{ij}(\omega) = \langle q_i(\omega)\, q_j(\omega') \rangle \ . \tag{11.32b}$$

Again we illustrate this with (11.21) which gives

$$q(\omega) = G(\omega)\, \zeta'(\omega)/\Gamma' \tag{11.33}$$

where $G(\omega)$ is given by (11.31). From (11.24) we obtain

$$\langle \zeta'(\omega)\, \zeta'(\omega') \rangle = 2\pi\,\delta(\omega+\omega')\, 2D' \ . \tag{11.34}$$

Therefore, (11.33) and (11.34) give the correlation function

$$\begin{aligned} C(\omega) &= 2D'\, G(\omega)\, G(-\omega)/\Gamma'^2 \\ &= (2D'/\Gamma'^2)(\omega^2/\Gamma'^2 + (K/T)^2) \\ &= (2/\omega)\ \mathrm{Im}\, G(\omega) \ , \end{aligned} \tag{11.35}$$

where we have used the relation $\Gamma' = D'$. The relation between $C(\omega)$ and the imaginary part of the response function $G(\omega)$ given by (11.35) is the "fluctuation-dissipation theorem."

6. THE VAN HOVE THEORY

In subsequent study of critical dynamics, we shall examine models described by kinetic equations like (11.8). The modes q_i will be the Fourier components σ_k and other quantities which are expected to have long relaxation times. Instead of two modes, we shall have virtually an infinite number of modes, and solutions to the kinetic equations will be very difficult to obtain. Let us start with the simplest model, the van Hove theory. More general discussions will follow.

The van Hove theory, often called "the conventional theory," is a simple model without mode-mode coupling. Let us consider a Ginzburg-Landau form of H in the Gaussian approximation (3.28) for $T > T_c$:

$$H[\sigma]/T \approx \sum_{k < \Lambda} (a_2 + ck^2)\, |\sigma_k|^2 \quad . \qquad (11.36)$$

Here we take $n = 1$ for simplicity. Then (11.8) has the form

$$\frac{\partial \sigma_k}{\partial t} = -\Gamma_k \, 2(a_2 + ck^2)\, \sigma_k + \zeta_k \ ,$$

$$\langle \zeta_k(t)\, \zeta_{k'}(t') \rangle = 2\Gamma_k \, \delta_{-kk'} \, \delta(t - t') \ . \tag{11.37}$$

For the moment, let us assume that $\Gamma_k = \Gamma$, independent of k. We can write (11.37) in terms of $\sigma(x, t)$:

$$\frac{\partial \sigma}{\partial t} = -\Gamma \, 2(a_2\sigma - c\nabla^2\sigma) + \zeta \ ,$$

$$\langle \zeta(x, t)\, \zeta(x', t') \rangle = 2\Gamma \, \delta(x - x') \, \delta(t - t') \ . \tag{11.38}$$

As before, it is understood that $\sigma(x, t)$ contains only Fourier components with $k < \Lambda$. Equation (11.38) explicitly describes the dynamics over a region of size Λ^{-1}. It is a "local" equation of motion, in the same spirit as model (5) discussed in Chapter II. It should be kept in mind that our purpose here is to derive critical dynamics, which concerns large-scale behaviors, from local equations of motion, which are based on dynamics over a much smaller scale. The phenomenological constant Γ and the random field ζ are supposed to simulate the effect of dynamical processes over a scale Λ^{-1}. Therefore, Γ is expected to be a smooth function of temperature and can be considered to be a constant within a small temperature range near T_c. This is the same argument as that in Chapter II leading to the conclusion that parameters in the block Hamiltonian are smooth functions of T.

Now let us return to the Fourier component representation (11.37). Clearly, each mode is independent of other modes and has a relaxation time given by

$$\tau_k = [2(a_2 + ck^2)]^{-1} \Gamma^{-1} . \qquad (11.39)$$

In the limit of small k ,

$$\tau_k \to (2a_2)^{-1} \Gamma^{-1}$$

$$= [2a_2'(T - T_c)]^{-1} \Gamma^{-1}$$

$$= \chi \Gamma^{-1} , \qquad (11.40)$$

where we have substituted $a_2'(T - T_c)$ for a_2 and χ is the static susceptibility as given by (3.25). As $T \to T_c$, χ and therefore τ_k diverges. This divergence of relaxation time is a crude description of the "critical slowing down" mentioned earlier. Intuitively, this result is rather obvious. As far as the mode σ_k is concerned, the effective Hamiltonian is just $(a_2 + ck^2)|\sigma_k|^2$. As a function of σ_k, this is a parabolic well. (See Fig. 11.1. Here σ_k plays the role of q_2.) For $k \to 0$, $a_2 = a_2'(T - T_c) \to 0$, this well becomes flat and it will take a very long time for $\langle \sigma_k \rangle$ to relax toward the center of the well. The flattening of the well was seen in Chapter III as the reason for large fluctuations of σ_k.

The relaxation time (11.40) can be expressed in terms of $\xi = (a_2/c)^{-1/2}$, the correlation length:

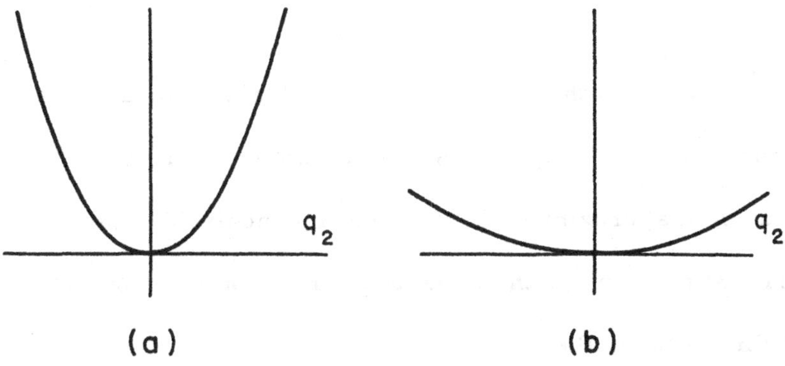

Figure 11.1. The potential $(1/2)Kq_2^2$ vs q_2.
(a) Large K. (b) Small K.

$$\tau_k = \xi^2 (1 + \xi^2 k^2)^{-1} (2c\Gamma)^{-1} . \tag{11.41}$$

The dynamic scaling hypothesis, which will be discussed later, says that, for $T \to T_c$

$$\tau_k = \xi^z f(k\xi) \tag{11.42}$$

where z is called the "dynamic exponent" or "characteristic time exponent, " and $f(y)$ is some function independent of $T - T_c$. Equation (11.41) says that $z = 2$, $f(y) = (1 + y)^{-1} (2c\Gamma)^{-1}$.

It is clear that (11.39) is of the form

$$\tau_k = G(k)\, \Gamma^{-1} \tag{11.43}$$

where $G(k)$ is the static correlation function in Gaussian approximation. Suppose that we do not treat H by Gaussian approximation. It is not inconceivable that (11.43) may hold with the exact correlation function $G(k)$. If this is true, then

$$\tau_k = \xi^{2-\eta} f(k\xi) ,$$

$$z = 2 - \eta \quad , \tag{11.44}$$

since we know that $G(k)$ is $\xi^{2-\eta}$ times a function of $k\xi$ for $T \to T_c$. We shall show later that in many cases (11.44) is not true.

We have so far taken Γ_k of (11.37) as a constant. More generally, we should consider the k dependence of Γ_k. Let us write Γ_k as an expansion:

$$\Gamma_k = \Gamma + \gamma k^2 + O(k^4) \quad . \tag{11.45}$$

Note that Γ_k must be a smooth function of k because it is based on the behavior of spins over a block, i.e., a small distance $b \sim \Lambda^{-1}$. As long as $\Gamma \neq 0$, the above discussions are unchanged when k is sufficiently small. However, there are systems in which certain conservation laws force Γ to zero. For example, for an isotropic Heisenberg ferromagnet, the total spin, i.e., the $k = 0$ mode, is a constant of motion as a consequence of spin rotational invariance. The noise cannot change σ_k for $k = 0$ at any temperature. We must then have

$$\Gamma_k = \gamma k^2 + O(k^2) \tag{11.46}$$

if total spin is conserved. In this case, (11.43) is modified to

$$\tau_k = G(k)\, k^{-2}/\gamma \tag{11.47}$$

and the dynamic exponent becomes

$$z = 4 \tag{11.48}$$

or perhaps $z = 4 - \eta$.

For an antiferromagnet (with σ now denoting the order parameter, the staggered magnetization), the σ_k for $k = 0$ is not conserved. Therefore, $\Gamma \neq 0$.

Clearly, equations (11.37) were chosen for their simplicity. The question is whether they, at least approximately, describe the critical dynamics of ferromagnets, or other realistic systems. The answer is no. They are oversimplified. The Gaussian approximation for H is an oversimplification. The omission of mode-mode coupling (the v_i term in (11.8)] and slowly varying modes other than the order parameter is in many cases more serious. However, (11.37), owing to its simplicity, is a convenient starting point for investigations.

XII. THE RENORMALIZATION GROUP IN DYNAMICS

SUMMARY

The extension of RG ideas to the study of dynamics is straightforward and is outlined in this chapter. The dynamic scaling hypothesis follows from RG arguments when certain simplicities in the RG are assumed.

1. DEFINITION OF THE RG IN DYNAMICS

In this chapter we extend the discussion of the RG in Chapters V and VI to dynamic models. The motivation and basic ideas of the RG approach to dynamics are the same as those to statics. The formulation needs to be extended to accommodate those features of dynamics which are absent in statics.

For the purpose of this discussion, we shall consider kinetic equations of the form

$$\frac{\partial \sigma_k}{\partial t} = v_k - \Gamma_k \frac{\partial \mathcal{H}}{\partial \sigma_{-k}} + \zeta_k , \quad k < \Lambda ,$$

$$\langle \zeta_k(t) \zeta_{k'}(t) \rangle = 2\Gamma_k \, \delta_{-kk'} \, \delta(t - t') \quad , \tag{12.1}$$

and ignore other modes. These are just (11.8), with σ_k replacing q_i. The phase space velocity v_k must satisfy (11.9) and (11.10), namely

$$\sum_k \frac{\partial \mathcal{H}}{\partial \sigma_k} v_k = 0 ,$$

$$\sum_k \frac{\partial v_k}{\partial \sigma_k} = 0 \quad . \tag{12.2}$$

The probability distribution for σ_k satisfies (11.6), namely

$$\frac{\partial P}{\partial t} + \sum_k \frac{\partial}{\partial \sigma_k} \left(v_k P - \Gamma_k \frac{\partial \mathcal{H}}{\partial \sigma_{-k}} P - \Gamma_k \frac{\partial P}{\partial \sigma_{-k}} \right) = 0 , \tag{12.3}$$

which has the stationary solution

$$P \propto e^{-\mathcal{H}} \tag{12.4}$$

Recall that in Chapter V we defined a parameter space on which the RG operates. Each point μ in the parameter space specifies a probability distribution $e^{-\mathcal{H}}$. Here we can easily generalize the concept of parameter space. We simply include in μ all the parameters specifying v_k and Γ_k in addition to those specifying $\mathcal{H}$.

The RG transformation

$$\mu' = R_s \mu \tag{12.5}$$

has two steps, (i) a Kadanoff transformation and (ii) a change of scale as in the static case defined in Sec. V.2, except that the first step here must be carried out very differently, as follows.

Step (i). Eliminate the modes σ_q with

$$\Lambda/s < q < \Lambda$$

from the kinetic equations. This means solving those

equations of (12.1) for σ_q, substituting the solutions in the remaining equations for σ_k, $k < \Lambda/s$, and then averaging over ζ_q, a procedure comparable to our elimination of one of the harmonic oscillators in the previous chapter. This step is the dynamic generalization of the static Kadanoff transformation which simply integrates out σ_q in $e^{-\mathcal{H}}$.

Step (ii). In the remaining equations for $\sigma_k(t)$, $k < \Lambda/s$, we make the replacement

$$\sigma_k(t) \rightarrow s^{1-\eta/2} \sigma_{sk}(ts^{-z}) \quad , \qquad (12.6)$$

and L by sL′. This is the same as Step (ii) in the static RG except for the additional feature $t \rightarrow ts^{-z}$, with the new exponent z, which plays a role similar to that of η. We shall need to adjust the values of η and z in order that R_s will have a fixed point.

The new equations of motion are then written in the form of (12.1) with new parameters, which are identified as entries in $\mu' = R_s \mu$.

Step (ii) is very easy to carry out, but Step (i) is more involved and will cause us difficulties similar to those we encountered in the static case. That is, after application of Steps (i) and (ii), the new kinetic equation will in general assume a more complicated form than (12.1). For example, Γ'_k may become dependent on σ_k, and $\partial^2 \sigma_k / \partial t^2$ terms may be needed. In other words, the form of the kinetic equation (12.1) is not sufficiently general for a precise formulation of the RG. Nevertheless, (12.1) will be sufficient for the calculations which we shall discuss under certain approximations.

2. TRANSFORMATION OF CORRELATION FUNCTIONS AND RESPONSE FUNCTIONS

Let us define the correlation function $C(k, \omega)$ for σ_k in the same manner as (11.32):

$$C(k, \omega)\, 2\pi\, \delta(\omega + \omega')\, \delta_{-kk'} = \langle \sigma_k(\omega)\, \sigma_{k'}(\omega') \rangle \ , \tag{12.7}$$

$$\sigma_k(\omega) = \int dt\ e^{i\omega t}\, \sigma_k(t) \ , \tag{12.8}$$

or equivalently

$$C(k, \omega) = \int dt\ e^{i\omega t} \langle \sigma_k(t)\, \sigma_{-k}(0) \rangle \ . \tag{12.9}$$

$C(k, \omega)$ is proportional to the cross section of inelastic neutron scattering of momentum transfer k and energy transfer ω. The argument leading to this conclusion is similar to that leading to (1.5), where the spin configuration was assumed to be fixed during the scattering process, i.e., the scattering was assumed to be instantaneous. When time variation of σ_k is taken into consideration, the average scattering rate is

$$\Gamma_{fi} \propto \lim_{\tau \to \infty} \frac{1}{\tau} \left\langle \left| \int_0^\tau dt\, d^d x\, e^{iE_f t - ip_f \cdot x} \sigma(x, t)\, e^{-iE_i t + ip_i \cdot x} \right|^2 \right\rangle$$

$$= C(k, \omega)\, V \tag{12.10}$$

instead of (1.5). Here $k = p_f - p_i$, $\omega = E_f - E_i$. $C(k, \omega)$ is often called the "dynamic structural factor."

The linear response function $G(k, \omega)$ is defined by changing $\mathcal{H}$ to

$$\mathcal{H}' = \mathcal{H} - \sum_k h_k \sigma_{-k} \tag{12.11}$$

as was done in (11.27). This changes the first equation of (12.1) to

$$\frac{\partial \sigma_k}{\partial t} = v'_k - \Gamma_k(\partial \mathcal{H}/\partial \sigma_{-k}) + \Gamma_k h_k + \zeta_k \ . \tag{12.12}$$

Note that v'_k is generally not the same as v_k. Now the response function $G(k, \omega)$ is defined by

$$\langle \sigma_k(\omega) \rangle = G(k, \omega)\, h_k(\omega) + O(h^2) \tag{12.13}$$

for small h .

$G(k, \omega)$ and $C(k, \omega)$ defined by (12.9) are simply related by

$$C(k, \omega) = \frac{2}{\omega} \operatorname{Im} G(k, \omega) \tag{12.14}$$

$$G(k, \omega) = - \int \frac{d\omega'}{2\pi} \frac{C(k, \omega')\omega'}{\omega - \omega' + io^+} \ , \tag{12.15}$$

$$G(k, 0) = \int \frac{d\omega'}{2\pi} C(k, \omega')$$

$$= C(k, t=0) = \langle \sigma_k(t') \sigma_{-k}(t') \rangle = G(k) \ . \tag{12.16}$$

Equation (12.16) says that the static correlation function $G(k)$ studied in previous chapters, which is the instantaneous correlation function, is the same as the zero frequency response function $G(k, 0)$. It is not easy to prove that (12.14), the fluctuation dissipation theorem, follows from (12.1) and (12.12). A proof by perturbation theory for some models will be given in Sec. XIV.3.

As far as those modes which are not eliminated in carrying out R_s are concerned, R_s is simply a change of name of the modes and no physical content of the kinetic equations is altered. Since $\sigma_k(t)$ is replaced by $s^{1-\eta/2}\sigma_{sk}(ts^{-z})$ in Step (ii), we must have

$$\langle \sigma_k(t)\,\sigma_{-k}(0)\rangle_\mu = s^{2-\eta}\langle \sigma_{sk}(ts^{-z})\,\sigma_{-sk}(0)\rangle_{\mu'} \,, \tag{12.17}$$

where the subscripts μ, μ' denote that the quantities are computed with kinetic equations specified, respectively, by μ and $\mu' = R_s\mu$. Equation (12.17) is simply a generalization of (5.16). In an obvious notation, the Fourier transform of (12.17) reads

$$C(k, \omega, \mu) = s^{2-\eta+z}\, C(sk, \omega s^z, R_s\mu) \,. \tag{12.18}$$

In view of (12.14) or (12.15), we have

$$G(k, \omega, \mu) = s^{2-\eta} G(sk, \omega s^z, R_s \mu) . \qquad (12.19)$$

Equations (12.18) and (12.19) are all that the RG can tell us about the correlation and response functions. As we have emphasized before, the RG is a set of transformations. It does not solve the model for us and cannot tell us what the explicit form of G or C is.

3. FIXED POINTS, CRITICAL BEHAVIOR, AND DYNAMIC SCALING

The application of the RG to critical dynamics is similar to its application to statics. The discussion of Chapter VI is easily extended to dynamics.

The fixed point μ^* is invariant under R_s. That is,

$$\mu^* = R_s \mu^* . \qquad (12.20)$$

The constants η and z in Step (ii) [see (12.6)] are adjusted so that (12.20) has a solution, i.e., one in general cannot find a fixed point unless η and z assume certain values.

Again, just as in (6.2), the critical surface of the fixed point μ^* is defined as the subspace of the parameter space on which any point μ has the property

$$\lim_{s \to \infty} R_s \mu = \mu^* . \tag{12.21}$$

The linearized transformation R_s^L and its general properties are formally derived and expressed by the same equations as (6.3) - (6.11). The fundamental hypothesis linking the RG to critical phenomena is still expressed by (6.12). The whole formal discussion from (6.12) through (6.23) is applicable to dynamics with some straightforward modification of results. We shall not repeat the discussion and the equations (the reader should review Chapter VI if necessary), but only note the modified results. Equation (6.18) now becomes

$$\begin{aligned} G(k, \omega, \mu(T)) &= s^{2-\eta} G(sk, s^z \omega, R_s \mu) \\ &= s^{2-\eta} G(sk, s^z \omega, \mu^* \pm (s/\xi)^{1/\nu} e_1 + O(s^{y_2})) \end{aligned} \tag{12.22}$$

for large s. Formally, the only new ingredients are the

new argument ω and the new exponent z. It must be noted, however, that μ^* and e_1 in (12.22) are entities defined in a larger parameter space than that in statics and thus do not mean the same quantities as the symbols μ^* and e_1 in (6.18). The exponent y_2 might not be the same as that in (6.18). Of course, (12.22) must say the same thing as (6.18) if we set $\omega = 0$, in view of (12.16). All parameters, except those specifying $\mathcal{H}$, must drop out when we set $\omega = 0$. This provides a nontrivial check in practical calculations. Here we assume that the exponent $1/\nu$, assumed to be the only positive exponent, is the same as that given by statics. Also $\xi \propto |T - T_c|^{-\nu}$ as before. Setting $s = \xi$ in (12.22), we get the generalization of (6.19):

$$G(k, \omega, \mu(T)) = \xi^{2-\eta} G(\xi k, \xi^z \omega, \mu^* \pm e_1 + O(\xi^{y_2})) \quad . \tag{12.23}$$

If $O(\xi^{y_2})$ can be ignored, then

$$G(k, \omega, \mu(T)) \approx \xi^{2-\eta} g(\xi k, \xi^z \omega) \quad , \tag{12.24}$$

i.e., G is $\xi^{2-\eta}$ times a function of ξk and $\xi^z \omega$. This is a statement of the dynamic scaling hypothesis, which is a direct generalization of the scaling hypothesis in statics. At a fixed value of ξk, the function g, as a function of ω, has a characteristic frequency scale ξ^{-z}. Of course, we cannot say more unless more information is provided. For example, suppose that we somehow knew that $G(k, \omega, \mu(T))$ as a function of ω is a single peak at $\omega = 0$. Then the dynamic scaling hypothesis given above says that the width of the peak is a function of ξk times ξ^{-z}. Equation (12.24) implies a characteristic time

$$\tau_k = \xi^z f(k\xi) \qquad (12.25)$$

where $f(x)$ is some function independent of temperature.

At $T = T_c$, we set $s = k^{-1}$ in (12.22) to obtain, for very small k,

$$G(k, \omega, \mu(T_c)) = k^{-2+\eta} G(1, \omega k^{-z}, \mu^* + O(k^{-y_2})) . \qquad (12.26)$$

If we can ignore $O(k^{-y_2})$, we obtain

$$G(k, \omega, \mu(T_c)) \approx k^{-2+\eta} \tilde{g}(\omega k^{-z}) \ . \tag{12.27}$$

This is the generalization of (6.20) and (6.21). Thus, at $T = T_c$, Eq. (12.27) says that the characteristic time is

$$\tau_k \propto k^z \ . \tag{12.28}$$

When additional information concerning the response function or the correlation function is supplied, for example, via hydrodynamic considerations, the dynamic scaling hypothesis (12.24) and also (12.27) enable one to obtain a great deal more information. A very nice discussion of dynamic scaling and its applications has been given by Halperin and Hohenberg (1969).

The ideas and conclusions discussed above are indeed plausible and straightforward extensions of the static RG. Their discussion, however, has been formal and without substantiation. So far there has been no serious numerical attempt to apply the dynamic RG, but have limited explicit realization of these ideas and conclusions to a few simple models where perturbation methods are applicable.

In the next chapter we shall discuss some such simple dynamic models and the results of applying the RG to them. We shall see that even though there is no new concept put into the dynamic RG, there is a great deal of new structure which was absent in the static RG.

XIII. SIMPLE DYNAMIC MODELS

SUMMARY

In this chapter we apply RG ideas to the study of a few selected simple models, including time-dependent Ginzburg-Landau models (TDGL models), a model with slow heat conduction, and a model of the Heisenberg ferromagnet. The physical basis of these models will be explained at length. The application of the RG and its results will be discussed, but extended calculations leading to some of the results are not included in this chapter. They will be included in the next chapter where technical information will be systematically presented.

1. THE TIME-DEPENDENT GINZBURG-LANDAU MODELS (TDGL)

A model described by (12.1) with $v_k = 0$, i.e., without mode-mode coupling, is often referred to as a "time-dependent Ginzburg-Landau model" (or a TDGL model), when $\mathcal{H}$ takes the Ginzburg-Landau form. Let us write out the kinetic equations explicitly:

$$\frac{\partial \sigma_{ik}}{\partial t} = -\Gamma_k \frac{\partial \mathcal{H}}{\partial \sigma_{i-k}} + \zeta_{ik}$$

$$= -\Gamma_k \left[(r_o + k^2)\sigma_{ik} + \frac{1}{2} u L^{-d} \sum_{jk'k''} \sigma_{jk'} \sigma_{jk''} \sigma_{ik-k'-k''} \right]$$

$$+ \zeta_{ik} \, , \tag{13.1a}$$

$$\langle \zeta_{ik}(t) \zeta_{jk'}(t') \rangle = 2\Gamma_k \delta_{ij} \delta_{-kk'} \delta(t - t') \, , \tag{13.1b}$$

where we have generalized (12.1) slightly to include multi-component spins ($i, j = 1, 2, \ldots, n$). As was mentioned in Chapter XI, if the total spin is conserved, we have $\Gamma_k \propto k^2$. Otherwise Γ_k can be taken as a constant. Let us distinguish these two cases by calling

$$\Gamma_k = \Gamma \qquad \text{model (a)}$$

$$\Gamma_k = \gamma k^2 \qquad \text{model (b)} \tag{13.2}$$

where Γ, γ are constants.

Let us restrict our attention to model (a) for the moment. This model can be viewed as describing a set of spins, each of which is in contact with a large thermal reservoir. This view can be understood more easily if we write (13.1) in terms of block spin variables $\sigma_i(x, t)$

$$\frac{\partial \sigma_i}{\partial t} = -\Gamma \frac{\delta \mathcal{H}}{\delta \sigma_i} + \zeta_i \ ,$$

$$\langle \zeta_i(x,t)\, \zeta_j(x',t') \rangle = \delta_{ij}\, \delta(x-x')\, \delta(t-t') \ . \tag{13.3}$$

Here $\delta(x-x')$ has a spatial resolution of $b \sim \Lambda^{-1}$, the block size and $\delta \mathcal{H}/\delta\sigma_i$ is $b^{-d}\, \partial \mathcal{H}/\partial \sigma_{ix}$ in terms of the discrete block spin variable σ_{ix}. Thus each block has an independent noise source. The dynamics is generated by these noise sources, or reservoirs. The condition of the reservoirs does not depend on the spin configuration. This means that heat conduction through unspecified processes must be sufficiently fast. Otherwise local energy build-up or depletion may invalidate the conclusions of (13.1), as we shall discuss in detail in the next section. We now turn to the application of the RG.

This model needs only one more parameter than those specifying $\mathcal{H}$, namely Γ. Since (13.1) is consistent with statics, the only additional formula we need for $\mu' = R_s \mu$ is the formula for the transformation of Γ. The rest of the RG is furnished by statics.

We define

$$\mu = (\Gamma, \bar{\mu}) \tag{13.4}$$

where $\bar{\mu}$ is the set of parameters specifying $\mathcal{H}$, namely $\bar{\mu} = (r_o, u)$ for the Ginzburg-Landau model.

For $d > 4$, we know that the static fixed point is at $\mu^* = 0$. For $\mu = (\Gamma, 0)$, the RG transformation $\mu' = R_s \mu$ is trivial to work out since Step (i), the elimination of σ_q with $\Lambda/s < q < \Lambda$, is just the dropping of equations for σ_q. Step (ii) is the replacement of $\sigma_k(t)$ by $s\sigma_{sk}(ts^{-z})$ in the remaining equations:

$$\frac{\partial}{\partial t} s\sigma_{isk}(ts^{-z}) = -\Gamma k^2 s\sigma_{isk}(ts^{-z}) + \zeta_{ik}(t) . \quad (13.5)$$

We have set $\eta = 0$. Now we want to write the transformed equation (13.5) in the old form. Let us define

$$sk = k' ,$$

$$s^{-z}t = t' ,$$

$$s^{-1+z} \zeta_{ik'/s}(t' s^z) = \zeta'_{ik'}(t') . \quad (13.6)$$

Then (13.5) becomes

$$\frac{\partial}{\partial t'} \sigma_{ik'}(t') = -\Gamma s^{z-2} k'^2 \sigma_{ik'}(t') + \zeta'_{ik'}(t') ,$$

$$\langle \zeta'_{ik'}(t') \zeta'_{jk''}(t'') \rangle = 2\Gamma s^{z-2} \delta(t'-t'') \delta_{ij} \delta_{-k'k''} . \quad (13.7)$$

Clearly, the transformation formula for Γ is

$$\Gamma' = \Gamma s^{z-2} . \quad (13.8)$$

If we want to have a fixed point, we must have

$$z = 2 \quad (13.9)$$

to keep $\Gamma' = \Gamma$. As long as $z = 2$, $\mu^* = (\Gamma, 0)$ is a fixed point regardless of the value of Γ. In fact, Γ can be changed by simply redefining the unit of time. The choice of a particular z is simply to keep Γ constant. This is similar to choosing the right η to keep the coefficient of $(\nabla\sigma)^2$ constant in the static case.

The fixed point $\mu^* = (\Gamma, 0)$ will be called the trivial fixed point. It is a stable fixed point for $d > 4$ for this model, but not necessarily so when additional terms are kept in the kinetic equations.

For $d < 4$, the trivial fixed point $\bar{\mu}^* = 0$ for the static RG is no longer stable. This necessarily implies that $\mu^* = (\Gamma, 0)$ [or $(\gamma, 0)$] is no longer stable for the dynamic RG.

For $d = 4 - \epsilon$ with small ϵ, one can use a perturbation expansion in ϵ to determine the RG, fixed points, and exponents. The determination of nontrivial fixed points requires some tedious calculations, which will be discussed in detail in the next chapter. Here we shall merely summarize the results.

To $O(\epsilon)$, we simply find the fixed point

$$\begin{aligned} \mu^* &= (\Gamma, \bar{\mu}^*) , \\ z &= 2 , \end{aligned} \tag{13.10}$$

where $\bar{\mu}^*$ is the nontrivial fixed point for the static RG discussed in Chapter VII.

To $O(\epsilon^2)$, calculation shows that

$$\begin{aligned} z &= 2 + c\eta , \\ c &= 6 \ln (4/3) - 1 . \end{aligned} \tag{13.11}$$

In the limit of $n \to \infty$, and $2 < d < 4$ the RG can also be worked out. Again one obtains (13.10) with $\bar{\mu}^*$ being the fixed point for the static RG in this limit. Of course, one needs to consider more general forms of $\mathcal{H}$ than the Ginzburg-Landau form in this case. To $O(1/n)$ one finds $z = 2 + c\eta$ with

$$c = \left(\frac{4}{4-d}\right)\left[\frac{B\left(\frac{d}{2}-1, \frac{d}{2}-1\right)d}{8\int_0^{1/2} dx\,[x(2-x)]^{d/2-2}} - 1\right] , \tag{13.12}$$

which is $\frac{1}{2}$ for $d = 3$ and 0 for $d = 2$, and the value

(13.11) for $d \to 4$ from below.

For model (b), i.e., the case of conserved total spin with $\Gamma_k = \gamma k^2$, the above analysis can be carried out the same way except for minor modifications. One defines $\mu = (\gamma, \bar{\mu})$. One easily finds a trivial fixed point $\mu^* = (\gamma, 0)$ with

$$z = 4$$

instead of $z = 2$. This is easily seen by replacing Γ by γk^2 in (13.5) and then repeating (13.6). For $d > 4$, the trivial fixed point is stable, while for $d < 4$ it is unstable.

For $d < 4$, the nontrivial fixed point again has the form (13.10) except that

$$z = 4 - \eta \quad . \tag{13.13}$$

This is a result mentioned in Chapter XI in connection with the van Hove theory.

The above results (13.10) - (13.13) [Halperin, Hohenberg and Ma (1972)] give an estimate of the correction from the σ^4 term in $\mathcal{H}$ to the van Hove theory discussed in Sec. XI.6. The next model is a simple illustration of the effects of additional modes.

2. EFFECTS OF SLOW HEAT CONDUCTION

We proceed to devise a model to study the effect of slow heat conduction. The physical basis for this model is explained as follows.

Again imagine a block spin $\sigma(x)$ which is the net spin density in the block located at x. The "reservoir" in direct contact with $\sigma(x)$ consists of all motions within the block, including lattice vibrations, spin fluctuations of wavelengths shorter than the block size, electronic motions, etc. We have approximated the net effect of the reservoir by a noise in the preceding models. The noise has been assumed to be rapidly varying and has zero correlation time. Now if the conduction of heat is slow, there will be slow accumulation and depletion of energy and thereby the effect can no longer be included as a part of the noise of short correlation time and must be singled out.

Let $\rho(x, t)$ be the energy in the reservoir at the block x. We write a phenomenological $\tilde{\mathcal{H}}$ including ρ as new modes:

$$\tilde{\mathcal{K}}[\sigma, \rho] = \frac{1}{2}\int d^d x \left[(\nabla\sigma)^2 + r_o\sigma^2 + \frac{1}{4}\tilde{u}\sigma^4 + w\rho^2 + g\rho\sigma^2\right] \tag{13.14}$$

where g, w are positive constants. Now $\exp(-\tilde{\mathcal{K}})$ is the joint equilibrium probability distribution for the spin configuration σ and the energy ρ in local reservoirs. We have assumed a simple Gaussian distribution $e^{-\frac{w}{2}\rho^2}$ for ρ in every block in the absence of σ. The coupling term $g\rho\sigma^2$ in $\tilde{\mathcal{K}}$ can be viewed as a local change of r_o by an amount $g\rho(x, t)$. Since r_o is roughly linear in temperature, we can view this coupling term as describing the effect of local fluctuations in temperature.

For static properties involving only σ, we do not need to keep track of ρ. We can integrate out ρ from the joint probability distribution $\exp(-\tilde{\mathcal{K}})$ to obtain

$$e^{-\mathcal{K}[\sigma]} = \int \delta\rho \; e^{-\tilde{\mathcal{K}}[\sigma, \rho]} \quad . \tag{13.15}$$

This means integrating over ρ for every block. The integrals are independent and Gaussian. We recover the Ginzburg-Landau $\mathcal{K}$

$$\mathcal{H}[\sigma] = \frac{1}{2}\int d^dx \left[(\nabla\sigma)^2 + r_o\sigma^2 + \frac{1}{4}u\sigma^4 \right]$$

$$u \equiv \tilde{u} - g^2/w \ . \qquad (13.16)$$

Now we can write down the kinetic equations for the Fourier components of σ and ρ following the general form of (11.8):

$$\frac{\partial \sigma_k}{\partial t} = -\Gamma_k \frac{\partial \tilde{\mathcal{H}}}{\partial \sigma_{-k}} + \zeta_{ik} \ , \qquad (13.17a)$$

$$\frac{\partial \rho_k}{\partial t} = -Dk^2 \frac{\partial \tilde{\mathcal{H}}}{\partial \rho_{-k}} + \varphi_k \ , \qquad (13.17b)$$

$$\langle \zeta_{ik}(t)\, \zeta_{jk'}(t') \rangle = 2\delta_{ij}\, \delta_{-kk'}\, \delta(t-t')\, \Gamma_k \ , \qquad (13.17c)$$

$$\langle \varphi_k(t)\, \varphi_{k'}(t') \rangle = 2\delta_{-kk'}\, \delta(t-t')\, Dk^2 \ . \qquad (13.17d)$$

All Fourier components are restricted to $k < \Lambda$. Equations (13.14) and (13.17) define a model, which is just the TDGL model with one more set of modes, ρ_k, added. In coordinate representation, (13.17b) reads

$$\frac{\partial \rho}{\partial t} = Dw \, \nabla^2 (\rho + g\sigma^2/2w) + \varphi \ . \qquad (13.18a)$$

This equation is somewhat more complicated than the usual diffusion equation for heat conduction, but its physical meaning is quite transparent. It says that the rate of change of the reservoir energy ρ in a block is due to the influx of energy from neighboring blocks and a random source within the block. The influx depends on the <u>total energy</u>, i.e., the reservoir energy ρ plus $g\sigma^2/2w$, interpreted as the energy of the block spin σ, in the neighboring blocks. This point can be made clearer by integrating (13.18a) over a finite region and applying Gauss's theorem:

$$\frac{\partial}{\partial t} \int d^d x \, \rho = - \int dS \cdot j + \int d^d x \ \varphi \ ,$$

$$j = -Dw \ \nabla (\rho + g\sigma^2/2w) \quad . \qquad (13.18b)$$

Here j is the flux of energy by diffusion and Dw is the diffusion coefficient. The integral $\int dS$ is taken over the surface enclosing the region. Note that the influx of energy j does not go directly into the energy of the block spin $g\sigma^2/w$, but only into the reservoir energy ρ, which in turn affects σ. Of course, σ is directly affected by neighboring spins via (13.17a).

Equation (13.17) defines the simplest extension of the previous models to include the effect of heat conduction. There is still no organized motion in this model since no mode-mode coupling term has been included. Compared to the TDGL models discussed earlier, this model simply has more complicated noises and dissipation.

For the modes ρ_k, we define the correlation function $K(k, \omega)$ in the same manner as (12.7) - (12.9):

$$\langle \rho_k(\omega)\, \rho_{k'}(\omega') \rangle = 2\pi\, \delta(\omega + \omega')\, \delta_{-kk'}\, K(k, \omega) \qquad (13.19)$$

$$K(k, \omega) = \int dt\; e^{i\omega t} \langle \rho_k(t)\, \rho_{-k}(0) \rangle \; . \qquad (13.20)$$

The response function $F(k, \omega)$ is defined in the same manner as (12.11) - (12.13). We add a small term $-\sum_k \beta_k \rho_{-k}$ to $\tilde{\mathcal{H}}$, compute $\langle \rho_k(\omega) \rangle$, and then define $F(k, \omega)$ by

$$\langle \rho_k(\omega) \rangle = \beta_k(\omega)\, F(k, \omega) + O(\beta^2) \; . \qquad (13.21)$$

$K(k, \omega)$ and $F(k, \omega)$ are related by

$$K(k, \omega) = \frac{2}{\omega} \text{ Im } F(k, \omega) \ . \qquad (13.22)$$

We now turn to the RG study of (13.17). The presence of new modes ρ_k requires some minor extensions of the definition of the RG given in Sec. XII. 1. In Step (i), we need to eliminate the modes ρ_q, $\Lambda/s < q < \Lambda$ as well as σ_q. In Step (ii), we need in addition the replacement

$$\rho_k(t) \rightarrow \rho_{sk}(ts^{-z})\, s^y \qquad (13.23)$$

where y is still another exponent like η to be adjusted to achieve a fixed point for the RG. This adjustment, however, will not be sufficient to achieve a fixed point. A further adjustment of ρ by including in it an additive term will be necessary. This additive term is simply the energy of the eliminated modes. Note that when we eliminate a mode, we push it into the reservoir. Since the new ρ is the energy in the new reservoir, it must include the contribution of the eliminated modes to assure an unchanged physical meaning.

These additional features brought in by ρ make the structure of the RG much more complicated. There will be a variety of fixed points as we shall see shortly.

The RG for this model has been worked out only for $d = 4 - \epsilon$ to $O(\epsilon)$. [See Halperin, Hohenberg and Ma (1974).] Let us go through some of the calculcations.

There are six parameters in the model. Each point in the parameter space is a set of six parameters:

$$\mu = (\Gamma, D, g, w, r_o, \tilde{u}) \quad . \tag{13.24}$$

The transformation R_s takes μ to

$$R_s\mu = \mu' = (\Gamma', D', g', w', r_o', \tilde{u}') .$$

Following the steps (i) and (ii), we obtain

$$\Gamma' = s^{z-2}\,\Gamma\left(1 - \frac{\Gamma}{\Gamma + Dw}\,\frac{g^2}{w}\,K_4 \ln s\right) , \tag{13.25a}$$

$$D'w' = s^{z-2}\,Dw\left(1 - \frac{n}{2}\,\frac{g^2}{w}\,K_4 \ln s\right) , \tag{13.25b}$$

$$\frac{g'}{w'} = s^{\epsilon/2-y}\,\frac{g}{w}\left(1 - \left(\frac{n}{2} + 1\right)\left(\tilde{u} - \frac{g^2}{w}\right) K_4 \ln s\right) , \tag{13.25c}$$

$$g' = s^{\epsilon/2+y} g \left(1 - \left(\left(\frac{n}{2}+1\right)\tilde{u} - \frac{g^2}{w}\right) K_4 \ln s\right) \qquad (13.25d)$$

$$r_o' = s^2 \left[r_o + \left(\frac{n}{2}+1\right)\left(\tilde{u} - \frac{g^2}{w}\right) K_4 \left(\frac{\Lambda^2}{2}(1-s^{-2}) - r_o \ln s\right)\right]$$
(13.25e)

$$\tilde{u}' = s^{\epsilon}\left[\tilde{u} - \left(\left(\frac{n}{2}+4\right) u^2 + 2g^4/w^2 - 6g^2\tilde{u}/w\right) K_4 \ln s\right] .$$
(13.25f)

Equations (a), (c), (e), and (f) are obtained from the kinetic equation for σ_{ik}, and (b) and (d) from that for ρ_k. The dynamics is contained only in (a) and (b). The rest can be obtained directly from the static RG. The additional adjustment made for ρ as mentioned earlier is

$$\rho + \frac{g}{2w} \langle \phi^2 \rangle \rightarrow \rho$$

$$\langle \phi^2 \rangle = nK_4 \int_{\Lambda/s}^{\Lambda} dq\, q^{d-1} (r_o + q^2)^{-1} . \qquad (13.26)$$

In other words, the new ρ now includes $g \frac{1}{2w} \langle \phi^2 \rangle$, which is the contribution to the energy from the eliminated modes to first order in g. In the above, D and Γ are assumed to be of $O(1)$, and g^2, r_o, u are assumed to be of $O(\epsilon)$.

Equation (13.25) can be put into a simpler form. Let us introduce the notation

$$\lambda = g/\sqrt{w} \ ,$$

$$u = \tilde{u} - g^2/w = \tilde{u} - \lambda^2 \ , \qquad (13.27)$$

$$\Gamma/(Dw) = f \ .$$

The meaning of u is the same as that in statics, as explained by (13.16). We obtain from (13.25)

$$f' = f\left(1 + \left(\frac{n}{2} - \frac{f}{1+f}\right)\lambda^2 K_4 \ln s\right) , \qquad (13.28a)$$

$$D' = s^{-2+z-2y} D \ , \qquad (13.28b)$$

$$w' = s^{2y} w \left(1 - \frac{n}{2}\lambda^2 K_4 \ln s\right) \ , \qquad (13.28c)$$

$$\lambda' = s^{\epsilon/2} \lambda\left(1 - \left(\left(\frac{n}{2}+1\right) u + \frac{n}{4}\lambda^4\right) K_4 \ln s\right) . \qquad (13.28d)$$

The equations for u' and r_o' become those given by the static RG [see (7.59), (7.69)], as expected. To search for fixed points, let us first set (r_o, u) equal to its stable fixed point value

$$r_o^* = -\left(\frac{n+2}{n+8}\right)\frac{1}{2}\,\epsilon\,\Lambda^2$$

$$u^* = \frac{2\epsilon}{(n+8)K_4} \tag{13.29}$$

as given by statics [see (7.71)], and then determine the fixed point values of the other parameters.

For a fixed point with $\lambda^* \neq 0$, we must have, in view of (13.28d),

$$\lambda^{*2} = \frac{2}{n}\left(\frac{\epsilon}{K_4} - (n+2)u^*\right)$$

$$= \frac{2}{n}\left(\frac{4-n}{n+8}\right)\frac{\epsilon}{K_4}\ , \quad n < 4\ , \tag{13.30}$$

which makes sense only if $n < 4$ because $\lambda^2 \propto g^2$ must be positive as defined by (13.14) and (13.27). Given (13.30), we obtain the exponent y from (13.28c) by setting $w = w'$:

$$2y = \frac{n}{2}\,\lambda^{*2}K_4 = \left(\frac{4-n}{n+8}\right)\epsilon\ . \tag{13.31}$$

This value happens to be that of the specific heat exponent α divided by ν. We shall elaborate upon this point later. Now suppose that f^* is finite and nonzero. Then (13.28a) tells us that

$$1/f^* = \frac{2}{n} - 1 \tag{13.32}$$

which makes sense only for $n < 2$. Since Γ or Dw can be used as a unit of inverse time, a finite D requires that

$$z = 2 + 2y = 2 + \alpha/\nu , \quad n < 2 . \tag{13.33}$$

So we have found a fixed point for $n < 2$ as given by (13.30) - (13.33). This fixed point is stable, as we easily obtain the linearized equations

$$\begin{aligned} \delta f' &= \delta f \; s^{y_f} , \\ \delta \lambda' &= \delta \lambda \; s^{y_\lambda} ; \end{aligned} \tag{13.34}$$

$$\begin{aligned} y_f &= -\left(\frac{2}{n} - 1\right)^2 ny , \\ y_\lambda &= -y , \end{aligned} \tag{13.35}$$

where $\delta f \equiv f - f^*$, $\delta\lambda \equiv \lambda - \lambda^*$; y_f and y_λ are both negative.

There is another fixed point, namely at $f^* = 0$, which means D is much larger than Γ near the fixed point. This is the situation of very large heat conduction described by the previous models. Since $D^* = \infty$, where Γ remains finite, we need to go back to (13.25a) to find that $z = 2$. This fixed point is unstable, since, from (13.28a) we get $\delta f' = \delta f \, s^{y_f}$ with

$$y_f = \frac{n}{2} \lambda^2 K_4 = 2y = \alpha/\nu > 0 \ . \qquad (13.36)$$

There is of course the fixed point at $f^* = \infty$. This is again unstable for $n < 2$, since

$$\delta f'^{-1} = \delta f^{-1} \, s^{(1 - n/2)\lambda^{*2}} \ , \qquad (13.37)$$

as obtained from (13.28a) for $1/f^* = 0$.

For $4 > n > 2$, this fixed point is evidently stable according to (13.37), and there is no fixed point for finite f^* [Eq. (13.32) breaks down if $n > 2$]. The $f^* = 0$ fixed point is still unstable since (13.36) holds for $n < 4$. The limit $f \to \infty$ describes the situation where the heat conduction is extremely slow compared to the spin fluctuation. The exponent z for $f^* = \infty$ is directly obtained from (13.25a) by setting $D = 0$:

$$z = 2 + 2\alpha/n\nu \tag{13.38}$$

Note that if $D \to 0$, then ρ_k with large k will also have long relaxation times whereas (13.25) assumes short relaxation times for large k modes. Thus the $D = 0$ fixed point has little meaning.

Finally, there is the fixed point with $\lambda^* = 0$, $y = 0$, $z = 2$. This fixed point is stable for $n > 4$ because (13.28d) gives

$$\delta\lambda' = \delta\lambda \ s^{\alpha/2\nu} ,$$

$$\alpha/\nu = \left(\frac{4 - n}{n + 8}\right)\epsilon \tag{13.39}$$

for $\lambda^* = 0$. This means that the coupling between the spins and the heat conduction is weakened by the transformation R_s and will not affect the critical behavior predicted by the time-dependent Ginzburg-Landau model discussed before.

The above results on fixed points are summarized in Table 13.1.

So far we have set $\Gamma_k = \Gamma$. For the case $\Gamma_k = \gamma k^2$, i.e., the case where total spin is conserved, the results are much simpler. Equation (13.25) remains the same except that (13.25a) is replaced by

$$\gamma' = s^{z-4} \gamma . \tag{13.40}$$

Thus for a fixed point we must have $z = 4$. This means $D^* = \infty$, i.e., infinite heat conduction. Therefore the previous model defined by (13.1), (13.2b) is adequate.

We conclude this section with a few remarks on the conspicuous role of the specific heat exponent α, as shown in Table 13.1.

As we have noted earlier, when we apply R_s, the transformed ρ will include energy fluctuations of the eliminated spin modes. If $\alpha > 0$, the energy fluctuation of spin modes is very large and will constitute the major part of the transformed ρ for large s. It is thus not surprising

Table 13.1

Fixed points and exponents to $O(\epsilon)$ for the model with slow heat conduction, $\Gamma_k = \Gamma$

Fixed Point	(i)	(ii)	(iii)	(iv)
λ^{*2}	$\frac{2\alpha}{n\nu K_4}$	$\frac{2\alpha}{n\nu K_4}$	$\frac{2\alpha}{n\nu K_4}$	0
f^*	$\left(\frac{2}{n} - 1\right)^{-1}$	∞	0	0
y_λ	$\frac{-\alpha}{2\nu}$	$\frac{-\alpha}{2\nu}$	$\frac{-\alpha}{2\nu}$	$\frac{\alpha}{2\nu}$
y_f	$-\left(\frac{2}{n} - 1\right)^2 \frac{n\alpha}{\nu}$	$\left(1 - \frac{n}{2}\right)\frac{2\alpha}{n\nu K_4}$	α/ν	0
y	$\frac{\alpha}{2\nu}$	$\frac{\alpha}{2\nu}$	$\frac{\alpha}{2\nu}$	0
z	$2 + \frac{\alpha}{\nu}$	$2 + \frac{2\alpha}{n\nu}$	2	2
Defined for	$n < 2$ only	$n < 4$ only	$n < 4$ only	Any n
Stable only for	$n < 2$	$2 < n < 4$	Unstable	$n > 4$

$$\alpha/\nu = \left(\frac{4 - n}{n + 8}\right)\epsilon$$

that the exponent for ρ, i.e., y, is just $\alpha/2\nu$ if $\alpha > 0$ [fixed points (i), (ii), (iii)]. If $\alpha < 0$, the spin energy fluctuation is not large and ρ will essentially be the original reservoir energy, and the exponent for ρ is zero [fixed point (iv)]. The fact that $y_f = 0$ for (iv) reflects the arbitrariness in choosing the unit of Γ.

The above results show that, if the spin energy fluctuation is large, i.e., $\alpha > 0$, slow heat conduction will modify the dynamics significantly, as expected. For the case $\Gamma_k = \gamma k^2$, the spin relaxation is so slow that even when $\alpha > 0$, the dynamics is not modified significantly.

The fact that $n = 4$ is the borderline for the sign of α applies only when d is very close to 4. From other calculations of α, we know that for $d = 3$ the borderline should be at an n less than 2 because α for $n = 2$ is already negative for $d = 3$. This suggests that the fixed point (ii), whose meaning is very dubious, may not exist for ϵ larger than some value less than 1. Just how large this value is remains unanswered. Of course, all results in Table 13.1 are correct to $O(\epsilon)$ and may not hold for $d = 3$. For more discussion, see Halperin,

Hohenberg, and Ma (1974).

3. THE ISOTROPIC FERROMAGNET

The above models did not include any regular or organized motion. There were no mode-mode coupling terms in the kinetic equations. The dynamics they described was strictly dissipative, generated by random noise. Now we examine how a model with mode-mode coupling transforms under the RG and how its critical behavior is affected by the presence of the mode-mode coupling terms. Consider a simple model of an isotropic ferromagnet constructed as follows.

The organized motion in a ferromagnet is the precession of spins. Let $\sigma_i(x, t)$ denote the spin configuration and $B_i(x, t)$ be the local magnetic field. Here σ and B are taken to be three-dimensional vectors, i.e., $i = 1, 2, 3$, $n = 3$. The equation describing the precession is well known:

$$\frac{\partial \sigma}{\partial t} = \lambda \sigma \times B \tag{3.41}$$

where λ is a constant (obviously unrelated to the λ in the previous model), and the cross means the usual cross product of three-dimensional vectors.

Of course, in addition to precession, there are still dissipation and noise. Now we take over the TDGL model (13.1) with $\Gamma_k = \gamma k^2$ and add a mode-mode coupling term given by (13.41):

$$\frac{\partial \sigma_{ik}}{\partial t} = v_{ik} - \gamma k^2 \frac{\partial \mathcal{H}}{\partial \sigma_{i\,-k}} + \zeta_{ik} , \qquad (13.42)$$

$$k < \Lambda, \quad i = 1, 2, 3,$$

where v_k is the Fourier component of $\lambda \sigma \times B$:

$$v_k = \lambda L^{-d} \sum_{k'} \sigma_{k+k'} \times B_{-k'} \quad . \qquad (13.43)$$

We need an expression for B, which is the sum of the external field h and the field provided by the neighboring spins. It is effectively given by

$$B_i = \nabla^2 \sigma_i + h_i \quad . \qquad (13.44)$$

Here $\nabla^2 \sigma(x)$ is proportional to the mean of spins in the

$2 \times d$ neighboring blocks minus the block spin itself at x. The part of the local field contributed by the block spin itself is proportional to $\sigma(x)$ and does not affect $\sigma \times B$ since $\sigma \times \sigma = 0$. More precisely, B is given by

$$B_i(x) = -\delta \mathcal{H} / \delta \sigma_i(x) \tag{13.45}$$

which is (13.44), apart from terms proportional to σ. Finally, we substitute (13.44) in (13.43) to obtain

$$v_k = \lambda L^{-d/2} \sum_{k'} \sigma_{k-k'} \times (h_{k'} - k'^2 \sigma_{k'}) . \tag{13.46}$$

Equations (13.42) and (13.46) define our model completely. We use $\Gamma_k = \gamma k^2$ in (13.42) since the total spin is a conserved quantity. The effect of slow heat conduction was shown to be unimportant for $\Gamma_k = \gamma k^2$ in the preceding models, so we do not include it here.

The only new parameter is λ. We write

$$\mu = (\lambda, \gamma, r_o, u) . \tag{13.47}$$

The fixed point found in the beginning of this chapter is for $\lambda^* = 0$:

$$\mu_o^* = (0, \gamma, r_o^*, u^*) ,$$

$$z = 4 , \tag{13.48}$$

valid to $O(\epsilon)$ for $d = 4 - \epsilon$, with (r_o^*, u^*) denoting the stable static fixed point, and also valid for $d > 4$ (with $r_o^* = u^* = 0$).

To check whether μ_o^* is stable, we need to find the linearized transformation law of λ under the RG. This is easily found since Step (i), the elimination of modes, plays no part in the calculation. Step (ii) makes the replacement

$$\frac{\partial \sigma_k}{\partial t} \rightarrow \frac{\partial \sigma_{k'}}{\partial t'} s^{-z+1} , \tag{13.49}$$

$$v_k \rightarrow v_{k'} s^{-d/2} \tag{13.50}$$

with $k' = sk$, and we must write $L = L's$. Equation (13.50) follows from (13.46). Substituting (13.49) and (13.50) in (13.42) we get

$$\lambda' = \lambda s^{z-1-d/2} = \lambda s^{3-d/2} , \tag{13.51}$$

since $z = 4$. Clearly $3 - d/2 > 0$ and μ_o^* is unstable for $d < 6$ and stable for $d > 6$. The borderline is at $d = 6$. The $\epsilon = 4 - d$ expansion around $d = 4$ therefore will not help in searching for a stable fixed point. As a first step in looking for and studying the stable fixed point for $d < 6$, we shall consider $d = 6 - \epsilon$ with small positive ϵ.

The stable static fixed point for $d > 4$ is the trivial fixed point. For the purpose of searching for dynamic fixed points, it is sufficient to work out the RG for $\mu' = R_s \mu$ with $r_o = u = 0$. Going through Steps (i) and (ii), we obtain the formulas via calculations to $O(\epsilon)$:

$$\lambda' = s^{z-4+\epsilon/2} \lambda \; , \tag{13.52a}$$

$$\gamma' = s^{z-4} \gamma \left(1 + \frac{1}{3} \tilde{\lambda}^2 K_6 \ln s\right) \; , \tag{13.52b}$$

where

$$K_6 = (64 \pi^3)^{-1} \; , \tag{13.53}$$

$$\tilde{\lambda} = \lambda / \gamma \; . \tag{13.54}$$

Combining the two equations in (13.52), we obtain

$$\tilde{\lambda}' = s^{\epsilon/2} \lambda \left(1 - \frac{1}{3} \tilde{\lambda}^2 K_6 \ln s\right) . \qquad (13.55)$$

Clearly, there is the unstable trivial fixed point μ_o^*, mentioned earlier, with $\lambda^* = 0$, $z = 4$. There are also two nontrivial fixed points given by

$$\tilde{\lambda}^* = \pm \left(\frac{3\epsilon}{2K_6}\right)^{1/2} = \pm (96\pi^3 \epsilon)^{1/2} ,$$

$$z = 4 - \epsilon/2 = 1 + d/2 . \qquad (13.56)$$

These two fixed points differ only by the sign of λ^*. The sign of λ can be reversed by simply reversing the signs of σ and h and has no special significance. We can simply ignore the fixed point with negative λ^*.

The nontrivial fixed point is stable, as one easily finds

$$\delta\tilde{\lambda}' = \delta\tilde{\lambda}\, s^{-\epsilon} \qquad (13.57)$$

by linearizing (13.55).

What we have just shown is that a mode-mode coupling term can make a great difference in the fixed point and the dynamic exponent z .

The above three models illustrate the corrections to the simple van Hove theory discussed near the end of Chapter XI. These corrections are due to the non-Gaussian nature of $\mathcal{H}$, the additional slow modes, and the organized motion, i.e., the mode-mode coupling. The RG approach allows us to treat these corrections as small when d is restricted to $4-\epsilon$ or $6-\epsilon$.

4. UNIVERSALITY IN CRITICAL DYNAMICS

From our study of the above models, it is evident that universality of critical behaviors is far more restricted in dynamics than it is in statics. That is to say, among the systems sharing the same static critical behaviors, there can be widely different dynamic critical behaviors. All the models studied above share the same fixed point and exponents under the static RG for fixed n and d. But when dynamics is taken into account, many fixed points with different exponents appear. The basic question is to what extent universality remains. So far it looks as if whenever we add something to the model, we get a new type of critical behavior. One might answer that there is not any universality. This is of course an overpessimistic answer. In the above discussions, we have devoted a great deal of attention to concrete physical pictures and details for each model with the result that it is easy to lose sight of the overall features. The most significant feature of these model studies is the strong evidence they present that the RG ideas given in Chapter XII are correct for dynamics as well as for statics. That is to say, universality of dynamic critical behaviors is determined by the properties of fixed points. Universality is observed for all systems sharing the same critical surface of a fixed point. What we need is to classify all fixed points of interest. This task is far from being accomplished. Let us summarize the main gross features by which we have learned so far to distinguish fixed points of the RG:

(a) Static fixed points

In our analysis above it is evident that a prerequisite for a fixed point for the full RG is that the static part of the RG have a fixed point. Thus, one has to classify all static fixed points first.

(b) Conservation laws

Conservation laws affect the transformation properties under the RG in a crucial manner. They determine whether a local disturbance can be randomized or relaxed locally or must spread out to an infinite region in order to disappear. Whenever large sizes enter into the picture, transformation properties under the RG are affected strongly. For the same static fixed point, different conservation laws can lead to different full fixed points.

(c) Mode-mode coupling

This is the regular or nonrandom aspect of dynamics and is probably very difficult to classify because of its variety. The random part of dynamics, i.e., dissipation and noise, is basically statistical in nature, limited only by conservation laws, and expected to depend only on the phase space considerations which determine static behaviors. The mode-mode coupling, however, does depend on details of the dynamics. The precession of spins in a ferromagnet is a very special feature, not shared by antiferromagnets, for example. The exponent $z = 1 + d/2$ for the ferromagnetic model is a consequence of the form of the precession term in the kinetic equation. Our task is thus to classify various features of mode-mode coupling terms which are relevant in determining the transformation properties under the RG.

XIV. PERTURBATION EXPANSION IN DYNAMICS

SUMMARY

We develop perturbation expansions for solving kinetic equations. Rules of calculation using graphs are systematically presented. The development runs parallel to that given in Chapter IX for statics. The main purpose of this chapter is to provide basic technical information to those readers who want to learn the hardware of perturbation calculations.

1. ITERATION SOLUTION OF KINETIC EQUATIONS

We begin with the time-dependent Ginzburg-Landau model given by (13.1). For the case $u = 0$, (13.1a) is simply

$$-i\omega\, \sigma^{o}_{ik}(\omega) = -\Gamma_k (r_o + k^2)\, \sigma^{o}_{ik}(\omega) + \zeta_{ik}(\omega) \;. \tag{14.1}$$

Define G_o as

$$G_o(k, \omega) = \left[r_o + k^2 - i\omega/\Gamma_k \right]^{-1} \;. \tag{14.2}$$

Then the solution of (14.1) is

$$\sigma^{o}_{ik}(\omega) = G_o(k, \omega)\, \zeta_{ik}(\omega)/\Gamma_k \;. \tag{14.3}$$

The correlation function for $u = 0$ is $C_o(k, \omega)$ defined by

$$\langle \sigma^{o}_{ik}(\omega)\, \sigma^{o}_{jk'}(\omega') \rangle = 2\pi\, \delta(\omega + \omega')\, \delta_{ij}\, \delta_{-kk'}\, C_o(k, \omega) \;. \tag{14.4}$$

Using (14.3) and the fact that

$$\langle \zeta_{ik}(\omega)\, \zeta_{jk'}(\omega') \rangle = 2\pi\, \delta(\omega + \omega')\, \delta_{ij}\, \delta_{-kk'}\, 2\Gamma_k \;, \tag{14.5}$$

which follows from (13.1b), we obtain

$$C_o(k, \omega) = \frac{2}{\Gamma_k}\, G_o(k, \omega)\, G_o(-k, -\omega) = \frac{2}{\omega}\, \mathrm{Im}\, G_o(k, \omega) \;. \tag{14.6}$$

The subscript or superscript o will always denote the solutions of the linear equation (14. 1), the "unperturbed equation."

For $u \neq 0$, (13. 1a) can be written in Fourier transformed form as

$$\sigma_{ik}(\omega) = \sigma^{o}_{ik}(\omega) - G_o(k, \omega)u$$

$$\times \frac{1}{2} L^{-d} \sum_{j, k', k''} \int \frac{d\omega'}{2\pi} \frac{d\omega''}{2\pi} \sigma_{jk'}(\omega') \sigma_{jk''}(\omega'') \sigma_{ik-k'-k''}(\omega - \omega' - \omega'')$$

$$+ G_o(k, \omega) h_{ik}(\omega) \quad . \qquad (14.7)$$

We have turned on an external magnetic field h, which gives rise to the last term in (14. 7). We can now iterate (14. 7) to obtain $\sigma_{ik}(\omega)$ in powers of u and h. This is the perturbation expansion. Note that since ζ_{ik} is assumed to be a Gaussian noise, σ^{o}_{ik} is, too, in view of (14. 3). The average of a product of σ^{o}'s equals the sum of products of all pairwise averages, just as in the static case. Each pairwise average is given by (14. 4) and (14. 6).

2. REPRESENTATION OF TERMS BY GRAPHS, RULES OF CALCULATION

As in the static case, the dynamic perturbation expansion can be represented by graphs. Here graphs represent terms generated by iterating (14.7), and then averaging over σ^o.

Equation (14.7) is represented by Fig. 14.1a. The thick short lines represent σ, the thin short lines represent σ^o. The longer thin lines represent G_o, and h is represented by a cross. The dashed line represents -u. The arrows keep track of the direction of iteration. Figure 14.1b shows the solution of (14.7) to first order in h and u, obtained by replacing σ, the thick short lines by σ^o, the thin short lines and $G_o h$, a thin line ending with a cross. Figure 14.1c gives second-order terms in u and first-order in h for σ.

It is easy to see that further iterations generate graphs like Fig. 14.2. Such graphs look like trees and will be called tree graphs. More complicated tree graphs are obtained from simpler ones by "growing more branches." Each tree graph begins with a G_o. All thin lines are G_o's

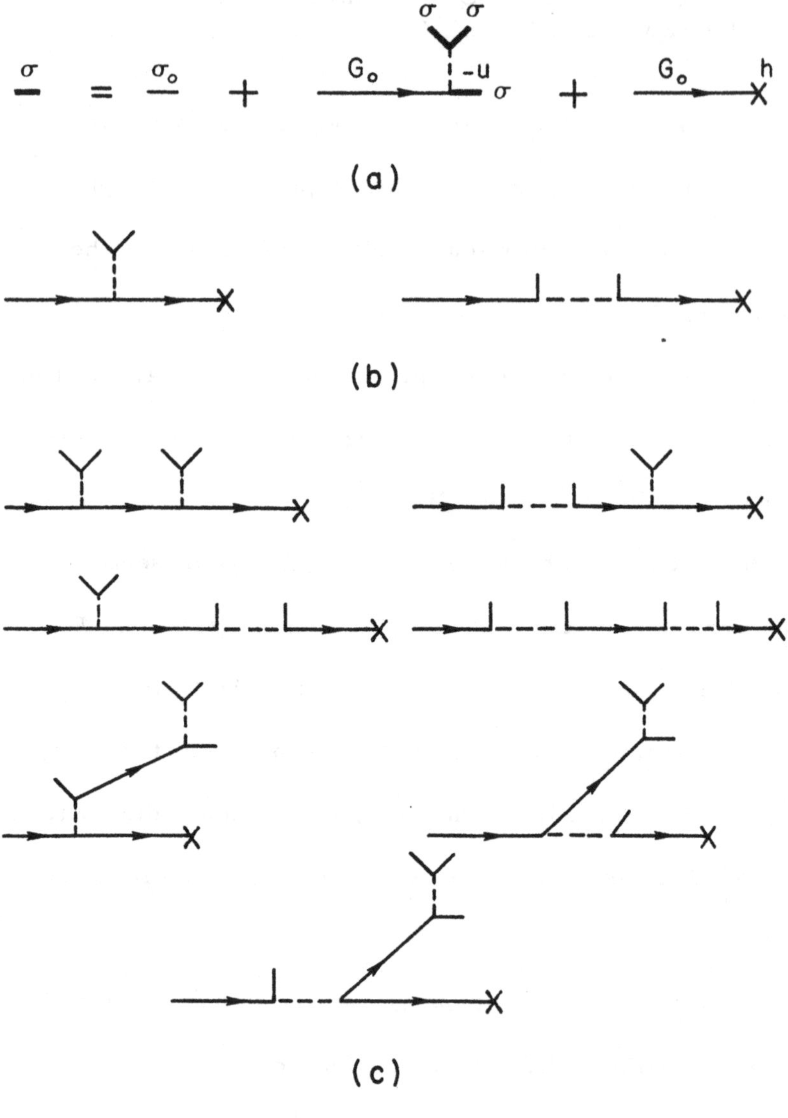

Figure 14.1. (a) Representation of (14.7) by graphs.
(b) Graphs for Gh of O(uh).
(c) Graphs for Gh of $O(u^2h)$.

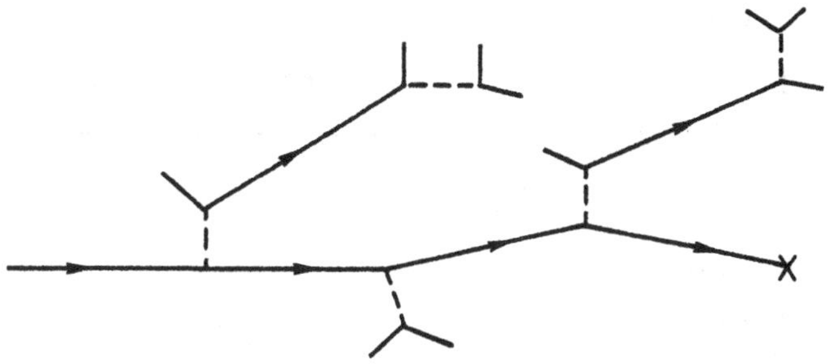

Figure 14.2. A tree graph for Gh.

except those short ones forming the ends of branches, which are σ^{o}'s.

When we want to take the average of the solution, we take pairwise averages of σ^{o}'s. For each pairwise average, we have a factor C_o, as we mentioned earlier. We can easily represent such averages graphically. We simply join the ends of branches in pairs in all possible ways. The lines formed by joining a pair of σ^{o}'s will be represented by a line with a small circle in the middle. Figure 14. 3 shows the averages of terms shown in Fig. 14. 1.

Clearly, the graphs look exactly the same as those in statics except that some of the lines now have circles on them. Intuitively, a plain line, which represents a G_o,

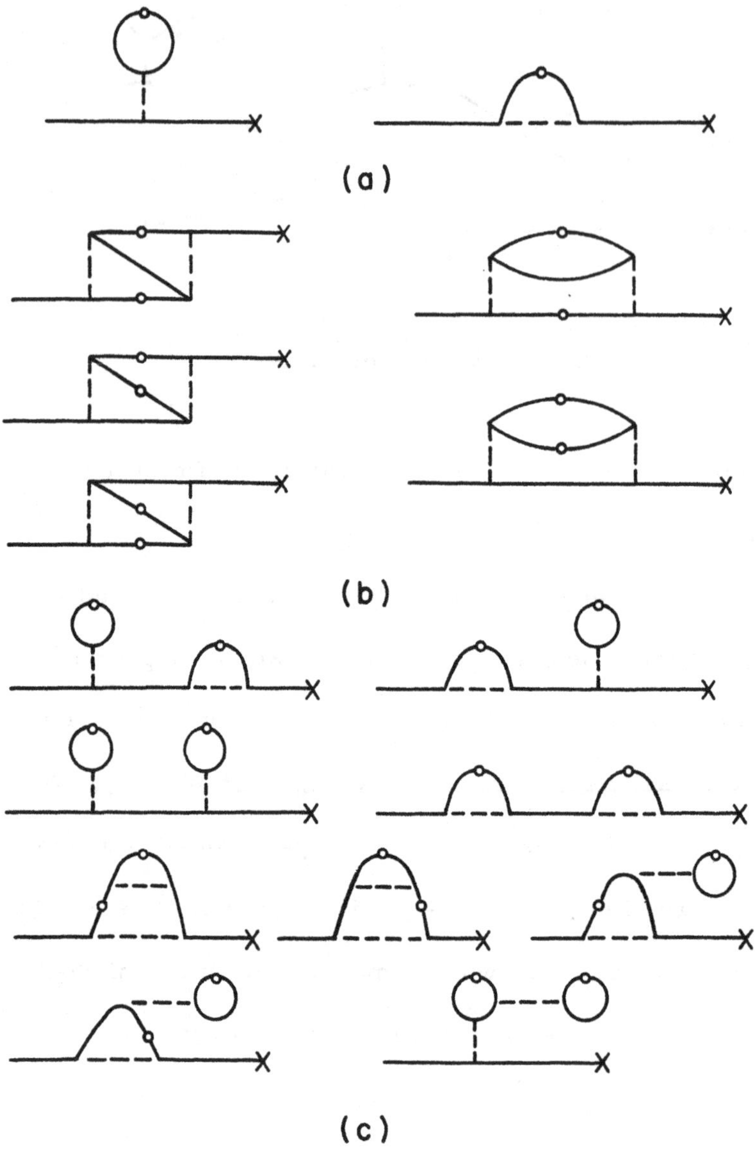

Figure 14. 3. Graphs for Gh.
(a) First order in u.
(b), (c) Second order in u.

describes the time evolution. It is the "propagator." A circled line represents the average of a pair of σ^{o}'s. It gives the zeroth-order "power spectrum" of spin fluctuations, and describes the statistical average.

The response function $G(k, \omega)$ is defined by (12. 13). If we sum all graphs with one cross, we get $G(k, \omega)\, h_k(\omega)$.

Let us summarize with the following rules for calculating perturbation terms by graphs. We restrict our statements to graphs for $G(k, \omega)\, h_k(\omega)$ for simplicity.

Rule (i): Draw a tree graph with one cross. Join all end points (the beginning of the tree and the end with a cross excepted) in pairs to form circled lines. Note that there is in general more than one way to pair up ends. Each way gives a separate term.

Rule (ii): Label each solid line with a wave vector and a frequency. The wave vectors and frequencies must follow conservation laws, i. e., the incoming one must equal the sum of the outgoing three at each vertex.

Rule (iii): Write $G_o(q, \nu)$ for each solid line without circle, $C_o(q, \nu)$ for each solid line with a circle, -u for each dashed line, and $h_k(\omega)$ for the cross.

Rule (iv): Integrate over all frequencies and wave vectors which are not fixed by the conservation laws in Rule (ii). A factor $(2\pi)^{-d-1}$ goes with each wave vector-frequency integral. For each closed loop of solid lines, write a factor n (which comes from the sum over spin components just as in the static case).

One may take advantage of the knowledge gained from static graph calculations and simplify Rule (i) and the counting of graphs. Rule (i) may be replaced by

Rule (i′): Draw a graph for G in statics. Keep the same factors of u, n, and other counting factors. Put small circles on the solid lines in all possible ways under the following constraint: When all circled lines are cut, the graph must become a tree graph.

The above rules are for the TDGL model. Generalizations to other models are straightforward and will be made as we proceed.

As a simple illustration of the rules, consider the graph Fig. 14. 3a. It gives a contribution

$$-G_o(k,\omega)\,\Sigma_a(k,\omega)\,G_o(k,\omega)\,h_k(\omega)$$

to $\langle\sigma_k(\omega)\rangle$ with

$$\Sigma_a(k,\omega) = \left(\frac{n}{2}+1\right) u(2\pi)^{-d-1}\int d^dq\; d\nu\; C_o(q,\nu)\;. \tag{14.8}$$

The ν integral is easily done using (14.6) and the explicit expression (14.2). We get

$$\Sigma_a(k,\omega) = \left(\frac{n}{2}+1\right) u(2\pi)^{-d}\int d^dq\,(r_o+q^2)^{-1} \tag{14.9}$$

which is independent of k and ω and is the same as the first-order self energy in statics.

As in the static case, the response function $G(k,\omega)$ can be expressed as

$$G(k,\omega) = (G_o^{-1}(k,\omega)+\Sigma(k,\omega))^{-1} \tag{14.10}$$

where the self energy $\Sigma(k,\omega)$ is the sum of graphs for $G(k,\omega)$ with isolated single $G_o(k,\omega)$ lines excluded.

As a second illustration, consider the graphs in Fig. 14.3b. We get an $O(u^2)$ contribution to $\Sigma(k,\omega)$:

$$\Sigma^{(2)}(k, \omega) = -3u^2 \left(\frac{n}{2}+1\right) (2\pi)^{-d-2} \int d^dq \; d\nu \; d^dq' \; d\nu'$$

$$\times \; C_o(q, \nu) \, C_o(q', \nu') \, G_o(k-q-q', \omega-\nu-\nu') \; . \tag{14.11}$$

There are very important differences between the static and the dynamic graphs, besides the frequency variables. As we mentioned earlier, the two kinds of lines, i.e., the circled and the plain, have different physical significance. The plain lines, the G_o's, describe time evolution while the circled lines, the C_o's, describe statistical averaging. In statics, there is only statistical averaging. In static graphs, one can consider a part of a graph separately, such as the vertex part or the self energy. One can easily join parts together to form new graphs. In dynamics, the fact that there are two kinds of lines and the condition that a graph must be a tree graph upon cutting all circled lines make the graph analysis much more complicated. These complications are all results of the necessity of keeping track of both time evolution and statistical averaging together. In quantum perturbation theory at a finite temperature, they are mixed together by using imaginary time variables. This is the Matsubara method, or the temperature Green's function method, where graphs can be conveniently handled like static graphs. One pays the price of having to do complicated analytic continuations in going from imaginary frequencies to real frequencies.

One can show that for $k, \omega \to 0$ the perturbation expansion will not converge near the critical point for $d < 4$ by the power-counting procedure used in the static graph expansion discussed in Chapter IX. There appears to be some complication here from the frequency integrals and the fact that each circled line gives a factor $C_o(q, \nu)$, which as -2 powers of q <u>and</u> -1 power of ν. [$G_o(q, \nu)$ has -2 powers of q as $G_o(q)$ in statics.] But the number of frequency integrals is the same as the number of circled lines because each circled line came from joining two ends of a tree, thereby forming a loop for a wave vector-frequency integral. Thus the -1 power of ν cancels the +1 power of ν of the ν-integral. We can thus count powers of wave vectors as we did in static graphs and forget about powers of ν. Consequently, as in statics, we find that for $d < 4$, the perturbation series diverges.

3. THE FLUCTUATION-DISSIPATION THEOREM

This theorem relates the correlation function to the imaginary part of the response function:

$$C(k, \omega) = \frac{2}{\omega} \text{ Im } G(k, \omega) \tag{14.12}$$

It can be trivially satisfied to zeroth order, as (14.6) shows, and easily proven in a microscopic theory. Also, in a formal way, it can be shown to follow from our general kinetic equations. Here let us illustrate how this theorem may be satisfied by studying graphs. The reader can gain some insight into the structure of graphs through this illustration.

Any graph for $G(k, \omega)$ has one continuous chain of G_o's and u's running from the beginning to the end to be hooked to h, as the previous examples showed. A general graph for $G(k, \omega)$ looks like Fig. 14.4a. All the ends of the branches growing out of the main chain will pair up to

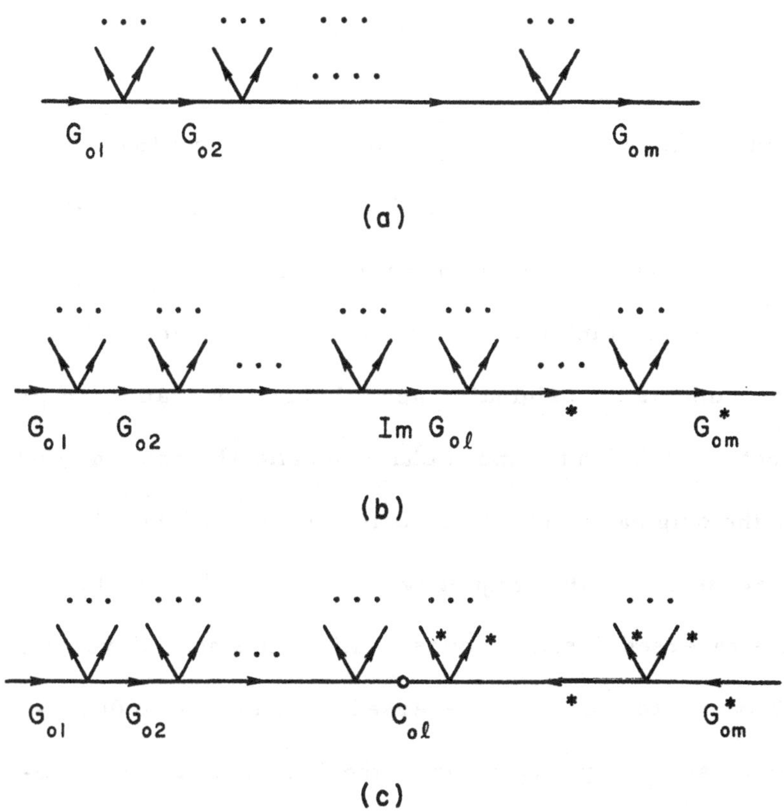

Figure 14.4. (a) A general graph for G (ignore the factor of h). Other lines are not shown explicitly.
(b) A term of (14.16) for Im G(k, ω).
(c) A term of (14.17) for C(k, ω).

form circle lines. Before we perform any integral, the graph gives

$$G_{o1} G_{o2} \cdots G_{om} A \tag{14.13}$$

where we have used the simplified notation

$$G_{o\ell} \equiv G_o(k_\ell, \omega_\ell) \tag{14.14}$$

and all other factors are included in A. Note that $k_1 = k_m = k$, $\omega_1 = \omega_m = \omega$. Now we turn the graph around and reverse the arrows on the main chain. We, of course, get another graph for $G(k, \omega)$, which may or may not be the same as the original graph. In the new graph, the product of G_o's on the main chain remains the same as that in the original graph, since only the order of the G_o's is reversed. But the frequency variables in the side branches are reversed to satisfy conservation laws at each vertex. (Wave vectors are also reversed but it makes no difference, since they all appear in the form of squares.) Reversing frequencies is the same as taking the complex conjugate. The sum of the original graph and the new one is therefore

$$G_{o1} G_{o2} \cdots G_{om} (A + A^*) . \tag{14.15}$$

The conclusion we want from (14.15) is that the imaginary part of A will not contribute. Therefore, if we take the imaginary part of a graph for G, we can regard A as real. In effect, the imaginary part of (14. 13) is

$$\begin{aligned}
&\mathrm{Im}(G_{o1} \cdots G_{om})\, A \\
&= [\,(\mathrm{Im}\, G_{o1})\, G^*_{o2} \cdots G^*_{om} \\
&+ G_{o1}\, (\mathrm{Im}\, G_{o2}) \cdots G^*_{om} \\
&+ \cdots \\
&+ G_{o1} \cdots G_{om-1}\, \mathrm{Im}\, G_{om}\,]\, A \\
&= \sum_{\ell=1}^{m} (G_{o1} \cdots G_{o\ell-1})\, \mathrm{Im}\, G_{o\ell}\, (G^*_{o\ell+1} \cdots G^*_{om})\, A
\end{aligned} \tag{14.16}$$

Figure 14. 4b shows a given term in this series.

Now we note that the contribution to $C(k, \omega)$, which is just $\langle \sigma_k(\omega)\, \sigma_{-k}(-\omega') \rangle = \langle \sigma_k(\omega)\, \sigma_k(\omega')^* \rangle$ apart from a δ function $\delta(\omega - \omega')$, can be represented by graphs generally of the form shown in Fig. 14. 4c. That is, we multiply two tree graphs, one for $\sigma_k(\omega)$ and another for $\sigma_k(\omega)^*$, and then average. In Fig. 14. 4c, only one circle line resulting from averaging is shown explicitly. The others are understood. The contribution from the chain explicitly shown to $C(k, \omega)$ is

$$(G_{o1} \cdots G_{o\ell-1})\, C_{o\ell}\, (G_{o\ell+1} \cdots G_{om})^* \tag{14.17}$$

which is just a term in the sum (14.16). Now for every graph for $\operatorname{Im} G(k, \omega)$, such as the one given in Fig. 14.4b, there is a graph for $C(k, \omega)$ given in Fig. 14.4c. The side branches have identical structure. On the right side of $C_{o\ell}$ in Fig. 14.4c, all factors including those from the branches have a star, i.e., are complex conjugated. Now we remove the stars from the factors on the right side branches and reverse all frequency variables in the branches on this side; we also reverse the arrows on the main chain to the right of $C_{o\ell}$. We then get a graph identical to Fig. 14.4b apart from the circle on the ℓth line. We get the contribution

$$(G_{o1} \cdots G_{o\ell-1})\, C_{o\ell}\, (G^*_{o\ell+1} \cdots G^*_{om})\, A \; . \tag{14.18}$$

Since $C_{o\ell} = (2/\omega_\ell) \operatorname{Im} G_{o\ell}$, the ℓth term in (14.16) for $\operatorname{Im} G(k, \omega)$ is $\omega_\ell/2$ times the corresponding term for $C(k, \omega)$ in (14.18). We have $\omega_\ell = \omega$ if the branches to the left of the ℓth line are not joined to those to the right. We represent this situation by Fig. 14.5a. This means that $\text{Im}\ G(k, \omega) = (\omega/2)\ C(k, \omega)$ for such graphs. Of course, there are also graphs in which they are joined, as represented by Fig. 14.5b. The additional circled lines give the factor $C_{o\ell'} C_{o\ell''} \ldots C_{o\ell'''}$. In the graph for $\text{Im}\ G(k, \omega)$, we need also to account for graphs in which the ℓ line changes places with one of the ℓ', ℓ'', ..., ℓ''' lines. This makes a total factor

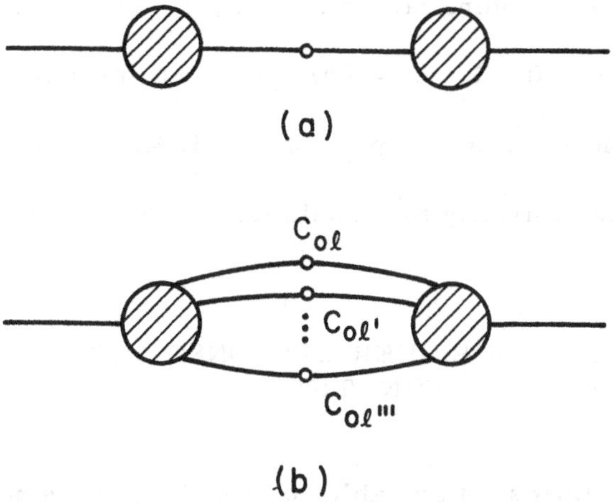

Figure 14.5. A graph for $C(k, \omega)$ is obtained by joining two trees. The joint can be a single circled line as shown in (a), or it can be several circled lines as shown in (b).

$$
\begin{aligned}
&\operatorname{Im} G_{o\ell}\, C_{o\ell'}\, C_{o\ell''} \cdots C_{o\ell'''} \\
&\quad + C_{o\ell} \operatorname{Im} G_{o\ell'}\, C_{o\ell''} \cdots C_{o\ell'''} \\
&\quad + \cdots \\
&\quad + C_{o\ell}\, C_{o\ell'} \cdots \operatorname{Im} G_{o\ell'''} \\
&\quad = C_{o\ell}\, C_{o\ell'} \cdots C_{o\ell'''}(\omega_\ell/2 + \omega_{\ell'}/2 + \cdots + \omega_{\ell'''}/2) \\
&\quad = C_{o\ell}\, C_{o\ell'} \cdots C_{o\ell'''}(\omega/2) \qquad (14.19)
\end{aligned}
$$

since $\omega_\ell + \omega_{\ell'} + \cdots + \omega_{\ell'''} = \omega$. Now $C_{o\ell} C_{o\ell'} \ldots C_{o\ell'''}$ is the corresponding factor in $C(k, \omega)$. This conclusion implies that $\operatorname{Im} G(k, \omega) = (\omega/2)\, C(k, \omega)$ for general graphs, and the theorem is thus proved. This kind of proof can be generalized trivially to models with more modes.

4. GRAPHS FOR HIGHER RESPONSE AND CORRELATION FUNCTIONS

In statics we are able to classify graphs according to the number of external lines or "legs." The sum of graphs of two legs gives $G(k)$, and the sum of m-leg graphs gives $G(k_1 \ldots k_m)$, as we have discussed in Chapter IX.

In dynamics, the classification needs to be generalized, because of the presence of the circled lines in addition to the plain lines. Let us define a <u>one-tree graph</u> to be that which becomes a tree graph upon cutting all circled lines. An <u>m-tree graph</u> is a graph which becomes m disconnected trees upon cutting all circled lines. Thus, $G(k, \omega)$ is the sum of two-leg one-tree graphs, and $C(k, \omega)$ is the sum of two-leg two-tree graphs. We can define generalized response functions $G_{ii_1 \ldots i_m}(k_1, \omega_1 \ldots k_m, \omega_m)$ by expanding $\langle \sigma_{ik}(\omega) \rangle$ in powers of the external field $h_{ik}(\omega)$:

$$\langle \sigma_{ik}(\omega) \rangle = \sum_{m=1}^{\infty} \sum_{i_1 \ldots i_m = 1}^{n} L^{(1-m)d/2}$$

$$\times \left[\prod_{\lambda=1}^{m} \sum_{k_\lambda} \int \frac{d\omega_\lambda}{2\pi} h_{i_\lambda k_\lambda}(\omega_\lambda) \right]$$

$$\times 2\pi \delta(\omega - \omega_1 - \cdots - \omega_m) G_{ii_1 \ldots i_m}(k_1, \omega_1, \ldots, k_m, \omega_m) \tag{14.12}$$

where $k_1 + k_2 + \cdots + k_m = k$ is understood. Of course,

$G_{ii_1}(k, \omega) = \delta_{ii_1} G(k, \omega)$. The other G's are the nonlinear response functions. In terms of graphs, $G_{ii_1 \ldots i_m}$ is the sum of all $m+1$ - leg one-tree graphs.

We can define correlation functions of more than two σ's at $h = 0$ as

$$C_{i_1 \ldots i_\lambda}(k_1, \omega_1, \ldots, k_\lambda, \omega_\lambda)\, 2\pi\, \delta(\omega_1 + \omega_2 + \cdots + \omega_\lambda)$$

$$= \langle \sigma_{i_1 k_1}(\omega_1) \ldots \sigma_{i_\lambda k_\lambda}(\omega_\lambda) \rangle\, L^{-d+\lambda d/2} \qquad (14.13)$$

where $k_1 + \cdots + k_\lambda = 0$ is understood. It is also understood that disconnected graphs are excluded, i.e., (14.13) is defined as a cumulant. $C_{i_1 \ldots i_\lambda}$ is the sum of all λ-leg λ-tree graphs.

Clearly, we can also define mixtures of correlation and response functions by considering the correlation functions like (14.13) in the presence of an external field h and expanding them in powers of h as we did in (14.12). We then get $\lambda + m$ - leg λ-tree graphs.

Some examples are shown in Fig. 14.6. Clearly, these graphs are very much like those in the analysis of

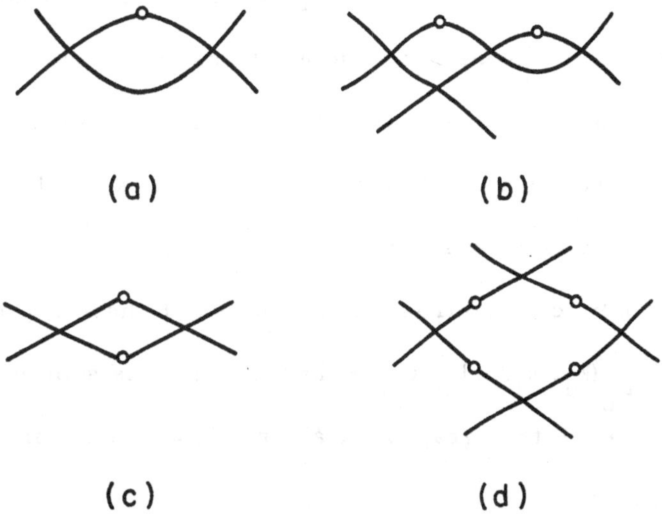

Figure 14.6. (a) Four-leg one-tree.
(b) Six-leg one-tree.
(c) Four-leg two-tree.
(d) Eight-leg four-tree.

quenched random systems in Chapter X. The quenched impurities act as sources of time-independent random noise.

As we have shown, $C(k, \omega)$ and $G(k, \omega)$ are simply related by the fluctuation dissipation theorem $C(k, \omega) = (2/\omega) \operatorname{Im} G(k, \omega)$. The higher response functions and correlation functions are also related by similar but much more complicated theorems. While $G(k, \omega)$ can be continued into the complex ω-plane and defined as an analytic function in the upper and lower plane and with a cut along the real axis, higher response functions can be continued into the space of several complex variables with complicated cut structures. Correlation functions are related to discontinuities across various cuts. We shall not study these analytic structures. Let us only note here that once $G_{ii_1 \ldots i_m}(k_1, \omega_1 \ldots k_m \omega_m)$ is known, then the sum of $m + 1$ - leg λ-tree graphs for all $m + 1 \geq \lambda \geq 1$ can be determined.

5. ADDITIONAL MODES AND MODE-MODE COUPLING TERMS

The above discussion has been restricted to time-dependent Ginzburg-Landau models. Extensions of graph rules to kinetic equations with additional modes and mode-mode coupling terms v_k are quite straightforward. As an illustration, we consider the model with heat conduction defined by (13. 14) and (13. 17), and the ferromagnet model defined by (13. 42) and (13. 46).

(a) Model with heat conduction

We have the additional modes ρ_k and additional coupling constant g. All we need to do is to introduce graph elements to account for ρ_k and g in the same manner as we did for σ_k and u.

Setting $\tilde{u} = g = 0$ in (13.14) we obtain from (13.17) the zeroth-order equation for σ_k^o, which is the same as (14.1), and that for ρ_k^o:

$$-i\omega\rho_k^o(\omega) = -Dk^2 \rho_k^o(\omega) + \varphi_k(\omega) \quad . \tag{14.20}$$

We have set $w = 1$ for simplicity. The zeroth-order correlation function K_o and response function F_o [see (13.20) and (13.21) for definitions of K and F] are easily obtained:

$$K_o(k, \omega) = \frac{2}{\omega} \operatorname{Im} F_o(k, \omega) \quad ,$$

$$F_o(k, \omega) = (1 - i\omega/(Dk^2))^{-1} \quad . \tag{14.21}$$

For nonzero $\tilde{u}$ and g, we get from (13.17a) an equation which is the same as (14.7) for $\sigma_k(\omega)$ except that u is replace by $\tilde{u}$ and the term

$$-g\, G_o(k, \omega)\, L^{-d/2} \sum_{k'} \int \frac{d\omega'}{2\pi}\, \rho_{k'}(\omega')\, \sigma_{ik-k'}(\omega - \omega') \tag{14.22}$$

is added to the right-hand side. There is also an additional equation obtained from (13.17b):

$$\rho_k(\omega) = \rho_k^o(\omega) - gF_o(k, \omega)\, L^{-d/2}$$

$$\times \sum_{jk'} \int \frac{d\omega'}{2\pi}\, \sigma_{jk'}(\omega')\, \sigma_{jk-k'}(\omega - \omega') \; . \tag{14.23}$$

Let us represent F_o by a wavy line and K_o by a wavy line with a small circle in the middle. The previous graph rules (i), (ii), and (iv) are unchanged provided wavy lines and straight lines are treated on equal footing. In rule (iii) we change u to $\tilde{u}$ and add the sentence "write $F_o(q, \nu)$ for each plain wavy line, $K_o(q, \nu)$ for each circled wavy line, and -g for each vertex joining two straight lines and a wavy line."

As an example, Fig. 14.7 gives the O(u) and $O(g^2)$ terms of self energy:

$$\Sigma(k, \omega) = -g^2 (2\pi)^{-d-1} \int d\nu \, d^d q \, [F_o(q, \nu) \, C_o(k-q, \omega-\nu)$$

$$+ K_o(q, \nu) \, G_o(k-q, \omega-\nu)]$$

$$+\left(\left(\frac{n}{2} + 1\right) \tilde{u} - ng^2 F_o(0, 0)\right) (2\pi)^{-d-1} \int d\nu \, d^d q \;\; C_o(q, \nu) .$$

Note that $F_o(q, 0) = 1$, as (14.21) shows. Let us write, following (13.16), $u = \tilde{u} - g^2$. Then (14.24) gives

$$\Sigma(k, \omega) = \Sigma_a(k, \omega) - g^2 (2\pi)^{-d-1} \int d\nu \, d^d q \, [F_o'(q, \nu) \, C_o(k-q, \omega-\nu)$$

$$+ K_o(q, \nu) \, G_o(k-q, \omega-\nu)] \tag{14.25}$$

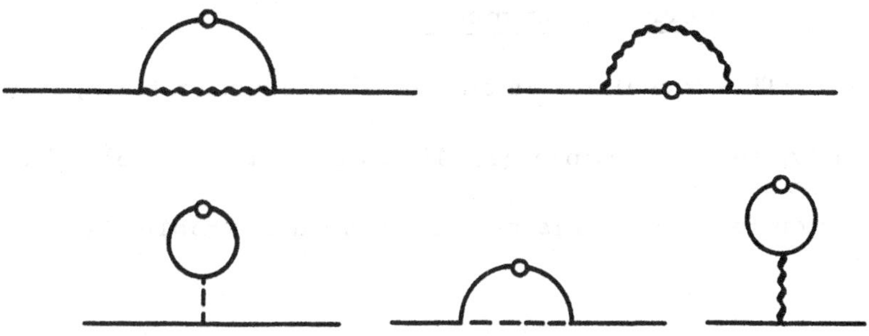

Figure 14.7. Self energy graphs to $O(\tilde{u})$ and $O(g^2)$.

where Σ_a is given by (14.8) and

$$F_o'(q, \nu) \equiv F_o(q, \nu) - F_o(q, 0)$$

$$= \frac{i\nu}{Dq^2}\left(1 - \frac{i\nu}{Dq^2}\right)^{-1} . \qquad (14.26)$$

We have combined $\tilde{u}$ with the $-g^2 F_o(q, 0)$, which is the static (zero frequency) value of the wavy line, to give u and subtracted the static value from $F_o(q, \nu)$. This in fact can be done in all graphs. We arrive at the new rule (iii) that

"Every dashed line still gives a factor $-u$ (not $-\tilde{u}$), and every wavy line gives a factor $F_o'(q, \nu)$ [instead of $F_o(q, \nu)$]."

(b) <u>Ferromagnet model</u>

This model has an additional interaction term, namely, the v_k term in (13.42) as given by (13.46). The equation for iteration is just (14.7) with one more term:

$$\sigma_{ik}(\omega) = \sigma^{o}_{ik}(\omega) + G_o(k, \omega)(\tilde{\lambda}/k^2)$$

$$\times L^{-d/2} \sum_{k', j, \ell} \int \frac{d\omega'}{2\pi} e_{ij\ell}\, \sigma_{jk-k'}(\omega-\omega')[-k'^2 \sigma_{\ell k'}(\omega')$$

$$+ h_{\ell k'}(\omega')] + G_o(k, \omega)\, h_{ik}(\omega) \qquad (14.28)$$

plus the u-term in (14.7), which we omit here. The symbol $e_{ij\ell}$ is the three-dimensional completely anti-symmetric tensor used here to express the cross product in component form, and $\tilde{\lambda} = \lambda/\gamma$.

The modification of the graph rules are

(i) Add graphs with three-line vertices. A three-line vertex has one line in and two lines out. It gives a factor

$$\lambda\, e_{ij\ell}(k'^2 - k''^2)/k^2 \quad , \qquad (14.29)$$

i and k being, respectively, the component label and the wave vector for the incoming line, while j, k′, ℓ, k″ are those for the outgoing lines.

(ii) Owing to the additional $h_{\ell k'}$ in the square bracket in (14.28), we need to add vertices with a cross, representing $h_{\ell k'}$, joining two lines and giving a factor

$$\tilde{\lambda}\, e_{ij\ell}\, h_{\ell k'} \quad . \tag{14.30}$$

Figure 14.8 shows $O(\lambda^2)$ graphs for $G(k, \omega)\, h_{ik}$. Let us take $i = 1$. Then j, ℓ must be 2.3, or 3.2. The contribution of Fig. 14.8a is

$$G_a h_k = 2\, G_o(k, \omega)(\tilde{\lambda}/k^2)(2\pi)^{-d-1} \int d^d q\, d\nu\, (q^2 - (q+k)^2)$$

$$\times\, G_o(q+k, \nu+\omega)\, C_o(q, \nu)\, [\tilde{\lambda}/(q+k)^2]\, (k^2 - q^2)\, G_o(k, \omega)\, h_k(\omega) \quad . \tag{14.31}$$

The factor 2 in front comes from the fact that $j\ell = 23$ and $j\ell = 32$ contribute equally. The contribution of Fig. 14.8b is

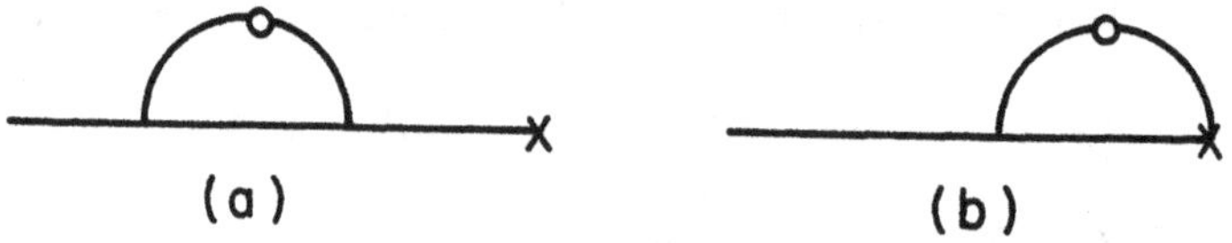

Figure 14.8. $O(\lambda^2)$ correction to $G_o h$ in the ferromagnet model.
(a) is given by (14.31) and (b) by (14.32).

$$G_b h_k = 2G_o(k, \omega)(\tilde{\lambda}/k^2)(2\pi)^{-d-1}\int d^d q\, d\nu\, (q^2 - (q+k)^2)$$

$$\times\, G_o(q+k, \nu+\omega)\, C_o(q, \nu)\, \tilde{\lambda}\, h_k(\omega) \quad . \qquad (14.32)$$

The power counting argument can be used to show that the perturbation expansion does not converge for $d < 6$ for $k, \omega \rightarrow 0$ at the critical point. This is left as an exercise for the reader.

APPENDIX

This appendix is supplementary to Chapters II and V, where a few technical points were skipped for the sake of continuity and simplicity.

1. AN ALTERNATIVE FORMULATION OF COARSE GRAINING, THE CLASSICAL FIELD CONFIGURATIONS

In Chapter II we have formulated the ideas of coarse graining and block Hamiltonians in terms of classical spins in a ferromagnet. These ideas apply equally well to other kinds o: variables such as staggered magnetization, Cooper pair amp:itude, and the density of a fluid. The formalism can be easily generalized. The following discussion has been found useful by theoretically oriented readers. It will be slightly more formal than what has gone before.

To be very general, we begin with a microscopic quantum mechanical Hamiltonian $\hat{H}$. Suppose that we expect the order parameter to be the average of the quantum mechanical operator

$$\int d^d x \ \hat{\phi}(x) \qquad \text{(A. 1)}$$

where $\hat{\phi}(x)$ could be the Boson field operator in HeII, or the density operator $\hat{n}(x)$ in a fluid, or the pair operator $\psi_\uparrow(x)\,\psi_\downarrow(x)$ in a superconductor, or the spin density operator, etc.

Let $\rho(x)$ be a smooth function which diminishes for x larger than a length b, for example, a Gaussian

$$\rho(x) = A\, e^{-x^2/b^2} \quad . \qquad \text{(A. 2)}$$

The constant A is a normalization factor such that $\int \rho(x)\, d^d x = 1$. Let us define the smeared operator $\hat{\Phi}(x)$ as

$$\hat{\Phi}(x) = \int d^d x' \ \rho(x - x')\ \hat{\phi}(x') \ , \qquad \text{(A. 3)}$$

which is clearly $\hat{\phi}(x)$ smeared over a volume of size b^d.

Now we want to obtain an effective Hamiltonian $H[\sigma]$ for a classical field $\sigma(x)$ which contains all the relevant information concerning $\hat{\Phi}$. We can define a classical Hamiltonian $H[\sigma]$ by

$$e^{-H[\sigma]/T} = \mathrm{Tr}\; e^{-\hat{H}/T} \rho[\sigma, \Phi] \qquad (A.4)$$

which is the generalized form of (2.16) with $\rho[\sigma, \hat{\Phi}]$ playing the role of the product of δ-functions and the trace playing the role of the integrals. We demand that

$$\int \delta\sigma \rho[\sigma, \hat{\Phi}] = 1 \qquad (A.5)$$

for any $\hat{\Phi}$ and that $\rho[\sigma, \hat{\Phi}]$ should peak at $\sigma = \hat{\Phi}$. A convenient choice is

$$\rho[\sigma, \hat{\Phi}] = B\, e^{-\frac{w}{2}\int d^d x(\sigma(x) - \hat{\Phi}(x))^2}, \qquad (A.6)$$

$$B = \int \delta\sigma\; e^{-\frac{w}{2}\int d^d x\, \sigma^2(x)}, \qquad (A.7)$$

where w is an arbitrary positive constant. If w is very

large, then $P[\sigma, \hat{\Phi}]$ resembles the product of δ-functions in (2.16). However, there is no need to choose a large w. Equation (A.5) is satisfied by (A.6), since we can change the variable of integral $\delta\sigma$ to $\delta\sigma'$ with $\sigma' = \sigma - \hat{\Phi}$. We have, by use of Eqs. (A.4) and (A.5),

$$\int \delta\sigma \, e^{-H[\sigma]/T} = \mathrm{Tr} \, e^{-\hat{H}/T} = Z \,. \tag{A.8}$$

Thus the partition function of the classical Hamiltonian $H[\sigma]$ is the same as that of the quantum mechanical $\hat{H}$. Consequently we can get the correct thermodynamical information from the classical $H[\sigma]$.

In general, one will also be interested in the average and correlation functions of $\hat{\Phi}(x)$,

$$\langle \hat{\Phi}(x) \rangle = \mathrm{Tr} \, e^{-\hat{H}/T} \, \hat{\Phi}(x)/Z \,, \tag{A.9}$$

$$\langle \hat{\Phi}(x) \, \hat{\Phi}(x') \rangle = \mathrm{Tr} \, e^{-\hat{H}/T} \, \hat{\Phi}(x) \, \hat{\Phi}(x')/Z \,, \tag{A.10}$$

etc.

How are these related to quantities calculated with the classical Hamiltonian $H[\sigma]$? The answer generally depends on the form of $\mathcal{P}[\sigma, \Phi]$, and can be seen easily via the moment generating functions as follows.

Define the moment generating function $\mathcal{M}[\lambda]$ for the classical field σ as

$$\mathcal{M}[\lambda] = \left\langle e^{\int d^d x \; \lambda(x)\, \sigma(x)} \right\rangle$$

$$= Z^{-1} \int \delta\sigma \; e^{-H[\sigma]/T + \int d^d x \; \lambda(x)\, \sigma(x)} \; . \tag{A.11}$$

Similarly, the moment generating function for the operator $\hat{\Phi}$ is defined as

$$\tilde{\mathcal{M}}[\lambda] = \left\langle e^{\int d^d x \; \lambda(x)\, \hat{\Phi}(x)} \right\rangle$$

$$= Z^{-1} \, \mathrm{Tr} \left(e^{-\hat{H}/T} \; e^{\int d^d x \; \lambda(x)\, \hat{\Phi}(x)} \right) . \tag{A.12}$$

Averages and correlation functions can be generated by differentiating the generating functions with respect to λ and then setting $\lambda = 0$:

$$\langle \hat{\Phi}(x) \rangle = (\delta \widetilde{\mathcal{M}}[\lambda] / \delta \lambda(x))_{\lambda=0} \quad ,$$

$$\frac{1}{2} \langle \hat{\Phi}(x) \hat{\Phi}(x') + \hat{\Phi}(x') \hat{\Phi}(x) \rangle = (\delta^2 \widetilde{\mathcal{M}}[\lambda] / \delta \lambda(x) \, \delta \lambda(x'))_{\lambda=0} \quad ,$$

$$\langle \sigma(x) \rangle = (\delta \mathcal{M}[\lambda] / \delta \lambda(x))_{\lambda=0} \quad , \tag{A. 13}$$

$$\langle \sigma(x) \sigma(x') \rangle = (\delta^2 \mathcal{M}[\lambda] / \delta \lambda(x) \, \delta \lambda(x'))_{\lambda=0} \quad ,$$

and so on. Note that $\hat{\Phi}(x)$, $\hat{\Phi}(x')$ may or may not commute. The correlation functions generated by (A. 12) are averages of the sum of products of $\hat{\Phi}$ in all possible orders divided by the total number of ways of ordering. If one is only interested in the large-scale behavior, i.e., the case where x, x′ are far apart, then the commutators will play no role.

Now $\mathcal{M}$ and $\widetilde{\mathcal{M}}$ can be related via the definition (A. 4) of $H[\sigma]$. Let us use (A. 6) for ρ and substitute (A. 4) in (A. 11) to obtain

$$\mathcal{M}[\lambda] = Z^{-1} \int \delta\sigma \, \mathrm{Te} \, e^{-\hat{H}/T} \, \mathrm{B} e^{\int d^d x \left(\lambda\sigma - \frac{w}{2} (\sigma - \hat{\Phi})^2 \right)} \tag{A. 14}$$

where the argument x for λ, σ, and $\hat{\Phi}$ is understood.

Now we replace σ by $\sigma' + \hat{\Phi}$ in (A. 14) to obtain

$$\mathcal{M}[\lambda] = Z^{-1} \operatorname{Tr} e^{-\hat{H}/T} e^{\int d^d x \, \lambda \hat{\Phi}} B \int \delta\sigma' \, e^{-\frac{w}{2}\sigma'^2 + \lambda\sigma'}$$

$$= \widetilde{\mathcal{M}}[\lambda] \, e^{\int d^d x (\lambda^2/2w)} \qquad \text{(A. 15)}$$

which gives the relationship between $\mathcal{M}$ and $\widetilde{\mathcal{M}}$. In view of (A. 13), (A. 14), and (A. 15), it is evident that

$$\langle \hat{\Phi}(x) \rangle = \langle \sigma(x) \rangle ,$$

$$\frac{1}{2} \langle \hat{\Phi}(x) \hat{\Phi}(x') + \hat{\Phi}(x') \hat{\Phi}(x) \rangle = \langle \sigma(x) \sigma(x') \rangle - \frac{1}{w} \delta(x - x')$$

(A. 16)

and so on. Thus the average of the operator $\hat{\Phi}(x)$ is the same as the average of the classical field $\sigma(x)$. The correlation functions of $\hat{\Phi}$ are in general different from those of σ but are simply related.

We have done two things in the above discussion. First, we coarse grained $\hat{\phi}(x)$ to get $\hat{\Phi}(x)$ via (A. 3). Second, we introduced a classical field $\sigma(x)$ and its effective Hamiltonian $H[\sigma]$ via (A. 4) - (A. 7). This classical field has the same thermodynamic behavior as the quantum mechanical system. Furthermore, once the correlation functions of the classical field are known, those of $\hat{\Phi}$ are simply obtained via Eq. (A. 16). We may regard $\sigma(x)$ as a classical approximation for $\hat{\Phi}(x)$.

The above discussion obviously applies also when $\hat{\phi}(x)$ and $\hat{\Phi}(x)$ are classical variables instead of quantum mechanical operators.

The replacement of δ-functions by a smooth function $\rho[\sigma, \hat{\Phi}]$ such as the Gaussian (A. 6) smears up the variable $\hat{\Phi}$ itself in contrast to smearing its spatial dependence as in (A. 3) . It removes some unpleasant mathematical features of δ-functions. It also produces extra terms like the $1/w$ in (A. 16) which must be carefully kept track of. Different forms of $\rho[\sigma, \hat{\Phi}]$ in general produce different extra terms.

2. SMOOTH CUTOFF

The sharp cutoff in q in (5. 19) produces very undesirable mathematical features. We saw in Chapter VII that long-range interactions in coordinate space were generated by such a sharp cutoff in q space. Such long-range interaction is analogous to the Friedel oscillation due to a sharp Fermi surface familiar in the physics of metals. Here the long-range interaction is of a purely mathematical nature and has no physical consequence, in contrast to the Friedel oscillation. To remove it, one needs to make the cutoff smooth. In other words, we need to smooth the transition from the integrated to the unintegrated. To accomplish this, we proceed in the same way as in the discussion of the previous section. Let us concentrate on the Kadanoff transformation $K_s \mathcal{H} = \mathcal{H}'$. We introduce a new definition

$$e^{-\mathcal{H}'[\sigma']} = \int \delta\sigma \; e^{-\mathcal{H}[\sigma]} \, \rho[\sigma', \sigma] \quad , \qquad \text{(A. 17)}$$

where

$$\rho[\sigma', \sigma] = B \, e^{-\frac{w}{2} \sum_{k,i} (\sigma'_{ik} - \rho_k \sigma_{ik})^2} \quad ,$$

$$\delta\sigma = \prod_{i,k} d\sigma_{ik} \quad , \qquad \text{(A. 18)}$$

B is a normalization constant such that $\int \delta\sigma' P[\sigma', \sigma] = 1$, w is a positive constant. The function ρ_k is defined by

$$\rho_k = 1 \quad \text{for } k < \Lambda/s - \Delta \ ,$$

$$= 0 \quad \text{for } k > \Lambda/s + \Delta \ , \tag{A.19}$$

and, in between $\Lambda/s + \Delta$ and $\Lambda/s - \Delta$, ρ_k increases smoothly from 0 to 1. The quantity $\rho_k \sigma_{ik}$ is the Fourier component of

$$\int d^d x' \ \rho(x - x') \ \sigma_i(x') \ ,$$

$$\int d^d x' \ \rho(x') = 1 \ , \tag{A.20}$$

where ρ_k is the Fourier component of $\rho(x)$. Equation (A.19) says that $\rho(x)$ is something like a smooth weighting function over a region of a size $s\Lambda^{-1}$, and (A.20) gives the coarse grained spin weighed by this weighting function. Thus $\rho(x)$ defines the new spin blocks in a smooth way. Instead of a δ-function, we have a Gaussian in (A.17). Thus σ'_{ik} is not quite equal to $\rho_k \sigma_{ik}$, but close to it roughly within a width $w^{-1/2}$. It is a nondeterministic relationship between σ' and σ. It can be viewed as a special form of nonlinearity.

Note that to achieve a smooth cutoff it is not necessary to have a finite w. We can set $w \to \infty$ and $P[\sigma', \sigma]$ would be a product of δ-functions. Here we keep w flexible to make the analysis more general.

The rest of the argument is almost identical to that in (A. 9) - (A. 16). We shall give it for completeness and continuity.

We note that $\int \delta\sigma P[\sigma', \sigma] = 1$. Integrating both sides of (A. 17), we obtain

$$\int \delta\sigma' \, e^{-\mathcal{H}[\sigma']} = \int \delta\sigma \, e^{-\mathcal{H}[\sigma]} \quad . \tag{A.21}$$

Therefore the free energy of $\mathcal{H}$ and that of $\mathcal{H}'$ are the same.

The correlation functions are slightly more complicated. Let us calculate the generating function

$$\left\langle e^{-\sum_{ik} \lambda_{ik} \sigma'_{ik}} \right\rangle \equiv \frac{\int \delta\sigma' \, e^{-\mathcal{H}'[\sigma'] - \sum_{ik} \lambda_{ik} \sigma'_{ik}}}{\int \delta\sigma' \, e^{-\mathcal{H}'[\sigma']}} \quad . \tag{A.22}$$

By (A. 17), the numerator is

$$\int \delta\sigma \int \delta\sigma' \, e^{-\mathcal{H}[\sigma] - \sum_{ik} \left(\frac{w}{2} (\sigma'_{ik} - \rho_k \sigma_{ik})^2 + \lambda_{ik}\sigma_{ik} \right)}$$

$$= \int \delta\sigma \, e^{-\mathcal{H}[\sigma] - \sum_{ik} \lambda_{ik} \rho_k \sigma_{ik}} \int \delta\sigma' \, e^{-\sum_{ik} \left(\frac{w}{2} \sigma'^2_{ik} + \lambda_{ik}\sigma'_{ik} \right)} \tag{A.23}$$

where we have displaced the variables, σ'_{ik} to $\sigma'_{ik} + \rho_k \sigma_{ik}$. When we perform the σ' integral and then divide by $\int \delta\sigma' \, e^{-\mathcal{H}'}$, we get

$$\left\langle e^{-\sum_{ik} \lambda_{ik} \rho_k \sigma_{ik}} \right\rangle = \left\langle e^{-\sum_{ik} \lambda_{ik} \sigma'_{ik}} \right\rangle' e^{-\frac{1}{2w} \sum_{ik} |\lambda_{ik}|^2} \; . \tag{A.24}$$

As far as correlation functions involving σ_{ik}, $k < \Lambda/s - \Delta$ are concerned, $\rho_k = 1$. They can be obtained by differentiating the generating function and then setting $\lambda = 0$. For example,

$$\langle \sigma_{ik} \rangle = \langle \sigma'_{ik} \rangle' \, , \tag{A.25}$$

$$G(k) = \langle |\sigma_{ik}|^2 \rangle = \langle |\sigma'_{ik}|^2 \rangle' + \frac{1}{w} \quad . \qquad (A.26)$$

Thus the finite w introduced in (A. 17) does produce a difference in the average values of 2nd and in fact all higher powers of spin, although the first power is not affected. Note that in (A. 26), $1/w$ is a constant. Thus the singularities of interest in $G(k)$ are completely contained in $\langle |\sigma'_{ik}|^2 \rangle'$. For higher correlation functions, the extra terms analogous to $1/w$ will be singular, but the leading singularity will be contained in the same correlation functions computed with $\mathcal{H}'[\sigma']$.

A warning is in order. If we define R_s by the above smooth cutoff procedure (together with the replacement $\sigma_k \rightarrow \lambda_s s^{d/2} \sigma_{sk}$), we shall find that the relation $R_s R_{s'} = R_{ss'}$ will not always hold. It will not hold whenever the transition region $\Lambda/s' - \Delta < k < \Lambda/s' + \Delta$ for $R_{s'}$ overlaps the region $k > \Lambda - s\Delta$. Roughly speaking, smoothing out the cutoff twice in $R_s R_{s'}$ is in general not the same as smoothing out the cutoff once in $R_{ss'}$. To avoid this difficulty, we can define R_s by first defining a generator, like R_2, or $R_{1.5}$ or $R_{1+\delta}$, usin the above procedure and then generating R_s by repeated applications of the generator, e.g.,

$$R_s = (R_2)^\ell \ , \quad s = 2^\ell \ ,$$

or
$$R_s = (R_{1.5})^\ell \ , \quad s = (1.5)^\ell \ ,$$

or
$$R_s = (R_{1+\delta})^\ell \ , \quad s = (1+\delta)^\ell \ , \tag{A.27}$$

with
$$\ell = 1, 2, 3, \ldots \ .$$

Then $R_s R_{s'} = R_{ss'}$ by construction.

REFERENCES

Abe, R., 1972. Prog. Theo. Phys. 48, 1414.

Abe, R., 1973a. Prog. Theo. Phys. 49, 113.

Abe, R., 1973b. Prog. Theo. Phys. 49, 1074.

Abe, R., 1973c. Prog. Theo. Phys. 49, 1877.

Abe, R., 1975. Prog. Theo. Phys. 53, 1836.

Abe, R. and S. Hikami, 1973a. Phys. Lett. 45A, 11.

Abe, R. and S. Hikami, 1973b. Prog. Theo. Phys. 49, 442.

Abe, R. and S. Hikami, 1974. Prog. Theo. Phys. 52, 1463.

Abe, R. and S. Hikami, 1975. Prog. Theo. Phys. 54, 325.

Aharony, A., 1973a. Phys. Rev. B8, 3342, 3349, 3358, 3363; B9, 3649(E).

Aharony, A., 1973b. Phys. Rev. B8, 4270.

Aharony, A., 1974. Phys. Rev. B10, 2834.

Aharony, A., 1975a. In Domb and Green, 1975.

Aharony, A., 1975b. Phys. Rev. B12, 1038.

Aharony, A. and A. D. Bruce, 1974a. Phys. Rev. Lett. 33, 427.

Aharony, A. and A. D. Bruce, 1974b. Phys.Rev. B10, 2973.

Aharony, A. and M. E. Fisher, 1973. Phys. Rev. B8, 3323.

Ahlers, G., 1973. Phys. Rev. A8, 530.

Ahlers, G., A. Kornblit and M. B. Salamon, 1974. Phys. Rev. B9, 3932.

Als-Nielsen, J. and O. Dietrich, 1966. Phys. Rev. 153, 711, 717.

Arajs, S., 1965. J. Appl. Phys. 36, 1136.

Baker, G. A. Jr. and D. L. Hunter, 1973. Phys.Rev. B7, 3377.

Baker, G. A. Jr., H. E. Gilbert, J. Eve and G. S. Rushbrooke, 1967. Phys. Rev. 176, 739.

Barenblatt, G. I. and Ya. B. Zel'dovich, 1972. Ann. Rev. Fluid Mechanics 4, 285.

Baxter, R. J., 1971. Phys. Rev. Lett. 26, 832.

Bell, T. L. and K. G. Wilson, 1974. Phys.Rev. B10, 3935.

Bell, T. L. and K. G. Wilson, 1975. Phys.Rev. B11, 3431.

Bergman, D. J. and B. I. Halperin, 1975. Preprint.

Berlin, T. H. and M. Kac, 1952. Phys. Rev. 86, 821.

Binder, K. and P. C. Hohenberg, 1972. Phys. Rev. B6, 3461.

Bray, A. J., 1974. Phys. Rev. Lett. **32**, 1413.

Brézin, E., D. J. Amit and J. Zinn-Justin, 1974. Phys. Rev. Lett. **32**, 151.

Brézin, E., J. C. LeGuillou and J. Zinn-Justin, 1974. Phys. Rev. Lett. **32**, 473.

Brézin, E., J. C. LeGuillou and J. Zinn-Justin, 1975. In Domb and Green, 1975.

Brézin, E. and D. J. Wallace, 1973. Phys. Rev. **B7**, 1967.

Brézin, E., D. J. Wallace and K. G. Wilson, 1972. Phys. Rev. Lett. **29**, 591.

Brézin, E., D. J. Wallace and K. G. Wilson, 1973. Phys. Rev. **B7**, 232.

Bruce, A. D. and A. Aharony, 1975. Phys. Rev. **B11**, 478.

Budnick, J. T. and M. P. Kawatra, 1972, editors. *Dynamical Aspects of Critical Phenomena* (Gordon and Breach).

Cohen, E. D. G., 1975, editor. *Fundamental Problems in Statistical Mechanics*, Vol. 3 (North Holland Publishing Co., Amsterdam).

Corliss, L. M., A. Delapalme, J. M. Hastings, H. Y. Lau and R. Nathans, 1969. J. Appl. Phys. **40**, 1278.

DeDominieis, 1975. Lettere al Nuovo Cimento **12**, 567.

DeDominieis, E. Brézin and J. Zinn-Justin, 1975, to be published.

DeGennes, P. G., 1972. Phys. Lett. **38A**, 339.

DeGennes, P. G., 1974. Physics of Liquid Crystals (Oxford).

DiCastro, C., G. Jona-Lasinio and L. Peliti, 1974. Ann. Phys. 87, 327.

Domb, C. and M. S. Green, 1976, editors. Phase Transitions and Critical Phenomena, VI (Academic Press, London, to be published).

Edwards, C., J. A. Lipa and M. J. Buckingham, 1968. Phys. Rev. Lett. 20, 496.

Ferrell, R. A. and D. J. Scalapino, 1972. Phys. Rev. Lett. 29, 413.

Fisher, M. E., 1965. J. Chem. Phys. 44, 616.

Fisher, M. E., 1967. Rep. Prog. Phys. 30, 615.

Fisher, M. E., 1974. Rev. Mod. Phys. 46, 597.

Fisher, M. E., 1975a. Phys. Rev. Lett. 34, 1634.

Fisher, M. E., 1975b. AIP Conf. Proc. 24, 273.

Fisher, M. E. and A. Aharony, 1974. Phys. Rev. B10, 2818.

Fisher, M. E., S. Ma and B. G. Nickel, 1972. Phys. Rev. Lett. 29, 917.

Fisher, M. E. and D. R. Nelson, 1974. Phys. Rev. Lett. 32, 1350.

Fisher, M. E. and P. Pfeuty, 1972. Phys. Rev. B6, 1889.

Freedman, R. and G. F. Mazenko, 1975a. Phys. Rev. Lett. 34, 1575.

Freedman, R. and G. F. Mazenko, 1975b. Preprint.

Gell-Mann, M. and F. E. Low, 1954. Phys. Rev. 95, 1300.

Ginzburg, V. L., 1960. Soviet Phys. Solid State 2, 1824 (Translated from Fiz. Tverd. Tela 2, 2031).

Ginzburg, V. L. and L. D. Landau, 1950, Collected Papers of Landau, edited by D. ter Haar (Gordon and Breach, New York, 1965), p. 546, translated from JETP 20, 1064 (1950).

Golner, G. R., 1973a. Phys. Rev. B8, 339.

Golner, G. R., 1973b. Phys. Rev. B8, 3419.

Golner, G. R. and E. K. Riedel, 1975a. Phys. Rev. Lett. 34, 856.

Golner, G. R. and E. K. Riedel, 1975b. AIP Conf. Proc. 24, 315.

Gonzalo, J. A., 1966. Phys. Rev. 144, 662.

Gorodetsky, G., S. Shtrikman and D. Treves, 1966. Solid State Phys. Com. 4, 147.

Griffiths, R. B., 1968. Phys. Rev. 176, 655.

Griffiths, R. B., 1969. Phys. Rev. Lett. 23, 17.

Griffiths, R. B. and J. C. Wheeler, 1970. Phys. Rev. A2, 1047.

Grover, M. K., 1972. Phys. Rev. B6, 3546.

Grover, M. K., L. P. Kadanoff and F. J. Wegner, 1972. Phys. Rev. B6, 311.

Gunton, J. D. and M. S. Green, 1973, editors. Proceeding of Conference on Renormalization Group in

Critical Phenomena and Quantum Field Theory (Temple University).

Gunton, J. D. and K. Kawasaki, 1975. J. Phys. A8, 69.

Gunton, J. D. and Th. Niemeijer, 1975. Phys. Rev. B11, 567.

Halperin, B. I., 1975. Phys. Rev. B11, 178.

Halperin, B. I. and P. C. Hohenberg, 1969. Phys. Rev. 177, 952.

Halperin, B. I., P. C. Hohenberg and S. Ma., 1972. Phys. Rev. Lett. 29, 1548.

Halperin, B. I., P. C. Hohenberg and S. Ma., 1974. Phys. Rev. B10, 139.

Halperin, B. I., P. C. Hohenberg and E. Siggia, 1974. Phys. Rev. Lett. 32, 1289.

Halperin, B. I., P. C. Hohenberg and E. Siggia, 1975. Preprint.

Halperin, B. I., T. C. Lubensky and S. Ma., 1974. Phys. Rev. Lett. 32, 292.

Harris, A. B. and T. C. Lubensky, 1974. Phys. Rev. Lett. 33, 1540.

Harris, A. B., T. C. Lubensky, W. K. Holcomb and C. Dasgupta, 1975. Phys. Rev. Lett. 35, 327.

Heller, P., 1967. Rep. Prog. Phys. 30, 731.

Hikami, S. and R. Abe, 1974. Prog. Theo. Phys. 52, 369.

Hohenberg, P. C., 1967. Phys. Rev. 158, 383.

Houghton, A. and F. J. Wegner, 1974. Phys. Rev. A10, 435.

Hsu, S. C., Th. Niemeijer and J. D. Gunton, 1975. Phys. Rev. B11, 2699.

Hubbard, J. and P. Schofield, 1972. Phys. Lett. 40A, 245.

Imry, Y., 1974. Phys. Rev. Lett. 33, 1304.

Imry, Y. and S. Ma., 1975. Phys. Rev. Lett. 35, 1399.

Jona-Lasinio, G., 1975. Nuovo Cimento 26B, 99.

Kadanoff, L. P., 1966. Physics (N. Y.) 2, 263.

Kadanoff, L. P., 1969. Phys. Rev. Lett. 23, 1430.

Kadanoff, L. P., 1975. Phys. Rev. Lett. 34, 1005.

Kadanoff, L. P., W. Götze, D. Hamblen, R. Hecht, E. A. S. Lewis, V. V. Palciauskas, M. Rayl, J. Swift, D. Aspens and J. Kane, 1967. Rev. Mod. Phys. 39, 395.

Kadanoff, L. P. and A. Houghton, 1975. Phys. Rev. B11, 377.

Kadanoff, L. P., A. Houghton and M. C. Yalabik, 1975. Preprint.

Kadanoff, L. P. and J. Swift, 1968. Phys. Rev. 165, 310.

Kadanoff, L. P. and F. J. Wegner, 1971. Phys. Rev. B4, 3989.

Kawasaki, K., 1970. Ann. Phys. (N. Y.) 61, 1.

Kawasaki, K., 1973. J. Phys. A6, 1289.

Kawasaki, K., 1975. J. Phys. A8, 262.

Kawasaki, K., 1975. J. Stat. Phys. In press.

Kawasaki, K. and J. D. Gunton, 1972. Phys. Rev. Lett. 29, 1661.

Kawasaki, K. and T. Yamada, 1975. Prog. Theo. Phys. 53, 111.

Ketley, J. J. and D. J. Wallace, 1973. J. Phys. A6, 1667.

Khmelnitski, D. E., 1975. Preprint.

Kouvel, J. S. and M. E. Fisher, 1964. Phys. Rev. 136A, 1626.

Landau, L. D. and E. M. Lifschitz, 1958. Course of Theoretical Physics: Statistical Physics (Pergamon Press, Oxford and New York).

Lederman, F. L., M. B. Salamon and L. W. Schaklette, 1974. Phys. Rev. B9, 2981.

Levelt Sengers, J. M. H., 1974. Physica 73, 73.

Levelt Sengers, J. M. H., W. L. Greer and J. V. Sengers, 1975. J. Phys. Chem. Ref. Data. In press.

Lubensky, T. C., 1975. Phys. Rev. B11, 3573.

Lubensky, T. C. and A. B. Harris, 1974. AIP Conf. Proc. 24, 311.

Lubensky, T. C. and R. G. Priest, 1974. Phys. Lett. 48A, 103.

Lubensky, T. C. and M. H. Rubin, 1975. Phys. Rev. B11, 4533; B12, 3885.

Ma. S., 1973a. Phys. Rev. A7, 2172.

Ma. S., 1973b. Rev. Mod. Phys. 45, 589.

Ma. S., 1974a. J. Math. Phys. 15, 1866.

Ma, S., 1974b. Phys. Rev. A10, 1818.

Ma, S., 1975. In Domb and Green, 1975.

Ma, S. and G. F. Mazenko, 1974. Phys. Rev. Lett. 33, 1384.

Ma, S. and G. F. Mazenko, 1975. Phys. Rev. 11, 4077.

Ma, S. and L. Senbetu, 1974. Phys. Rev. A10, 2401.

Martin, P. C., 1975. Preprint.

McLaughlin, J. B. and P. C. Martin, 1975. Phys. Rev. A12, 186.

McMillan, W. L., 1972. Phys. Rev. A6, 936.

McMillan, W. L., 1973, Phys. Rev. A8, 1921.

Mattis, D. C., 1965. The Theory of Magnetism (Harper and Row, New York).

McCoy, B. M. and T. T. Wu, 1969. Phys. Rev. 188, 982.

McCoy, B. M. and T. T. Wu, 1973. The Two-Dimensional Ising Model (Harvard University Press, Cambridge, Mass.).

Mermin, N. D. and H. Wagner, 1966. Phys. Rev. Lett. 17, 1133.

Mills, D. L., 1971. Phys. Rev. B3, 3887.

Mitter, P. K., 1973. Phys. Rev. D7, 2927.

Moldover, M. R., 1969. Phys. Rev. 182, 342.

Mukamel, D., 1975. Phys. Rev. Lett. 34, 481.

Mukamel, D. and S. Krinsky, 1975. Preprint.

Myerson, R., 1975. Preprint.

Nauenberg, M. and B. Nienhuis, 1974a. Phys. Rev. Lett. 33, 344.

Nauenberg, M. and B. Nienhuis, 1974b. Phys. Rev. Lett. 33, 1598.

Nelkin, M., 1974. Phys. Rev. A9, 388.

Nelkin, M., 1975. Phys. Rev. A11, 1737.

Nelson, D. R., 1975. Phys. Rev. B11, 3504.

Nelson, D.R. and M.E.Fisher, 1975. Phys.Rev.B11, 1030.

Nelson, D. R., M. E. Fisher and J. M. Kosterlitz, 1974. Phys. Rev. Lett. 33, 813.

Nelson, D. R. and M. E. Fisher, 1975. Ann. Phys. (N.Y.) 91, 226.

Nickel, B. G., 1974. Preprint.

Niemeijer, Th. and J. M. J. van Leeuwen, 1973. Phys. Rev. Lett. 31, 1411.

Niemeijer, Th. and J. M. J. van Leeuwen, 1974. Physica 71, 17.

Niemeijer, Th. and J. M. J. van Leeuwen, 1975. In Domb and Green, 1975.

Nienhuis, B. and M.Nauenberg, 1975. Phys.Rev.B11, 4152.

Opperman, R., 1974. J. Phys. A7, L137.

Passel, L., K. Blivowski, T. Bron and P. Nielson, 1965. Phys. Rev. 139A, 1866.

Pfeuty, P., D. Jasnow and M. E. Fisher, 1974. Phys. Rev. B10, 2088.

Riedel, E. K., 1972. Phys. Rev. Lett. 28, 675.

Riedel, E. K., H. Meyer and R. P. Behringer, 1976. J. Low Temp. Phys. 22. In press.

Riedel, E. K. and F. Wegner, 1972a. In Budnick (1972).

Riedel, E. K. and F. Wegner, 1972b. Phys. Rev. Lett. 29, 349.

Riedel, E. K. and F. Wegner, 1974. Phys. Rev. B9, 294.

Rudnick, J., 1975. Phys. Rev. B11, 363.

Rudnick, J., D. J. Bergman and Y. Imry, 1974. Phys. Lett. A46, 449.

Sak, J., 1973. Phys. Rev. B8, 281.

Sak, J., 1974. Phys. Rev. B10, 3957.

Schroer, B., 1972. Phys. Rev. B8, 4200.

Shirane, G. and J. D. Axe, 1971. Phys. Rev. B4, 2957.

Singh, S. and D. Jasnow, 1975. Phys. Rev. B11, 3445; B12, 493.

Splittorff, O. and B. N. Miller, 1974. Phys. Rev. A9, 550.

Stanley, H. E., 1968. Phys. Rev. 176, 718.

Stanley, H. E., 1971. Introduction to Phase Transitions and Critical Phenomena (Oxford University Press, New York and Oxford).

Stanley, H. E., 1973, editor. Cooperative Phenomena near Phase Transitions, a bibliography with selected readings (MIT Press, Cambridge, Mass.).

Stell, G., 1972. Phys. Rev. B5, 981.

Stephen, M. J., E. Abrahams and J. P. Straley, 1975. Phys. Rev. B12, 256.

Stephen, M. J. and J. McCauly, 1973. Phys. Lett. A44, 89.

Suzuki, M., 1973a. Prog. Theo. Phys. 49, 424.

Suzuki, M., 1973b. Prog. Theo. Phys. 49, 1106.

Suzuki, M., 1973c. In Gunton and Green (1973).

Sykes, M. F., D. S. Gaunt, P. D. Roberts and J. A. Wyles, 1972. J. Phys. A5, 624, 640.

Thompson, C. J., 1972. Mathematical Statistical Mechanics (Macmillan, New York).

Toulouse, G. and P. Pfeuty, 1975. Introduction au Groupe de Rénormalisation et a ses Applications (Presses Universitaires de Grenoble).

Tracy, C. A. and B. M. McCoy, 1975. Phys. Rev. B12, 368.

Tuthill, G. F., J. F. Nicoll and H. E. Stanley, 1975. Phys. Rev. B11, 4579.

van Leeuwen, J. M. J., 1975. Phys. Rev. Lett. 34, 1056.

Wallace, B. and H. Meyer, 1970. Phys. Rev. A2, 1563.

Wallace, D. J., 1973. J. Phys. C6, 1390.

Wegner, F. J., 1971. J. Math. Phys. 12, 2259.

Wegner, F. J., 1972a. Phys. Rev. B5, 4529.

Wegner, F. J., 1972b. Phys. Rev. B6, 1891.

Wegner, F. J., 1974a. J. Phys. C7, 2098.

Wegner, F. J., 1974b. J. Phys. C7, 2109.

Wegner, F. J., 1975a. In Domb and Green (1975).

Wegner, F. J., 1975b. Lecture Notes in Physics 37, 171.

Wegner, F. J., 1975c. J. Phys. A8, 710.

Wegner, F. J. and A. Houghton, 1975. Phys. Rev. A8, 401.

Wegner, F. J. and E. K. Riedel, 1973. Phys. Rev. B7, 248.

Widom, B., 1965. J. Chem. Phys. 43, 3989.

Wilson, K. G., 1969. Phys. Rev. 179, 1499.

Wilson, K. G., 1971. Phys. Rev. B4, 3174, 3184.

Wilson, K. G., 1972, Phys. Rev. Lett. 28, 540.

Wilson, K. G., 1973. Phys. Rev. D7, 2911.

Wilson, K. G., 1975. Rev. Mod. Phys. 47, 773.

Wilson, K. G. and M. E. Fisher, 1972. Phys. Rev. Lett. 28, 248.

Wilson, K. G. and J. Kogut, 1974. Phys. Report C12, 75.

Wu, T. T., B. M. McCoy, C. A. Tracy and E. Barouch, 1975. Preprint.

Wortis, M., 1973. In Gunton and Green (1973), p. 96.

INDEX

(References given parenthetically are in addition to those cited in the text.)